Flow Hydrodynamic in Open Channels: Interaction with Natural or Man-Made Structures

Flow Hydrodynamic in Open Channels: Interaction with Natural or Man-Made Structures

Editor

Mouldi Ben Meftah

MDPI • Basel • Beijing • Wuhan • Barcelona • Belgrade • Manchester • Tokyo • Cluj • Tianjin

Editor
Mouldi Ben Meftah
Department of Civil,
Environmental, Land,
Building Engineering and
Chemistry - DICATECh
Polytechnic University of Bari
Bari
Italy

Editorial Office
MDPI
St. Alban-Anlage 66
4052 Basel, Switzerland

This is a reprint of articles from the Special Issue published online in the open access journal *Water* (ISSN 2073-4441) (available at: www.mdpi.com/journal/water/special_issues/flow_hydrodynamic_in_open_channels).

For citation purposes, cite each article independently as indicated on the article page online and as indicated below:

LastName, A.A.; LastName, B.B.; LastName, C.C. Article Title. *Journal Name* **Year**, *Volume Number*, Page Range.

ISBN 978-3-0365-7663-3 (Hbk)
ISBN 978-3-0365-7662-6 (PDF)

Cover image courtesy of Mouldi Ben Meftah

Contents

About the Editor

Mouldi Ben Meftah

Mouldi BEN MEFTAH is currently an Associate Professor of Hydraulics at the Polytechnic University of Bari - Civil, Environmental, Land, Building Engineering, and Chemistry Department (DICATECh). His research interests include flow hydrodynamics in open channels, vegetated channels, dynamics of pollutant discharge in natural receiving water bodies, turbulent flows, modeling of turbulent jets, hydraulic jump in extensive channels, scouring processes at hydraulic structures, solid transport, maritime hydraulics, and the integrated monitoring of coastal areas.

Preface to "Flow Hydrodynamic in Open Channels: Interaction with Natural or Man-Made Structures"

Natural and man-made watercourses play a vital role in our way of life and in sustaining biodiversity. However, with population growth, increased economic activity, and climate change, waterways are under severe pressure and, thus, require consistent monitoring, proactive management, and timely remediation. Improved knowledge and further understanding of flow hydrodynamics in open channels are crucial to effectively achieving these goals.

Flow hydrodynamic structures in open channels are very complex, persistently challenging phenomena. Despite the numerous studies conducted over decades on this topic, many aspects still require improvement. Many new research dimensions on open-channel flow dynamics can be explored by exploiting today's availability of sophisticated experimental tools, advanced theoretical methods, ample data storage, and the high computational capacity of computers. This Special Issue represents an essential contribution to the state of the art of open-channel flow fields, extending our knowledge and providing valuable references for hydraulic engineers, experts, and those interested in the environment.

Many thanks are extended to all the authors who contributed to the SI, which would not have seen the light of day without their input.

Mouldi Ben Meftah
Editor

water

MDPI

Editorial

Flow Hydrodynamic in Open Channels: A Constantly Evolving Topic

Mouldi Ben Meftah

Department of Civil, Environmental, Land, Building Engineering and Chemistry, Polytechnic University of Bari, Via E. Orabona 4, 70125 Bari, Italy; mouldi.benmeftah@poliba.it; Tel.: +39-080-5963508

1. Introduction

Streams and riverbeds are subject to considerable hydromorphological alterations due to the interaction of their flow with natural or man-made structures found throughout them, i.e., natural vegetation, grade control structures, piers and abutments, discharge/suction systems, seepage, and movable/fixed boundaries. Flow interactions with natural or artificial obstacles lead to the development of complex hydrodynamic structures, which require further studies in order to fully understand them. Thanks to advances in technological devices (i.e., sensors and software), the measurements and numerical modeling of turbulent flows become both easier to conduct and more accurate, giving rise to a better understanding of complex flow dynamics. The significant number of studies conducted on open channel flows certainly helps readers to understand the turbulent flow involved in open channels, as well as to apply this understanding to the design and practice of hydraulic engineering and river management.

Many previous studies have extensively investigated flow dynamics in open channels and their interactions with cross-sectional features. Some, for example, focused on the vegetation effects on flow structures (e.g., [1–9]), characterizing the flow evolution process in vegetated channels. Several studies focused on how to predict the flow discharge in open channels of different cross-sectional shapes (e.g., [10–14]). Others have dealt with estuaries, the interconnection dynamics between rivers and seas, and the relative transport processes of sediment, pollutants, nutrients, and many other substances (e.g., [15–20]). The presence of natural or man-made structures on riverbeds, i.e., natural vegetation, riverbed debris, sills, weirs, sluice gates, spillways, spur dikes, bridge piers, abutments, etc., produce additional complex effects on the flow hydrodynamic characteristics, affecting riverbed evolution and causing scouring processes. Many studies have been conducted on this topic, especially on scour development at hydraulic structures (e.g., [21–28]). Many other studies (e.g., [29–36]) have looked at flow hydrodynamic structures in open channels, i.e., current distributions, turbulence properties, vortex structures, shear/boundary layer development, hydraulic jumps, discharge of turbulent jets and mixing processes, etc.

Despite the numerous studies that have been conducted in recent decades on open channel flows, many aspects still need to be improved. The availability of sophisticated experimental tools, theoretical methods, large data storage, and the high capacity of computers has added many new dimensions to the study of open channel flows. To summarize the results of several previous studies, to improve or propose new and more accurate predictors of many characteristic parameters, and to encourage the development of further knowledge in the field of open channel flows, this Special Issue (SI) has been proposed.

The SI, entitled "Flow Hydrodynamic in Open Channels: Interaction with Natural or Man-Made Structures", comprises eleven original articles [37–47] and two reviews [48,49] on flow hydrodynamic structures in open channel flows.

The next section presents a short descriptive summary of each manuscript.

Citation: Ben Meftah, M. Flow Hydrodynamic in Open Channels: A Constantly Evolving Topic. *Water* **2023**, *14*, 4120. https://doi.org/10.3390/w14244120

Received: 12 December 2022
Accepted: 15 December 2022
Published: 17 December 2022

2. Short Descriptive Summary of the Different Manuscripts

In [37], the authors dealt with understanding the hydraulic characteristics of lateral deflectors with different geometries in gentle-slope free-surface tunnels. Indicating their great importance for many engineering applications, such as in gentle-slope tunnels of high dams, in this study, the authors analyzed the results of both experimental and numerical modeling of two new deflector geometries. The hydraulic characteristics of the newly designed deflectors and the traditional triangular deflector were analyzed and compared. Particular attention was focused on the flow pattern improvement and the energy dissipation characteristics of the three deflectors.

Estuaries are one of the most important interconnections between land and sea; they are often important areas for leisure and economic activities. Estuaries are also ecosystems which are highly vulnerable to human impacts and environmental change; thus, they must be adequately preserved. In [38], the authors attempted to comprehend the hydrodynamic impact of waves on a river estuary in a case study of the Oued Sebou on Morocco's north Atlantic coast. Specifically, the study focused on the dredging effect (caused by sand extraction) on the wave motion and its impact on the estuary environment. Different scenarios of wave-propagation simulations were carried out, using different significant wave heights and considering bathymetry both before and after two dredging cases of 2 and 4 m depths.

In [39], the authors carried out field measurements in a river estuary, a case study of the Kampar River estuary in Indonesia, to determine the real-time properties of tidal bore generation, hydraulic jump, and transport mechanisms. Using a least-square-based tidal model, the tidal harmonic and its range were analyzed. The tidal bore's hydraulic jump properties were determined using survey datasets of the height and flow velocities of the tidal bore in the estuary. In addition, an acoustic-based approach was employed to quantify the suspended sediment concentration and flux during the passage of the tidal bore.

In [40], the authors were interested in the impact of long waves on hydraulic structures, such as bridge piers. An experimental study was carried out on the propagation of two different long waves, released on a steady current, and their interactions with a bottom-mounted rigid emergent cylinder. The flow velocity fields around the emergent cylinder were measured using a Particle Image Velocimetry (PIV) system, and the flow turbulence structures were analyzed. The measured data were also used to verify analytical theories or to approximate the methods that best predicted the flow properties generated by the wave–cylinder interaction.

In [41], the authors sought to solve the scouring problem at hydraulic structures. They specifically proposed a methodology to estimate the scour hazard, considering both the scour–fill interaction and the Monte Carlo simulation method. The general extreme value probability distribution was used to characterize the intensity of the scouring events. The lognormal distribution was used to characterize the sedimentation process (fill), and a homogeneous Poisson process was employed to forecast the occurrence of both the sedimentation and scouring events. Based on this approach, several histories of scour–fill depths were performed and used to develop time-dependent scour hazard curves. In this study, the authors proposed several hazard curves associated with different time intervals for a bridge located in Oaxaca, Mexico. The presented approach takes into account both missing data and outliers.

In [42], the authors aimed to experimentally investigate the effects of bed sediments on flow in a vegetated channel. Since in previous laboratory studies, most of the experiments were conducted in vegetated channels with smooth beds, the effect of bed sediment was certainly neglected. In this study, the authors focused on the bed roughness impact on the flow turbulence characteristics. A large experimental series was carried out in a vegetated channel composed of randomly arranged rigid emergent vegetation and beds with different sizes of uniform bed sediments. Based on flow velocity measurements, using an Acoustic Doppler Velocimeter (ADV) profiler, a refined analysis of the bed sediment impact on flow turbulence structures within vegetation arrays was performed.

In [43], the authors raised the problem of the complex mechanism of head loss in open channel flow and its accurate prediction. Unlike classic methods such as the Darcy–Weisbach equation, in this study, the authors proposed a new theoretical friction coefficient equation based on the entropy concept and hydraulic properties. The proposed equation was determined using different cross-section shapes of open channels, i.e., rectangular, trapezoidal, parabolic, round-bottomed triangle, and parabolic-bottomed triangle. To validate the proposed equation, the estimated values of friction coefficients were compared with the measured values for both the smooth and rough flow conditions. The author claimed that this new equation was able to calculate friction coefficients with very high accuracy, without the use of energy slopes.

In [44], the authors performed laboratory experiments and numerical simulations to investigate the flow-field structures in a curved channel. In a channel with a U-shaped cross-section and 180-degree curvature, extensive velocity measurements were carried out using an ADV at different cross-sections along the channel. The bed roughness effect was also considered, using sediment particles with different grain sizes. The primary current, turbulent kinetic energy, and turbulent bursting were analyzed in depth. Flow numerical simulations were also conducted using the Renormalization-Group (RNG) turbulence model in the FLOW-3D software. The bed roughness effects on the secondary current distribution were also provided.

Flow turbulence properties and secondary currents strongly affect the sediment transport processes, and, in turn, suspended particle motion influences flow turbulence, such as Reynolds shear stress and velocities. Since most previous studies on scouring processes did not consider secondary currents, in [45], the authors attempted to investigate the generation and turbulence properties of secondary currents across a scour hole under equilibrium conditions. Based on comprehensive measurements of flow velocities, using an ADV, the development of secondary current cells was proven in a clear water equilibrium scour downstream of a grade control structure, in a channel with non-cohesive sediments. A detailed analysis of the turbulence intensities and Reynolds shear stresses was implemented and compared with previous studies. The results showed that the anisotropy term of normal stresses has exhibited significant magnitudes, dominating the other terms in the mean streamwise vorticity equation, which may reflect its potential effect in generating the secondary current motion in straight non-circular channel flows.

In [46], the authors presented the results of flume experiments conducted to investigate flow velocity and concentration distributions within debris bodies, using high-resolution images. The data analysis showed that the concentration and mobility of sediment grains vary according to depth. A linear law to interpret the grain concentration distribution, starting from the knowledge of the packing concentration and the surface concentration, Cs, was established. Based on this finding (linear law), the authors proposed modified expressions of Bagnold's number and velocity in stony-type debris flows. The use of these expressions led to the identification of three motion regimes along the depth, and the velocity profile within the debris body was determined as a function of the parameter Cs. The comparison of the predicted velocity profiles, determined using the modified equation, with measured profiles from the literature confirmed the credibility of the proposed approach.

In [47], the authors attempted to understand the erosion and deposition behaviors in meandering watercourses. A series of experiments were conducted in a sinuous channel in order to investigate both the bed morphological changes and the flow hydrodynamic structures. High-order velocity fluctuation moments were analyzed at the outer and inner banks to explain the morphological variations of the bed. The magnitude values of the flow velocity fluctuation moment, in both the outer and inner bend regions, were determined and related to the erosion rate. The authors claimed that the higher-order moments, in association with the morphological changes of the bed, contributed to a new approach that helped them to understand the flow turbulent characteristics of a meandering channel.

Recently, possibly due to climate change, extreme rainfall events have become more frequent, causing floods. These floods threaten the stability of many hydraulic structures due to significant scouring effects. Predicting scour dimensions of hydraulic structures continues to be a concern for engineers and researchers. In [48], the authors provided a comprehensive and relevant review of scour prediction at bridge piers. Several empirical methods from previous studies for predicting scour at bridge piers were reviewed and analyzed. A summary of relevant scour predictors was provided, which will be useful for practitioners. A discussion on future research directions that could contribute to the field was also proposed. Some recommendations for future research to improve the knowledge in this field were also proposed.

Finally, in [49], the authors provided an interesting review on the topic of vegetated channel flows. They particularly focused on the different methodologies and approaches used in previous studies to evaluate the resistance to flow caused by the presence of vegetation in open channels. Different expressions of the drag coefficient and Manning/Chézy coefficient were reviewed and analyzed for submerged and emergent, rigid, and flexible vegetation. Furthermore, new developments in the field of 3D numerical methods, currently used for the evaluation of the turbulence characteristics and the transport of sediments and pollutants, were also briefly reviewed. The authors also proposed several recommendations for future research to improve many aspects related to the characterization of hydrodynamic structures in vegetated channel flows.

3. Conclusions

This Special Issue consists of thirteen manuscripts relating to the topic of flow hydrodynamics in open channels. Various arguments are addressed herein, including flow interaction with hydraulic structures [37,40–42,45,48], flow dynamic in estuaries [38,39], flow–vegetation interactions [42,49], bed sediment effects on flow structures [42,43,45,46], and the effect of channel curvature on flow behaviors and sediment transport [44,47]. This Special Issue represents a significant contribution to the state of the art of the open channel flow fields, extending our knowledge and providing useful references for hydraulic engineers, experts, and those interested in the environment.

Acknowledgments: I would like to extend my thanks to all the authors who have contributed to the SI, which, without them, would not have seen the light of day.

Conflicts of Interest: The author declares no conflict of interest.

References

1. Nezu, I.; Onitsuka, K. Turbulent structures in partly vegetated open-channel flows with LDA and PIV measurements. *J. Hydraul. Res.* **2001**, *39*, 629–642. [CrossRef]
2. White, B.L.; Nepf, H.M. Shear instability and coherent structures in shallow flow adjacent to a porous layer. *J. Fluid Mech.* **2007**, *593*, 1–32. [CrossRef]
3. De Serio, F.; Ben Meftah, M.; Mossa, M.; Termini, D. Experimental investigation on dispersion mechanisms in rigid and flexible vegetated beds. *Adv. Water Resour.* **2018**, *120*, 98–113. [CrossRef]
4. Ben Meftah, M.; Mossa, M. Turbulence measurement of vertical dense jets in crossflow. *Water* **2018**, *10*, 286. [CrossRef]
5. Mossa, M.; Ben Meftah, M.; De Serio, F.; Nepf, H. How vegetation in flows modifies the turbulent mixing and spreading of jets. *Sci. Rep.* **2017**, *7*, 6587. [CrossRef]
6. Ben Meftah, M.; De Serio, F.; Mossa, M. Hydrodynamic behavior in the outer shear layer of partly obstructed open channels. *Phys. Fluids* **2014**, *26*, 065102. [CrossRef]
7. Huai, W.; Xue, W.; Qian, Z. Large-eddy simulation of turbulent rectangular open-channel flow with an emergent rigid vegetation patch. *Adv. Water Resour.* **2015**, *80*, 30–42. [CrossRef]
8. Ben Meftah, M.; Mossa, M. Partially obstructed channel: Contraction ratio effect on the flow hydrodynamic structure and prediction of the transversal mean velocity profile. *J. Hydrol.* **2016**, *542*, 87–100. [CrossRef]
9. Ben Meftah, M.; De Serio, F.; Malcangio, D.; Mossa, M.; Petrillo, A.F. A modified log-law of flow velocity distribution in partly obstructed open channels. *Environ. Fluid Mech.* **2016**, *16*, 453–479. [CrossRef]
10. Aricò, C.; Corato, G.; Tucciarelli, T.; Ben Meftah, M.; Mossa, M.; Petrillo, F.A. Discharge estimation in open channels by means of water level hydrograph analysis. *J. Hydraul. Res.* **2010**, *48*, 612–619. [CrossRef]
11. Huthoff, W.; Roos, P.C.; Augustijn, C.M.A.; Hulscher, S.J.M.H. Interacting divided channel method for compound channel flow. *J. Hydraul. Eng.* **2008**, *134*, 1158–1165. [CrossRef]

12. Tang, X. An improved method for predicting discharge of homogeneous compound channels based on energy concept. *Flow Meas. Instrum.* **2017**, *57*, 57–63. [CrossRef]
13. Yang, Z.; Li, D.; Huai, W.; Liu, J. A New method to estimate flow conveyance in a compound channel with vegetated floodplains based on energy balance. *J. Hydrol.* **2019**, *575*, 921–929. [CrossRef]
14. Ben Meftah, M.; Mossa, M. Discharge prediction in partly vegetated channel flows: Adaptation of IDCM method with a curved interface and large-scale roughness elements. *J. Hydrol.* **2023**, *616*, 128805. [CrossRef]
15. Kvočka, D.; Žagar, D.; Banovec, P. A review of river oil spill modeling. *Water* **2021**, *13*, 1620. [CrossRef]
16. Eke, D.C.; Anifowose, B.; Van De Wiel, M.J.; Lawler, D.; Knaapen, M.A.F. Numerical modelling of oil spill transport in tide-dominated estuaries: A case study of Humber Estuary, UK. *J. Mar. Sci. Eng.* **2021**, *9*, 1034. [CrossRef]
17. Guo, Y. Hydrodynamics in estuaries and coast: Analysis and modeling. *Water* **2022**, *14*, 1478. [CrossRef]
18. Sohrt, V.; Hein, S.S.V.; Nehlsen, E.; Strotmann, T.; Fröhle, P. Model based assessment of the reflection behavior of tidal waves at bathymetric changes in estuaries. *Water* **2021**, *13*, 489. [CrossRef]
19. Eyre, B.; Hossain, S.; McKee, L. A suspended sediment budget for the modified subtropical Brisbane River estuary, Australia. *Estuar. Coast. Shelf Sci.* **1998**, *47*, 513–522. [CrossRef]
20. Chi, Y.; Rong, Z. Assessment of extreme storm surges over the Changjiang river estuary from a wave-current coupled model. *J. Mar. Sci. Eng.* **2021**, *9*, 1222. [CrossRef]
21. Manes, C.; Brocchini, M. Local scour around structures and the phenomenology of turbulence. *J. Fluid Mech.* **2015**, *779*, 309–324. [CrossRef]
22. Ben Meftah, M.; Mossa, M. Scour holes downstream of bed sills in low-gradient channels. *J. Hydraul. Res.* **2006**, *44*(4), 497–509. [CrossRef]
23. Wang, L.; Melville, B.W.; Whittaker, C.N.; Guan, D. Temporal evolution of clear-water scour depth at submerged weirs. *J. Hydraul. Eng.* **2020**, *146*, 06020001. [CrossRef]
24. Ben Meftah, M.; De Serio, F.; De Padova, D.; Mossa, M. Hydrodynamic structure with scour hole downstream of bed sills. *Water* **2020**, *12*, 186. [CrossRef]
25. Papanicolaou, A.N.T.; Bressan, F.; Fox, J.; Kramer, C.; Kjos, L. Role of structure submergence on scour evolution in gravel bed rivers: Application to slope-crested structures. *J. Hydraul. Eng.* **2018**, *144*, 03117008. [CrossRef]
26. Amini, A.; Melville, B.W.; Ali, T.M.; Ghazal, A.H. Clear-water local scour around pile groups in shallow-water flow. *J. Hydraul. Eng.* **2012**, *138*, 177–185. [CrossRef]
27. Ben Meftah, M.; Mossa, M. New approach to predicting local scour downstream of grade-control structure. *J. Hydraul. Eng.* **2020**, *146*, 04019058. [CrossRef]
28. De Padova, D.; Ben Meftah, M.; Mossa, M.; Sibilla, S. A multi-phase SPH simulation of hydraulic jump oscillations and local scouring processes downstream of bed sills. *Adv. Water Resour.* **2022**, *159*, 104097. [CrossRef]
29. Njenga, K.J.; Kioko, K.J.; Wanjiru, G.P. Secondary current and classification of river channels. *Appl. Math.* **2013**, *4*, 70–78. [CrossRef]
30. Albayrak, I.; Lemmin, U. Secondary currents and corresponding surface velocity patterns in a turbulent open-channel flow over a rough bed. *J. Hydraul. Eng.* **2011**, *137*, 1318–1334. [CrossRef]
31. Ben Meftah, M.; Mossa, M.; Pollio, A. Considerations on shock wave/boundary layer interaction in undular hydraulic jumps in horizontal channels with a very high aspect ratio. *Eur. J. Mech. - B/Fluids* **2010**, *29*, 415–429. [CrossRef]
32. Ben Meftah, M.; De Serio, F.; Malcangio, D.; Mossa, M.; Petrillo, A.F. Experimental study of a vertical jet in a vegetated crossflow. *J. Environ. Manag.* **2015**, *164*, 19–31. [CrossRef]
33. Chen, K.; Zhang, Y.; Zhong, Q. Wavelet coherency structure in open channel flow. *Water* **2019**, *11*, 1664. [CrossRef]
34. Ben Meftah, M.; Malcangio, D.; De Serio, F.; Mossa, M. Vertical dense jet in flowing current. *Environ. Fluid Mech.* **2018**, *18*, 75–96. [CrossRef]
35. Yang, Z.; Bai, F.; Xiang, K. A lattice Boltzmann model for the open channel flows described by the Saint-Venant equations. *R. Soc. Open Sci.* **2019**, *6*, 190439. [CrossRef]
36. Ben Meftah, M.; De Serio, F.; Mossa, M.; Pollio, A. Experimental study of recirculating flows generated by lateral shock waves in very large channels. *Environ. Fluid Mech.* **2008**, *8*, 215–238. [CrossRef]
37. Da, J.; Wang, J.; Dong, Z.; Du, S. Hydraulic characteristics of lateral deflectors with different geometries in gentle-slope free-surface tunnels. *Water* **2022**, *14*, 2689. [CrossRef]
38. Iouzzi, N.; Mouakkir, L.; Ben Meftah, M.; Chagdali, M.; Loudyi, D. SWAN modeling of dredging effect on the Oued Sebou Estuary. *Water* **2022**, *14*, 2633. [CrossRef]
39. Wisha, U.J.; Wijaya, Y.J.; Hisaki, Y. Real-time properties of hydraulic jump off a tidal bore, its generation and transport mechanisms: A case study of the Kampar River Estuary, Indonesia. *Water* **2022**, *14*, 2561. [CrossRef]
40. Basile, R.; De Serio, F. Flow field around a vertical cylinder in presence of long waves: An experimental study. *Water* **2022**, *14*, 1945. [CrossRef]
41. Flores-Vidriales, D.; Gómez, R.; Tolentino, D. Stochastic assessment of scour hazard. *Water* **2022**, *14*, 273. [CrossRef]
42. Coscarella, F.; Penna, N.; Ferrante, A.P.; Gualtieri, P.; Gaudio, R. Turbulent flow through random vegetation on a rough bed. *Water* **2021**, *13*, 2564. [CrossRef]

43. Choo, Y.-M.; Kim, J.-G.; Park, S.-H. A Study on the friction factor and Reynolds number relationship for flow in smooth and rough channels. *Water* **2021**, *13*, 1714. [CrossRef]
44. Lin, Y.-T.; Yang, Y.; Chiu, Y.-J.; Ji, X. Hydrodynamic characteristics of flow in a strongly curved channel with gravel beds. *Water* **2021**, *13*, 1519. [CrossRef]
45. Ben Meftah, M.; De Padova, D.; De Serio, F.; Mossa, M. Secondary currents with scour hole at grade control structures. *Water* **2021**, *13*, 319. [CrossRef]
46. Termini, D.; Fichera, A. Experimental analysis of velocity distribution in a coarse-grained debris flow: A modified Bagnold's equation. *Water* **2020**, *12*, 1415. [CrossRef]
47. Taye, J.; Barman, J.; Kumar, B.; Oliveto, G. Deciphering morphological changes in a sinuous river system by higher-order velocity moments. *Water* **2020**, *12*, 772. [CrossRef]
48. Harasti, A.; Gilja, G.; Potočki, K.; Lacko, M. Scour at bridge piers protected by the riprap sloping structure: A review. *Water* **2021**, *13*, 3606. [CrossRef]
49. D'Ippolito, A.; Calomino, F.; Alfonsi, G.; Lauria, A. Flow resistance in open channel due to vegetation at reach scale: A review. *Water* **2021**, *13*, 116. [CrossRef]

water

MDPI

Article

Hydraulic Characteristics of Lateral Deflectors with Different Geometries in Gentle-Slope Free-Surface Tunnels

Jinrong Da [1], Junxing Wang [1,*], Zongshi Dong [1] and Shuaiqun Du [2]

1 State Key Laboratory of Water Resources and Hydropower Engineering Science, Wuhan University, Wuhan 430072, China
2 Power China Guiyang Engineering Corporation Limited, Guiyang 550081, China
* Correspondence: jxwang@whu.edu.cn; Tel.: +86-137-0718-2138

Abstract: The gentle-slope tunnel has been adopted in many high dams, and aerators are usually required for high operating heads. For such tunnels, the lateral deflector is superior to the traditional bottom aerator, which loses its efficiency due to cavity blockage and fails to aerate the sidewalls. However, unfavorable flow patterns such as water-wings and shock waves are induced by the lateral deflectors. To address this problem, two novel lateral deflectors are proposed, and their hydraulic characteristics are comparatively investigated together with the triangular deflector by means of model test and numerical simulation. The triangular deflector was revealed to form a wide cavity that allows for the free rise up of the water-wings inside the cavity, leading to the development of a buddle-type shock wave, whereas the two-arc deflector yields a jet with a fluctuating surface, which induces water-wings and further develops into diamond-type shock waves. In contrast, the cavity formed behind the two-arc deflector with a straight downstream guiding line is stabler and shorter, thereby restricting the development of the rising flow and preventing the formation of water-wings and shock waves. Moreover, the two-arc deflector with a straight guiding line exhibits higher energy dissipation capacities and thus is recommended in practical engineering design.

Keywords: lateral deflectors; gentle-slope tunnel; water-wing; shock wave; energy dissipation

Citation: Da, J.; Wang, J.; Dong, Z.; Du, S. Hydraulic Characteristics of Lateral Deflectors with Different Geometries in Gentle-Slope Free-Surface Tunnels. *Water* **2022**, *14*, 2689. https://doi.org/10.3390/w14172689

Academic Editor: Mouldi Ben Meftah

Received: 16 July 2022
Accepted: 26 August 2022
Published: 30 August 2022

1. Introduction

Due to complex topographic and geological conditions [1] as well as material transport difficulties, high dams usually adopt arch concrete or local material types [2,3], the dam-body flood-discharge capacity of which is significantly limited [4]. Consequently, complementary flood release structures are of great importance for a sound design of the entire project [4]. In the past century, the tunnel spillway has stood out from other alternatives [4,5], such as chute spillways and shaft spillways owing to its high flood-release capacity, good terrain compatibility, and decent construction cost [6]. To avoid potential cavitation and denudation damage [1,4,5,7], the flow pattern and velocity in the spillway tunnel have to be strictly controlled. The ideal flow pattern should at least meet the following two requirements: (i) the flow is free-surface flow without shock waves and with adequate space between the aerated surface and the tunnel soffit [8]; (ii) the flow velocity is within 25 m/s [8] and the structure surface is carefully smoothed [8,9]. Under such conditions, technically no aerator is required. In the past decades, the sagging dragon tail tunnel has been proven to effectively fulfill the above requirement and thus has gained popularity in many high dams (e.g., Jinping, Baihetan, Xiaowan [5], Wudongde [10], Shuangjiangkou [11]) in China. The underlying reason for this kind of tunnel's satisfactory hydraulic behavior lies in its longitudinal layout, which is reflected in its name 'sagging dragon tail'. It comprises an overflow ogee weir (dragon head) close to the reservoir, a long gentle-slope free-surface tunnel in the middle (dragon body) and a short open and steep chute connected to a ski-jump bucket (the sagging dragon tail) at the end, as shown in Figure 1. With such a design, the flow acceleration is negligible inside the gentle-slope tunnel [4,5] and the high-velocity flow only occurs at the short steep chute, where the flow is usually highly aerated and the potential local structure destructions can be easily found and repaired once it occurs.

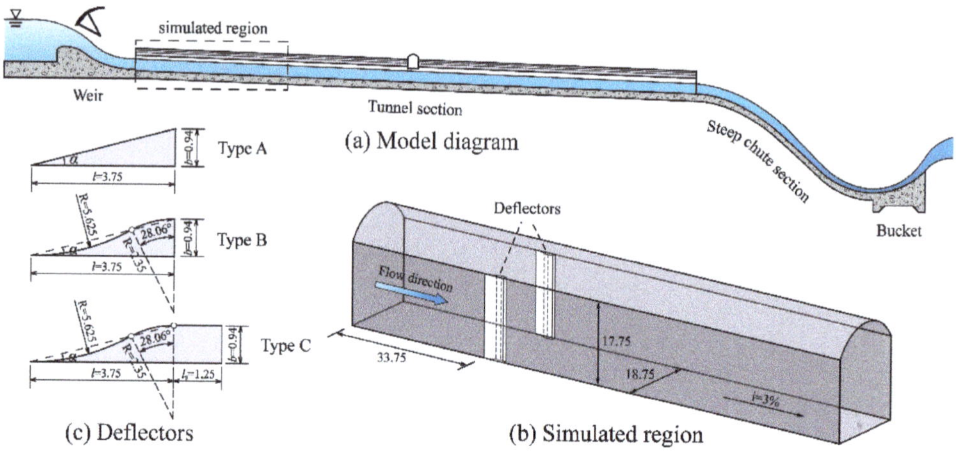

Figure 1. (**a**) Schematic diagram of the experimental spillway tunnel and details of (**b**) the simulated region and (**c**) the three lateral deflectors (unit is cm).

However, as the desired maximum operation head further increased, the flow velocity in the gentle-slope tunnel could exceed the threshold value of some 25 m/s [8], which consequently increases the cavitation risk of the tunnel [5,7,12] and the energy dissipation burden of the downstream chute. Under such conditions, aerators have to be installed to reduce cavitation risk [9,13,14], and restrictions resulting from flow pattern control have to be adequately considered. The main challenges of designing bottom aerators for gentle-slope tunnels are the low aeration efficiency caused by cavity blockage [6,12,15,16] and the cavitation protection of tunnel sidewalls [17,18]. Moreover, Hager discovered that shock waves could also be introduced by local changes in the bottom profile [14], such as offset aerators and bottom deflectors [16,19]. Therefore, to improve the aeration efficiency, lateral deflectors are usually installed together with the bottom aerators [15,17,18,20–23], of which a typical application is the fall-expansion aerators that constructed at the inlet of the free-surface tunnel behind the gate chamber [1]. It has been discovered that the lateral deflectors could not only serve as effective aerators [17,18,24] independent of the tunnel slope, but also help dissipate a certain amount of kinetic energy [24–26] and reduce the energy dissipation pressure of downstream structures. From this point, the hydraulic characteristics of the standalone lateral deflector (SLD, i.e., without bottom aerators) deserve further investigation and it is expected that the SLD could achieve decent hydraulic performance with regard to flow pattern and aeration.

To date, the commonest lateral deflector design is the triangular deflector whose up- and downstream endpoints are connected to the tunnel sidewall abruptly, and without vertical variation. The most noteworthy two of the few exceptions are probably the right-angled tetrahedrons and vertical plate deflectors investigated by Hager [26]. With such designs, water-wings and shock waves are usually induced by the lateral deflectors and thus introducing additional denudation risks. For example, Nie et al. (2006) discovered the jet formed behind the deflector impacts the sidewall and then is deflected upstream, resulting in partial blockage of the lateral cavity and the unstable swirling flow inside the lateral cavity plays a vital role in causing surface erosion [27]. The water-wings [18,20,23,24,28–30] formation was observed downstream of the impact region, which further rises up to the tunnel soffit [31] and could possibly develop into air pocket flow [16]. Shock waves [14,16,20,24,26,29,32] were also found to be induced by lateral deflection, which could propagate far downstream, mainly in a diamond shape with water crowns colliding alternatively in the middle and on the sidewalls; thereby, causing sidewall erosion with repeating impact and pressure fluctuation. Under highly aerated situations, shockwaves could cause overtopping or trigger conduits choke [14], which dramatically impede the air transportation and thereby harm for the aeration protection [23]. Existing studies mostly focus on the optimization of the size [21,24,26,28,30] and layout [1,26,29–31] of the traditional triangular deflector, and few attempts can be found to improve the flow pattern and energy dissipation by means of modifying the deflector geometries.

In this paper, two new deflector geometries are proposed for the sake of both flow control and energy dissipation. The hydraulic characteristics of the newly proposed deflectors and the traditional triangular deflector are comparatively investigated using the hybrid approach of model test and numerical simulation. Special focus is put on the flow pattern improvement. The energy dissipation characteristics of these three deflectors are also compared based on the simulation results.

2. Experiment Setup

The experiments were carried out with the physical model of a sagging dragon tail tunnel constructed at the State Key Laboratory of Water Resources and Hydropower Engineering Science, Wuhan University. In the experiment, water was supplied by a circulating system composed of an underground reservoir, pumps, and pipelines. The non-pressured section of the tunnel was a B = 18.75 cm wide, h = 17.75 cm deep flume inclined with a i = 3% bottom slope. The corresponding scale fell into the range of 1/24~1/80 with regard to the typical width being 4.5~15 m of the high unit-width-discharge free-surface tunnels (e.g., Yele, Xiaowan, Xiluodu) in China [8]. Experimental studies adopting similar geometric scale or model tunnel width can be found in [33–36]. The scale effect is considered acceptable as the main focus of this paper are the flow pattern and other macroscopical flow characteristics such as energy dissipation behaviors [36,37], which are much less sensitive to the model scale compared to two-phase flow characteristics such as air concentration, bubble size and air vent discharge [19,23]. The lateral deflectors were installed symmetrically (i.e., with identical geometry and streamwise location) on the sidewalls with our bottom aerators. The flume and deflectors are both made of plexiglass to provide a transparent view of the flow pattern. Three types of deflectors (Figure 1c) were investigated, the width b of which were all 0.94 cm and the streamwise length of deflectors A and B was 3.75 cm, while deflector C was further extended 1.25 cm downstream with a straight guiding line parallel to the side walls. The detailed geometries of the three types of deflectors are sketched in Figure 1. Type A is the traditional triangular deflector, deflector B is composed of two arcs tangent to each other: one being negative and has a radius of 5.625 cm and the other being positive and features a radius of 2.35 cm. As for deflector C, it is the same as deflector B except for the aforementioned additional straight line. Thus, the deflectors feature an identical contraction ratio of 10% ($2b/B$) and a tangent value of 1/4, and the dimensionless tail extension of deflector C with regard to the streamwise length of the curved section l is 1/3.

In the experiment, the inflow conditions were controlled at the cross-section 33.75 cm upstream of the deflector. All the measurements in this paper were conducted under the situation of inflow depth h_{in} = 11.25 cm and volume flow rate Q_{in} = 57.65 L/s, featuring a Reynolds number $Re = \frac{Q_{in}}{h_{in} R \upsilon} = 1.2 \times 10^5$ (R represents the hydraulic diameter calculated as $\frac{bh_{in}}{b+2h_{in}}$ and $\upsilon = 1 \times 10^{-6}$ m^2/s is the kinematic viscosity of water). This Re value being larger than 1×10^5 indicates the scale effect arising from viscous stress can be neglected according to [38]. The corresponding depth-width ratio of 0.6 in the experiment is a representative value for the real-world high-discharge flood tunnels [15,22,23,33–36] in China. In this study, the flow depth was measured using a fluviograph (accuracy $\pm$ 0.18 mm). The water discharge was monitored using an electromagnetic flowmeter (IFM4080K, Jiangsu Runyi Instrument Co., Ltd., Huaian, China), featuring accuracy of 0.1 L/s. The flow velocity was measured with a propeller-type flow meter (LS300-A, Fuzhou Lesida Information Technology Co., Ltd., Fuzhou, China) featuring accuracy of 0.01 m/s.

3. Numerical Models and Simulation Setup

Numerical simulations were performed in this study to obtain full-field velocities and turbulence properties for the analysis of the underlying mechanisms of flow pattern improvement and energy dissipation characteristics. The simulations were conducted using the commercial software FLOW-3D, which is claimed to have advantages over other opponents for free-surface flows and has been widely used for spillway and tunnel flows [39–41].

3.1. Governing Equations

FLOW-3D utilizes the one-fluid framework for free-surface flow modeling. This was achieved by using the Tru-VOF [42] technique to dynamically track the interface and, in the meantime, impose proper boundary conditions at the free surface. In this way, the computational cost is significantly reduced. The continuity and momentum equations are:

$$\frac{\partial u_i}{\partial x_i} = 0 \tag{1}$$

$$\frac{\partial u_i}{\partial t} + u_j\frac{\partial u_i}{\partial x_j} = -\frac{1}{\rho}\frac{\partial p}{\partial x_i} + f_i + v\frac{\partial^2 u_i}{\partial x_j \partial x_j} - \frac{\partial}{\partial x_j}(\frac{2}{3}k\delta_{ij} - 2v_t S_{ij}) \tag{2}$$

Here, $\boldsymbol{u}$ is velocity, p is the pressure, ρ and v are the fluid density and kinematic viscosity, respectively. $\boldsymbol{f}$ stands for the body force, k and μ_t stand for the turbulent kinetic energy, and turbulent viscosity. δ_{ij} is the Kronecker delta, and S_{ij} is the mean rate of strain tensor calculated as $\frac{1}{2}(\frac{\partial u_i}{\partial x_j} + \frac{\partial u_j}{\partial x_i})$.

The VOF equation for interface tracking reads:

$$\frac{\partial \alpha}{\partial t} + u_i\frac{\partial \alpha}{\partial x_i} = 0 \tag{3}$$

where α is the volume fraction of the simulated fluid.

The RNG k-ε turbulence model [43] was adopted to account for the turbulence contribution to the time-averaged momentum transport, in which a transport equation was solved for the turbulent kinetic energy k and the turbulent dissipation rate ε, respectively:

$$\frac{\partial k}{\partial t} + \frac{\partial}{\partial x_i}(ku_i) = \frac{\partial}{\partial x_j}\left((v + \sigma_k v_t)\frac{\partial k}{\partial x_j}\right) + P_k - \varepsilon \tag{4}$$

$$\frac{\partial \varepsilon}{\partial t} + \frac{\partial}{\partial x_i}(\varepsilon u_i) = \frac{\partial}{\partial x_j}\left((v + \sigma_\varepsilon v_t)\frac{\partial \varepsilon}{\partial x_j}\right) + C_{1\varepsilon}P_k\frac{\varepsilon}{k} - C^*_{2\varepsilon}\rho\frac{\varepsilon^2}{k} \tag{5}$$

where $v_t = C_\mu\frac{k^2}{\varepsilon}$, $P_k = 2v_t S_{ij}S_{ij}$, $C^*_{2\varepsilon} = C_{2\varepsilon} + \frac{C_\mu\eta^3\left(1-\frac{\eta}{\eta_0}\right)}{1+\beta\eta^3}$, $\eta = \left(2S_{ij}\cdot S_{ij}\right)^{1/2}\frac{k}{\varepsilon}$, and model parameters are: $\sigma_k = \sigma_\varepsilon = 1.39$, $C_\mu = 0.085$, $C_{1\varepsilon} = 1.42$, $C_{2\varepsilon} = 1.68$, $\eta_0 = 4.38$, and $\beta = 0.012$.

3.2. Simulation Setup

The computational domain adopted the same coordinate system as the experiment and extends from $x = -33.75$ to $x = 135$ cm in the streamwise direction. A grid convergence study involving three mesh schemes was carried out in advance to select a proper mesh for the simulation. The details of the grid convergence study were shown in Appendix A. One structured mesh block was used to discretize the domain into 10.25 million cuboid cells, the majority of which feature an average size of 0.34 cm × 0.125 cm × 0.2 cm ($x \times y \times z$). To ensure the spatial resolution of the deflector geometry using the FAVOR® technique [44,45], the computational domain was rotated by the flume slope of 3% to be aligned with the x-coordinate and the gravity vector g was adjusted accordingly. Moreover, the domain was locally refined to 0.188 cm × 0.094 cm × 0.188 cm in the region of $-7.5 < x < 30$ cm, $-10 < y < -6.25$ cm and $6.25 < y < -10$ cm, and $0 < z < 12.5$ cm.

The outlet boundary was configured with the outflow boundary condition (BC), implying a zero-gradient condition since the outflow is supercritical. The bottom and sidewalls were set as non-slip walls with equivalent roughness k_s = 0.015 mm according to the plexiglass surface properties. The top boundary is specified with fixed relative pressure $p = 0$ (i.e., atmosphere pressure). As for the inlet boundary, a pre-simulation of the entire physical model (i.e., from the reservoir to the downstream cushion pool) was conducted first, and then all the flow parameters at $x = -33.75$ cm were mapped onto the inlet of the short domain using the grid overlay BC. The illustrative diagram of the simulation setup is shown in Figure 2.

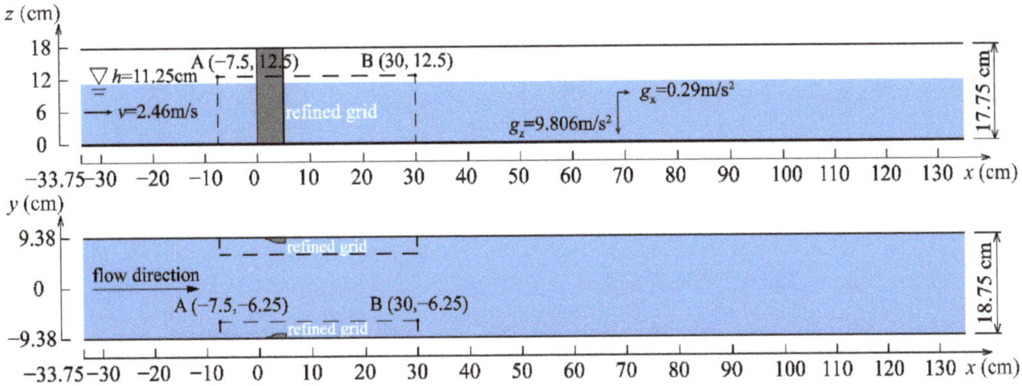

Figure 2. Schematic diagram of the simulation setup.

4. Results and Discussion

4.1. Model Validation

The simulated and measured water surface profiles at two longitudinal slices are comparatively shown in Figure 3a. It can be seen that the simulation results are in good agreement with the experimental data. Only small deviations can be found near the deflectors ($-5 < x < 10$ cm at $y = -8.375$ cm). Moreover, vertical and horizontal profiles of the measured and calculated streamwise velocity at $x = 112.5$ cm are presented in Figure 3b and c respectively. The simulated vertical velocity profile overlaps well with the measured data in the middle of the flume (i.e., $y = 0$ cm), whereas the calculated velocities near the sidewall ($y = -8.38$ cm) are approximately 10% smaller than the experimental data, which can be attributed to the wall function effects. As for the horizontal velocity distribution, the simulated velocities are also generally consistent with the measured data, with the bottom values ($z = 1$ cm) noticeably lower than those in the middle ($z = 6$ cm) and near the surface ($z = 11$ cm). While the bottom and middle velocities exhibit typical U-shape horizontal profiles, the surface velocities show a remarkable attenuation from the middle to both sides due to influence of shock waves, which are also reflected in the fluctuant measured data. Nevertheless, the overall accuracy of the simulation is considered acceptable to support the analysis of the flow pattern and energy dissipation characteristics.

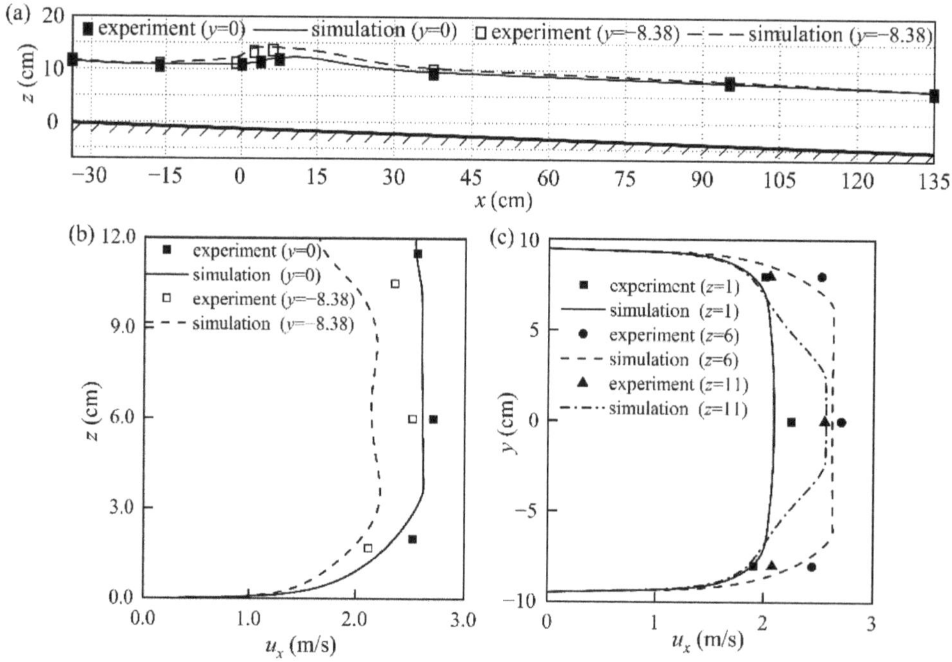

Figure 3. Comparison of the experimental and numerical results: (**a**) water surface profile; (**b**) vertical; and (**c**) horizontal distribution of the streamwise velocity u_x at x = 112.5 cm. The unit of the coordinates is cm.

4.2. Flow Pattern

Figure 4 shows the side view of the flow pattern downstream of three different deflectors. Lateral jets can be found to form behind the deflectors, and the jets travel further at higher elevations, leading to a non-uniform lateral cavity that is consequently longer and wider at higher elevations. The jet impacting the sidewall is deflected immediately and keeps rising, clinging to the sidewall, associated with noticeable turbulence and aeration. In contrast, the water below the impacting region remains non-aerated. The three deflectors also form different boundary lines between the white and black water regions. With type A, the boundary line is smooth, above which the deflected flow develops into water-wings and continually rises over the sidewalls. With type B, the boundary line is fluctuant, and the deflected flow develops into water-wings that intermittently jump over the sidewall, and a few bubbles could be found in the lower clear water. As for type C, neither distinguished boundary line nor water-wings can be found since the aeration in the lower region is also noticeable. It appears that the deflected flow is submerged in the lateral jet and induces stronger turbulence and intensive aeration since a larger amount of air bubbles can be seen at the impacting region throughout the flow depth. This is probably caused by the rapid close of the lateral cavity and the stronger interaction between the jet and the deflected flow.

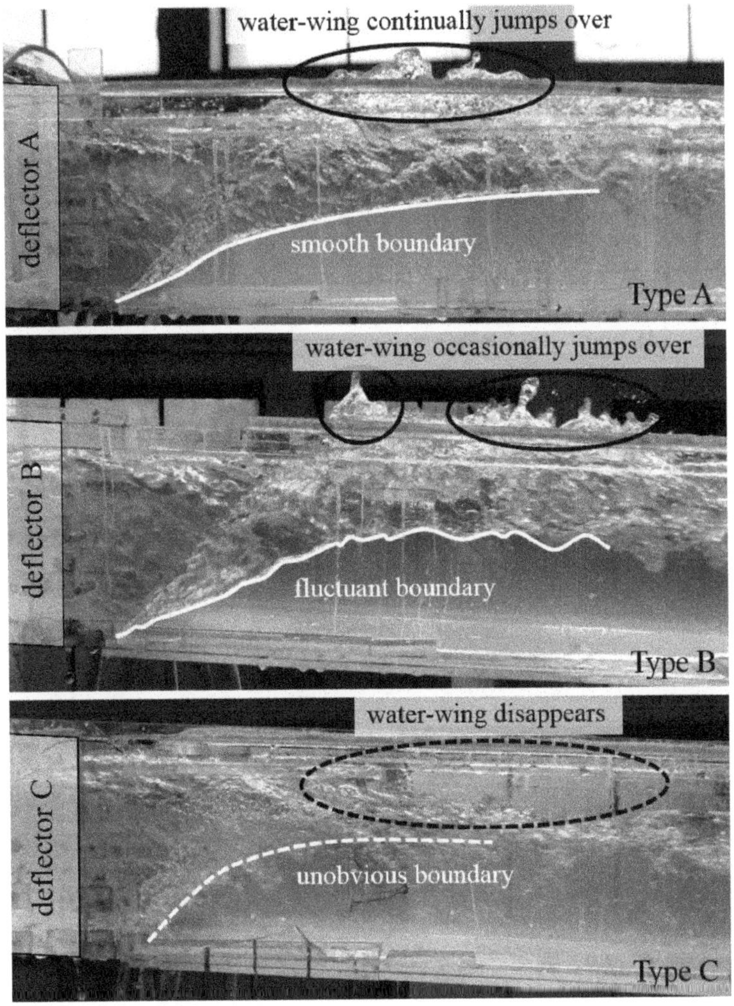

Figure 4. Side view of the flow pattern downstream of the deflectors.

The jet trajectories behind three different deflectors are marked with white lines and shown in Figure 5. The detaching jet behind deflector A moves in the direction tangent to the hypotenuse, leading to the widest and largest lateral cavity, which further leaves enough space for the lower deflected flow to develop to the upper region without contacting the jet trajectory. However, the jet behind deflector B exhibits a fluctuating phenomenon and forms an unstable cavity smaller and narrower than that behind deflector A. The fluctuating phenomenon can be mainly ascribed to the continuously varying edge slope of the deflector since it could yield non-uniform velocity distribution that exacerbates jet surface fluctuations. A benefit from the additional straight line in comparison to deflector B, the flow at the tail of deflector C has developed to a more uniform condition before detaching away from the deflector. It exhibits a trajectory almost parallel to the sidewall right behind the deflector, forming the narrowest and smallest lateral cavity. The rapidly closed cavity blocks the rising passage of the underneath deflected flow, preventing the formation of water-wings.

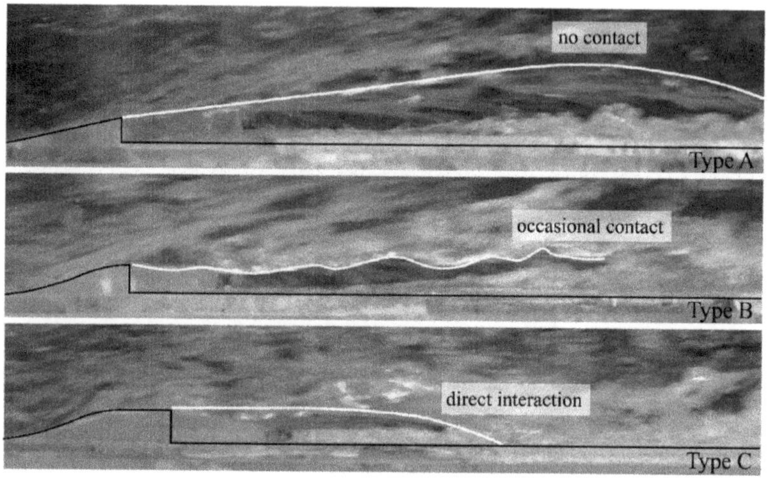

Figure 5. Top view of the flow pattern near the lateral cavity.

The development of water-wings and the subsequent shock waves downstream of the three deflectors are shown in Figure 6. It can be discovered that deflectors A and B form shock waves in bundle shape and diamond shape, respectively, whereas no shock wave is observed downstream of deflector C. With deflector A, the water-wings from both sides quickly rise up and detach from the main flow, jetting to the middle and colliding with each other in the air, and then fall back to the main flow with a certain vertical velocity and the lateral momentum dissipated. The returned water further produces disturbance on the surface and thereby inducing a bundle-shaped shock wave. With deflector B, the lateral momentum of the water-wings partially counteracts the jet in the cavity, and thus the water-wings fail to contact each other before falling back to the main flow. The returned water still contains certain lateral momentum and moves together with the main flow, producing two repeated reverse oblique developing lines on the flow surface and finally developing into a diamond-shaped shock wave. As for deflector C, the deflected flow is restricted in the relatively small cavity and thus only forms small water crowns, which are restricted in the vicinity of the sidewall and disappear further downstream.

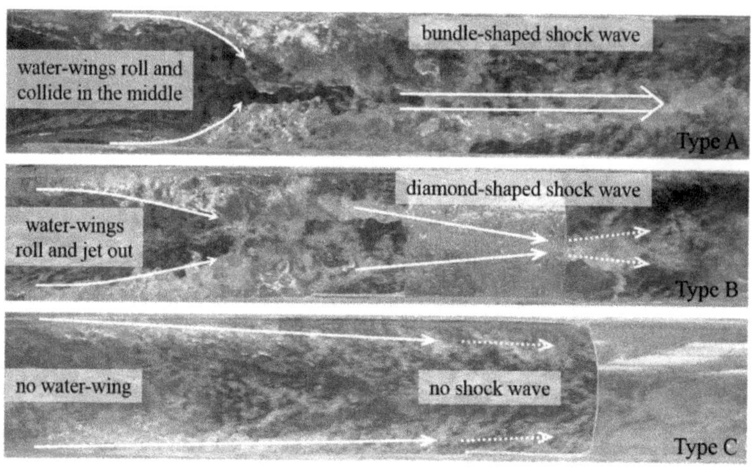

Figure 6. Top view of the flow pattern downstream of the deflectors.

Based on the above observations, it can be concluded that a continuous variation of the lateral deflector surface at the tail with an additional flow guiding extension is the key to the elimination of the water-wings and shock waves.

4.3. Velocity Distribution

The near-bottom (z = 1 cm) velocity distribution around three deflectors is comparatively shown in Figure 7. The jets behind deflectors A and B are firstly contracted and reach the maximum cavity width at approximately 1/3 of the cavity length. After that, the jets restart to spread to the sidewall. In comparison, the jet behind deflector C continuously moves close to the sidewall. Consequently, deflector A exhibits the largest cavity size, followed by deflectors B and C, which is in agreement with the experimental observation. It is worth emphasizing that the velocity vector distribution around deflector B is apparently more non-uniform compared to those around deflectors A and C, which is the reason for the fluctuating phenomenon of the jet surface. Moreover, the maximum velocity behind deflector C is about 0.2 m/s lower than those behind the other two deflectors, and the near-wall low-velocity region (i.e., the purple dashed rectangle in Figure 7) behind deflector C is obviously larger than those behind deflectors A and B. These phenomena all imply more effective energy dissipation of deflector C compared to A and B.

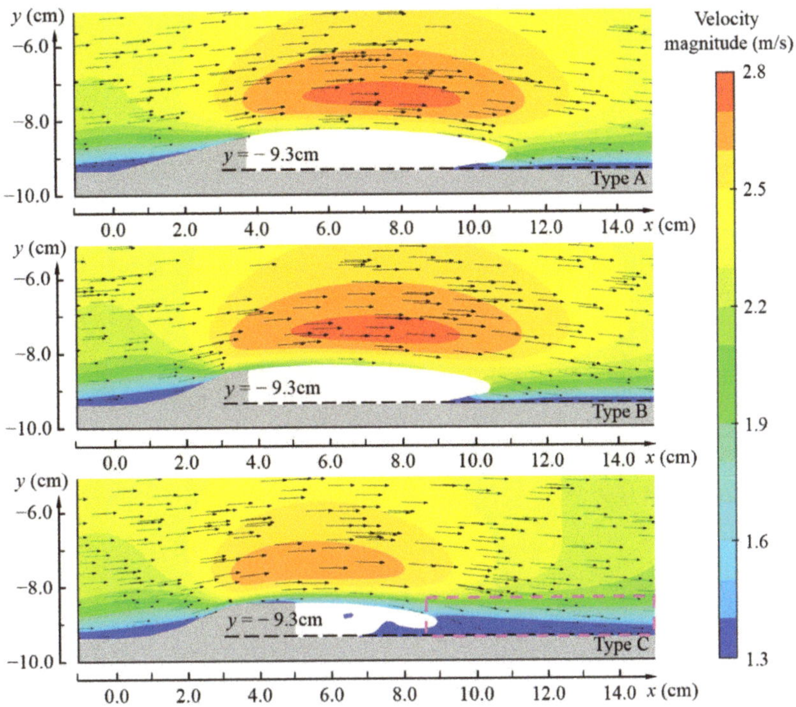

Figure 7. Horizontal velocity distribution around the deflectors at z = 1.0 cm. The vectors are generated using the same density and scale for the three types of deflectors.

The near-wall ($y = -9.3$ cm) vertical distribution of velocity behind the deflectors is comparatively shown in Figure 8 to illustrate the kinematic characteristics of the deflected flow. Distinguished by the velocity magnitude and vector direction, the longitudinal flow behind deflectors A and B can be divided into three distinct regions: the main flow region that flows downstream, the water-wing region that rises up while flowing downstream and the impacting region that exhibits the highest velocities and separating the above two regions. The rising water-wing regions are in agreement with the experimentally observed boundary lines. As for deflector C, the flow direction is almost not affected by the jet since only a small region can be found that has upward velocities and therefore is the underlying reason for the fast recovery of the rising surface and the elimination of shock waves. Moreover, the velocities behind deflector C are lower than those behind deflectors A and B, especially dropping about 50% at the impacting region. This again confirms the more effective energy dissipation of deflector C.

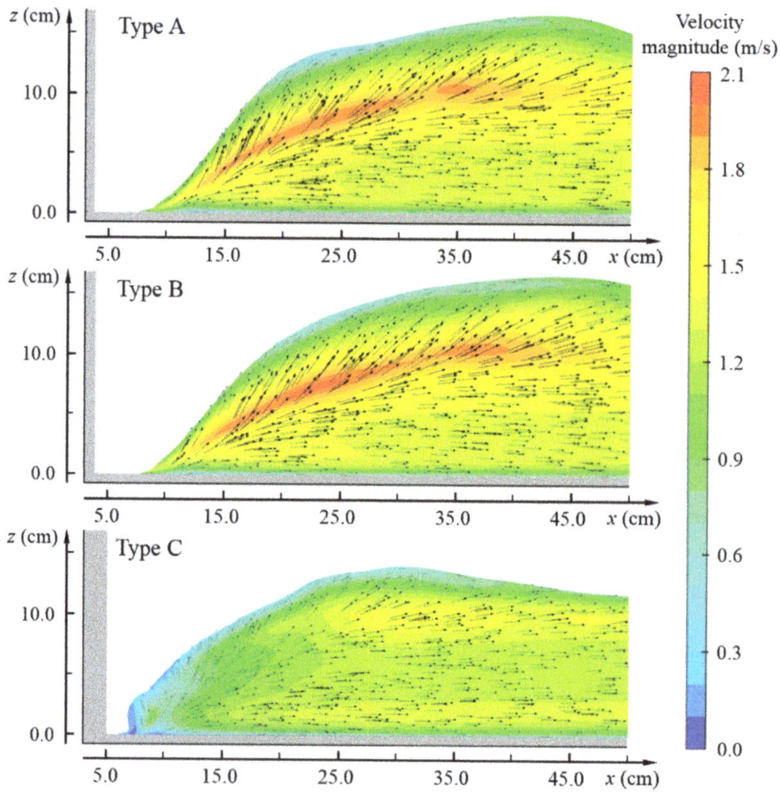

Figure 8. Velocity distribution near the sidewall ($y = -9.3$ cm). The vectors are generated using the same density and scale for the three types of deflectors.

Figure 9 shows the streamwise development of the cross-sectional velocity (CSV) with three different and without deflectors, which is computed from 17 flux surfaces with a spacing of 7.5 cm. The no-deflector scheme shows a linearly increased CSV, whereas for the other three schemes, the CSVs increase at the deflector region and decrease at the cavity region because of lateral flow contractions and expansions. With deflectors A and B, the CSVs first increase in the front of the cavity region and then decrease, whereas the CSVs of the flow behind deflector C exhibits a consistent decreasing trend due to continuous flow expansion. Affected by the shock waves, the CSVs of the downstream flow with deflectors A and B exhibit some fluctuations, and the fluctuating amplitude of deflector B is higher than that of deflector A. In contrast, the CSVs of deflector C feature a linear increase behavior similar to that of the no-deflector scheme in this region, although with lower values. Moreover, it can be found that the efficient energy dissipation mainly occurs at the impacting region (i.e., approximately $15 < x < 40$ cm).

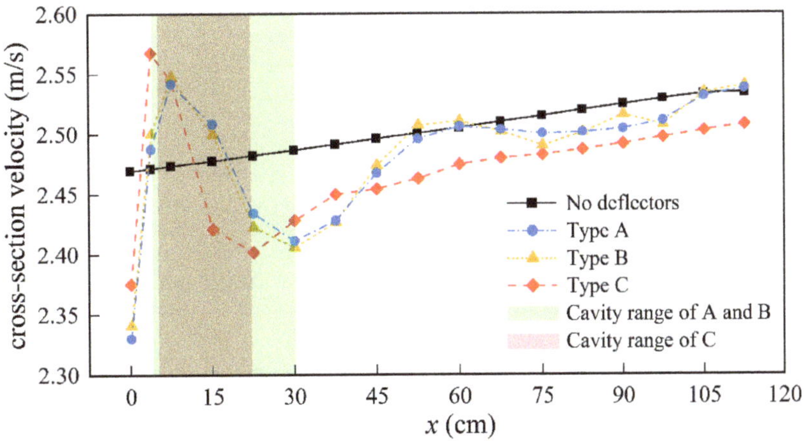

Figure 9. Streamwise development of the cross-sectional velocity along the flume with three different deflectors and without deflectors. The brown color indicates regions where the cavity range of C (pink) overlaps those of A and B (green).

4.4. Energy Dissipation Characteristics

Figure 10 shows the streamwise development of the flux-averaged hydraulic head (FAHH) along the flume with three different deflectors and without deflectors. In FLOW-3D, the FAHH is recommended to evaluate the hydraulic head where significant changes occur. It is calculated using the following equation:

$$h = \frac{\int \left(\frac{u^2}{2g} + \frac{p}{\rho g} + z \right) ds}{\int \varphi ds} \tag{6}$$

here, h is the FAHH, φ is the flux across a cell surface, and ds is the open area of the cell face.

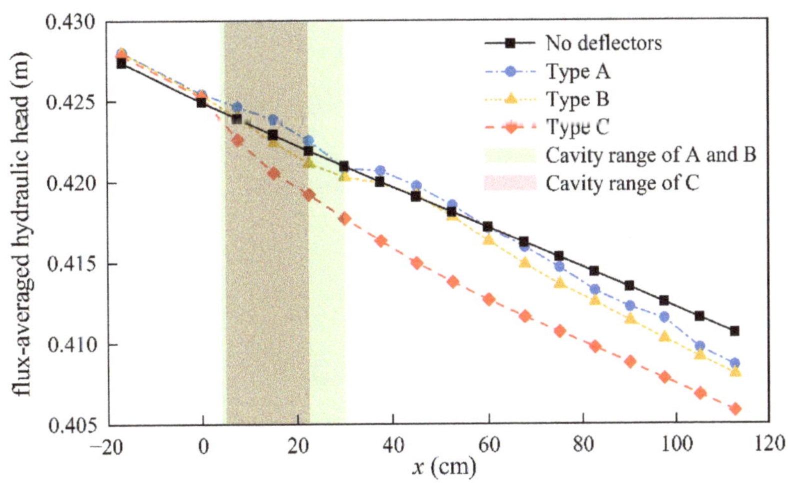

Figure 10. Streamwise development of the flux-averaged hydraulic head (FAHH) along the flume with three different and without deflectors. The brown color indicates regions where the cavity range of C (pink) overlaps those of A and B (green).

The FAHH decreases linearly in the flume without lateral deflectors. The FAHH of the flow underwent some certain fluctuations around the no-deflector values when passing deflectors A and B. These fluctuations continue until they are twice the cavity length downstream (i.e., to approximately $x = 60$ cm). Further downstream, the FAHH values of the flow behind deflectors A and B turn out to be lower than the no-deflector values. In contrast, a rapid decrease in FAHH can be found at the deflector and cavity region ($0 < x < 20$) for the flume equipped with deflector C, and further downstream ($x \geq 45$), the values of FAHH decrease steadily with a similar rate as those of the no-deflector scheme. A quantitative comparison of the energy dissipation effect of the three deflectors can be achieved using the local head loss coefficient $\zeta = (\mathrm{FAHH}_1 - \mathrm{FAHH}_2) \cdot 2g/v^2$ between $x = -17$ cm and $x = 45$ cm with v being a reference CSV calculated at $x = -17$ cm. The local head loss coefficients for deflectors A, B and C are 2.72%, 2.90% and 4.2%, respectively. This behavior is in agreement with the streamwise development of the CSV and confirms the kinetic energy is mainly dissipated at the deflector and jet region.

To further analyze the energy dissipation mechanism of the lateral deflectors, the distribution of turbulent kinetic energy k and turbulence dissipation rate ε in the deflector and jet region are comparatively shown in Figure 11. The most noticeable difference in Figure 11 is the larger k and ε values at the impacting region behind deflector C compared to those behind deflectors A and B. This implies more mean kinetic energies are turned into turbulent energies and are then dissipated for the flow behind deflector C. This is consistent with the intensive aeration observed in the flow photos shown in Figure 6, as higher turbulent levels induce more intensive aeration [44–46].

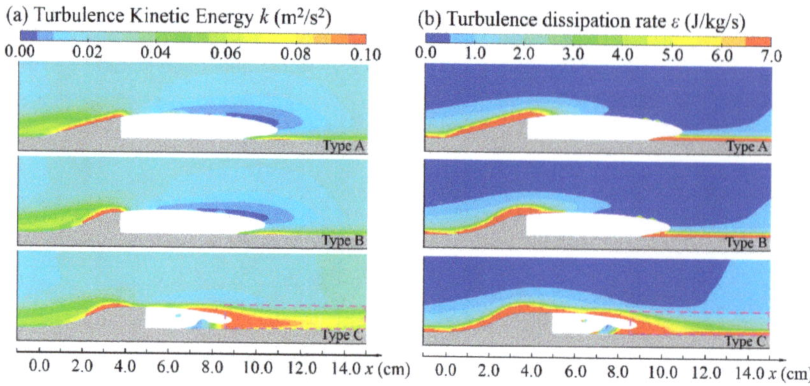

Figure 11. Horizontal distribution of (**a**) turbulent kinetic energy k; (**b**) turbulence dissipation rate ε in deflector and cavity region at $z = 1$ cm.

5. Conclusions

In this paper, the hydraulic characteristics of lateral deflectors with three different geometries in gentle-slope free-surface tunnels were investigated using the hybrid approach of model test and numerical simulation. Special focus was placed on the flow pattern and energy dissipation features. The main findings are:

1. The cavity formed behind lateral deflectors usually features a right-angled trapezoid shape with a larger streamwise length at higher elevations because of non-uniform velocity distributions. This makes the deflected flow rise up along the impacting region inside the cavity and potentially induced shock waves depending on the interaction of the rising up water-wings and the jet surfaces.
2. The traditional triangular deflector forms an adequately wide cavity that allows for the free rising up of the water-wings inside the cavity, which further contributes to the development of the buddle-type shock wave, whereas the two-arc deflector yields a jet with fluctuating surface, resulting from the non-uniform planar velocity distribution

caused by the continuously varying curvature of the arcs. Water-wings also develop inside the cavity and eventually produce a diamond-type shock wave downstream. In contrast, the jet behind the two-arc deflector with a straight guiding line at the tail is stabler and travels a shorter distance before impacting the side wall. The jet could thus restrict the development of the rising flow, and thereby eliminate the formation of water-wings and shock waves. Based on these observations, it is concluded that a continuous variation of the lateral deflector surface at the tail with an additional flow guiding extension is the key to the elimination of the water-wings and shock waves.
3. Compared to the triangular deflector and the two-arc deflector, the two-arc deflector with a straight line exhibits more effective energy dissipation, as reflected in the local energy loss coefficient. The underlying reason for its effective energy dissipation is the more intensive turbulence introduced by the stronger interaction between the deflected flow and the jet surface, which also leads to more intensive aeration.

Compared to real-world engineering problems, the present study has some limitations. The physical model is roughly 1/24~1/80 of real-world free-surface tunnels, and thus scale effect is expected, in particular for air–water two-phase flow characteristics [47–49]. However, the scale of the physical model is considered acceptable with regards to the main concern of the current paper, which is the comparative evaluation of the different lateral deflectors in terms of the flow pattern and the energy dissipation behaviors. Particularly, the jet trajectory behind the lateral deflector and its interaction with the deflected flow could be decently reproduced in the physical model. Moreover, considering that the numerical simulation approach remains immature for highly aerated flows but much more reliable for non-aerated flow [13,50,51], the hybrid approaches used in this paper could offer valuable insight into the underlying reason for the flow pattern improvement and higher energy dissipation observed with deflector type C. Aeration is another key concern for lateral deflector designs but is not investigated in detail in this paper due to facility reasons. Nonetheless, the findings from the study provide a novel and preferable lateral deflector design for gentle-slope free-surface tunnels, which could, if not resolve, significantly improve the unwanted flow patterns of water-wings and shock waves. Meanwhile, it could also achieve improved energy dissipation compared to traditional alternatives. The aeration characteristics of the novel deflector will be investigated in future research.

Author Contributions: The conceptualization of this research was directed by J.W. and the experimental data were from J.D. and S.D.; the numerical modeling was performed by J.D. under the supervision of J.W. and Z.D.; the validation is carried by J.D. and S.D.; the manuscript was mainly finished by J.D., Z.D. and J.W. All authors have read and agreed to the published version of the manuscript.

Funding: This research was funded by the National Key Research and Development Program of China (Grant No. 2021YFC3090105), the China Postdoctoral Science Foundation (No. 2021M692754), and Guizhou Science and Technology Joint Support (2019) No. 2890.

Institutional Review Board Statement: Not applicable.

Informed Consent Statement: Not applicable.

Data Availability Statement: Not applicable.

Conflicts of Interest: The authors declare no conflict of interest.

Appendix A

A grid convergence study involving three mesh schemes was carried out to evaluate the grid quality. The cell size ($x \times y \times z$) in the refined region of the three mesh schemes was 0.625 cm × 0.313 cm × 0.625 cm (coarse), 0.375 cm × 0.188 cm × 0.375 cm (medium), and 0.188 cm × 0.094 cm × 0.188 cm (fine), respectively.

The grid spacing of the 3D grids is defined as $h_k = \sqrt[3]{h_{x,k} \cdot h_{y,k} \cdot h_{z,k}}$. The computational solution f for a grid k is defined as the computed maximum streamwise velocities $u_{\max}$ in the middle longitudinal section at x = 112.5 cm. The *GCI* for each grid k is computed as $GCI_k = F_s|E_k|$, where a value of 1.25 is chosen for the factor of safety F_s, following the recommendations of Roache [52].

Table A1 shows the grid convergence index (GCI) calculated using the approach of Roache and Salas [52]. The observed order of accuracy is p = 1.84, and the *GCI* shows a convergence trend. The fine grid takes a very satisfactory value of 0.224%, which is an order of magnitude higher than the former and therefore is considered fine enough to resolve the flow field.

Table A1. Parameter of grid convergence calculation.

Mesh	h/cm	r	$p(u_{max})$	$f(u_{max})/(m \cdot s^{-1})$	$\varepsilon(u_{max})$/%	$GCI(u_{max})$/%
coarse	0.496	-		2.607	-	-
medium	0.297	1.67	1.84	2.633	0.997	2.040
fine	0.149	1.99		2.621	0.456	0.224

The vertical distributions of streamwise velocity u_x and turbulent kinetic energy k in the middle longitudinal section at x = 112.5 cm are comparatively shown in Figure A1. From Figure A1, it can be found that the main deviations of the u_x profiles lie in the region close to the water surface and the bottom wall, whereas in the middle, all three meshes return almost identical velocity values. Moreover, the difference between the medium and fine meshes is less obvious compared to that between the coarse and the medium meshes. As for the k profile, the result from the coarse mesh is strikingly different from those from the medium and fine meshes, the difference between which is relatively negligible. Therefore, the fine grid was selected for the simulation.

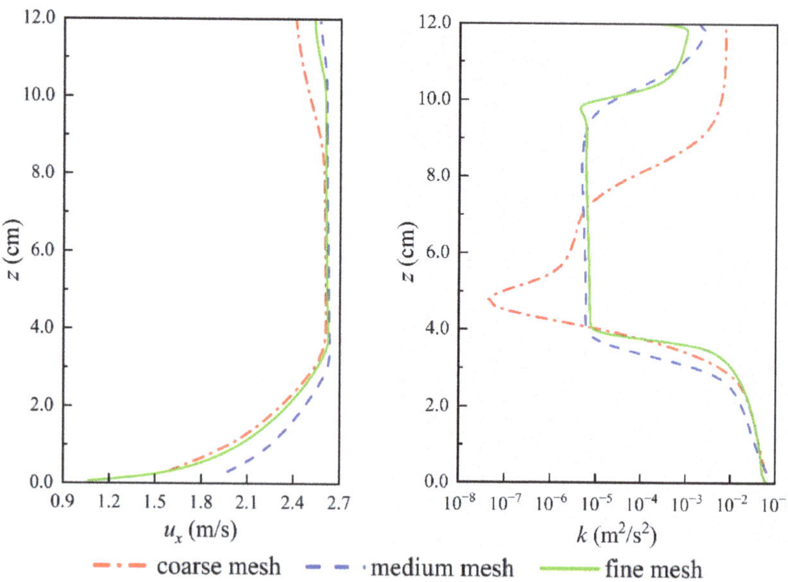

Figure A1. Vertical distributions of streamwise velocity u_x and turbulent kinetic energy k in the middle longitudinal section at x = 112.5 cm.

References

1. Xie, Z.S.; Wu, Y.H.; Chen, W.X. New technology and innovation on flood discharge and energy dissipation of high dams in China. *J. Hydraul. Eng.* **2016**, *47*, 324–336. (In Chinese) [CrossRef]
2. Ma, H.; Chi, F. Major technologies for safe construction of high earth-rockfill dams. *Engineering* **2016**, *2*, 498–509. [CrossRef]
3. Ma, H.; Chi, F. Technical progress on researches for the safety of high concrete-faced rockfill dams. *Engineering* **2016**, *2*, 332–339. [CrossRef]

4. Fan, L.; Zhang, H.W.; Liu, Z.P.; Zhang, D. Numerical study on hydraulic characteristic of free surface flow in spillway tunnel with different configuration. *J. Hydroelectr. Eng.* **2009**, *28*, 126–131. (In Chinese)
5. Guo, J.; Zhang, D.; Liu, Z.P.; Fan, L. Achievements on hydraulic problems in large spillway tunnel with a high head and large discharge flow and its risk analysis. *J. Hydraul. Eng.* **2006**, *37*, 1193–1198. (In Chinese) [CrossRef]
6. Pang, C.J.; Yuan, Y.Z. Research on type of aerator in large gentle-slope free-surface 'dragon-head-up' spillway tunnel. *J. Hydraul. Eng.* **1993**, *6*, 61–66. (In Chinese) [CrossRef]
7. Bai, R.; Liu, S.; Wang, W.; Zhang, F. Experimental investigation of the dissipation rate in a chute aerator flow. *Exp. Therm. Fluid Sci.* **2019**, *101*, 201–208. [CrossRef]
8. Suo, L.S.; Liu, N. *Handbook of Hydraulic Structure Design*, 2nd ed.; China Waterpower Press: Beijing, China, 2014. (In Chinese)
9. Chanson, H. Study of air entrainment and aeration devices. *J. Hydraul. Res.* **1989**, *27*, 301–319. [CrossRef]
10. Du, L.Q.; Tang, Y.Q.; Wang, J.S. Construction technology of slip form for bottom plate of turning point section of spillway tunnel of Wudongde Hydropower Station. *Yangtze River* **2020**, *51*, 193–196. (In Chinese) [CrossRef]
11. Suo, H.M.; Zhou, Z.J.; He, L.; Zhang, J.M. Study on type of tunnel spillway in Shuangjiangkou Hydropower Station. *Des. Hydroelectr. Power Stn.* **2021**, *37*, 9–12. (In Chinese) [CrossRef]
12. Wang, H.Y.; Dai, G.Q.; Yang, Q.; Liu, C. Experimental study on V-type aerator for spillway tunnel with inlet raised. *J. Hydraul. Eng.* **2005**, *36*, 102–105. (In Chinese) [CrossRef]
13. Yang, J.; Teng, P.; Zhang, H. Experiments and cfd modeling of high-velocity two-phase flows in a large chute aerator facility. *Eng. Appl. Comp. Fluid Mech.* **2019**, *13*, 48–66. [CrossRef]
14. Hager, W.H.; Boes, R.M. Hydraulic structures: A positive outlook into the future. *J. Hydraul. Res.* **2014**, *52*, 299–310. [CrossRef]
15. Li, S.; Zhang, J.; Xu, W.; Chen, J.; Peng, Y.; Li, J.; He, X. Simulation and experiments of aerated flow in curve-connective tunnel with high head and large discharge. *Int. J. Civ. Eng.* **2016**, *14*, 23–33. [CrossRef]
16. Juon, R.; Hager, W.H. Flip bucket without and with deflectors. *J. Hydraul. Eng.* **2000**, *126*, 837–845. [CrossRef]
17. Wang, H.Y.; Yang, Q.; Dai, G.Q.; Liu, C. Study on hydraulic characteristic of aerator with side-walls sudden enlargement in free-flow spillway tunnel. *Adv. Mater. Res.* **2012**, *614*, 613–616. [CrossRef]
18. Ma, X.; Dai, G.; Yang, Q.; Li, G.; Zhao, L. Analysis of influence factors of cavity length in the spillway tunnel downstream of middle gate chamber outlet with sudden lateral enlargement and vertical drop aerator. *J. Hydrodyn.* **2010**, *22*, 657–663. [CrossRef]
19. Pfister, M.; Hager, W.H. Chute aerators. II: Hydraulic design. *J. Hydraul. Eng.* **2010**, *136*, 360–367. [CrossRef]
20. Nie, M.; Wang, X.; Wu, G. Effect of lateral deflector on outlet cavity lengths. *J. Hydraul. Eng.* **2003**, *129*, 536–540. [CrossRef]
21. Xu, J.; Wu, J.; Peng, Y.; Ma, F. Fin performance of 3-d aerator devices with backward lateral deflectors. *J. Hydrodyn.* **2020**, *32*, 410–413. [CrossRef]
22. Siying, W.; Jinbo, Y.; Caihuan, W. Study on the aerator of spillway tunnel in one large-scale project. In Proceedings of the 35th World Congress of the International-Association-for-Hydro-Environment-Engineering-and-Research (IAHR), Chengdu, China, 8–13 September 2013.
23. Ye, F.; Xu, W.; Wei, W. Experimental study of the air concentration diffusion in aerated chute flows downstream of lateral and bottom aerators. *AIP Adv.* **2022**, *12*, 25221. [CrossRef]
24. Nie, M.X.; Wu, G.G. Effect of the lateral deflector on cavity design. *J. Tsinghua Univ.* **2002**, *42*, 546–550. (In Chinese) [CrossRef]
25. Nie, M. Comparison of energy dissipation with and without aerators. *Tsinghua Sci. Technol.* **2000**, *5*, 222–225.
26. Lucas, J.; Hager, W.H.; Boes, R.M. Deflector effect on chute flow. *J. Hydraul. Eng.* **2013**, *139*, 444–449. [CrossRef]
27. Nie, M.X.; Duan, B.; Li, L.L. Hydraulic characteristics along the sidewall of a sluice downstream of a sudden lateral enlargement and vertical drop. *J. Tsinghua Univ.* **2006**, *46*, 1969–1972. (In Chinese) [CrossRef]
28. Liu, C.; Zhang, G.K.; Li, N.W.; Zhang, G.B.; Wang, H.Y. The effect of base aerator on the length of lateral cavity. *Adv. Eng. Sci.* **2008**, *40*, 1–4. (In Chinese)
29. Wu, G.L.; Ma, F.; Wu, J.H.; Xu, J.R.; Peng, Y.; Wang, Z. Experimental investigation of flow regime control for dragon-drop-tail section of the spillway tunnel. *Chin. J. Hydrodyn.* **2015**, *30*, 140–145. (In Chinese) [CrossRef]
30. Wu, J.; Li, D.; Ma, F.; Qian, S. Fin characteristics of aerator devices with lateral deflectors. *J. Hydrodyn.* **2013**, *25*, 258–263. [CrossRef]
31. Zhang, H.W.; Liu, Z.P.; Zhang, D.; Wu, Y.H. Study on hydraulic characteristics of sidewall aerators in high-head large-discharge spillway tunnel. *J. Hydroelectr. Eng.* **2015**, *34*, 111–116. (In Chinese) [CrossRef]
32. Li, F.T.; Liu, P.Q.; Xu, W.L.; Tian, Z. Experimental study on effect of flaring piers on weir discharge capacity in high arch dam. *J. Hydraul. Eng.* **2003**, *11*, 43–47. (In Chinese) [CrossRef]
33. Zheng, X.; Zhang, L.; Wu, S.; Song, K. Study on the shape of the aerator of high-head discharge tunnel with mild bottom slope. *Water-Sui.* **2021**, *13*, 2128. [CrossRef]
34. Li, S.; Zhang, J.; Xu, W.; Chen, J.; Peng, Y. Evolution of pressure and cavitation on side walls affected by lateral divergence angle and opening of radial gate. *J. Hydraul. Eng.* **2016**, *142*, 05016003. [CrossRef]
35. Yang, B.; Wang, Z.; Wu, M. Numerical simulation of naval ship's roll damping based on cfd. *Procedia Eng.* **2012**, *37*, 287–293. [CrossRef]
36. Zhang, H.; Zhang, L.; Wu, S.; Wang, F.; Zhan, Z.; Zheng, X.; Zhang, H.; Bao, W. Research on flood discharge and energy dissipation of a tunnel group layout for a super-high rockfill dam in a high-altitude region. *Water* **2021**, *13*, 3408. [CrossRef]
37. Zhang, W.; Wang, J.; Zhou, C.; Dong, Z.; Zhou, Z. Numerical simulation of hydraulic characteristics in a vortex drop shaft. *Water* **2018**, *10*, 1393. [CrossRef]

38. Pfister, M.; Hager, W.H. Chute aerators. I: Air transport characteristics. *J. Hydraul. Eng.* **2010**, *136*, 352–359. [CrossRef]
39. Feurich, R.; Rutschmann, P. Efficient design optimization and investigation by combining numerical and physical models. In Proceedings of the Korea Water Resources Association Conference, Korea Water Resources Association, Seoul, Korea, 11–16 September 2005.
40. Johnson, M.C.; Savage, B.M. Physical and numerical comparison of flow over ogee spillway in the presence of tailwater. *J. Hydraul. Eng.* **2006**, *132*, 1353–1357. [CrossRef]
41. Stamou, A.I.; Chapsas, D.G.; Christodoulou, G.C. 3-d numerical modeling of supercritical flow in gradual expansions. *J. Hydraul. Res.* **2008**, *46*, 402–409. [CrossRef]
42. Hirt, C.W.; Nichols, B.D. Volume of fluid (vof) method for the dynamics of free boundaries. *J. Comput. Phys.* **1981**, *39*, 201–225. [CrossRef]
43. Yakhot, V.; Orszag, S.A.; Thangam, S.; Gatski, T.B.; Spezialea, C.G. Development of turbulence models for shear flows by a double expansion technique. *Phys. Fluids A Fluid Dyn.* **1992**, *4*, 1510–1520. [CrossRef]
44. Dong, Z.; Bürgler, M.; Hohermuth, B.; Vetsch, D.F. Density-based turbulence damping at large-scale interface for reynolds-averaged two-fluid models. *Chem. Eng. Sci.* **2022**, *247*, 116975. [CrossRef]
45. Hirt, C.W. *Modeling Turbulent Entrainment of Air at a Free Surface*; Technical Note FSI-01-12; Flow Science, Inc.: Santa Fe, NM, USA, 2003.
46. Dong, Z.; Wang, J.; Vetsch, D.F.; Boes, R.M.; Tan, G. Numerical simulation of air–water two-phase flow on stepped spillways behind x-shaped flaring gate piers under very high unit discharge. *Water* **2019**, *11*, 1956. [CrossRef]
47. Pfister, M.; Hager, W.H. Deflector-jets affected by pre-aerated approach flow. *J. Hydraul. Res.* **2012**, *50*, 181–191. [CrossRef]
48. Pfister, M.; Lucas, J.; Hager, W.H. Chute aerators: Preaerated approach flow. *J. Hydraul. Eng.* **2011**, *137*, 1452–1461. [CrossRef]
49. Pfister, M.; Chanson, H. Scale effects in physical hydraulic engineering models. *J. Hydraul. Res.* **2011**, *49*, 293–306. [CrossRef]
50. Hohermuth, B.; Schmocker, L.; Boes, R.M.; Vetsch, D.F. Numerical simulation of air entrainment in uniform chute flow. *J. Hydraul. Res.* **2021**, *59*, 378–391. [CrossRef]
51. Hohermuth, B.F. Aeration and Two-Phase Flow Characteristics of Low-Level Outlets. Ph.D. Thesis, ETH Zurich, Zürich, Switzerland, 2019.
52. Roache, P.J. Quantification of uncertainty in computational fluid dynamics. *Annu. Rev. Fluid Mech.* **1997**, *29*, 123–160. [CrossRef]

water

Article

SWAN Modeling of Dredging Effect on the Oued Sebou Estuary

Nisrine Iouzzi [1,*], Laila Mouakkir [1], Mouldi Ben Meftah [2,*], Mohamed Chagdali [1] and Dalila Loudyi [3]

1 Faculty of Sciences Ben M'Sik, Hassan II University of Casablanca, Casablanca 20450, Morocco
2 Department of Civil, Environmental, Land, Building Engineering and Chemistry, Polytechnic University of Bari, Via E. Orabona 4, 70125 Bari, Italy
3 Faculty of Sciences and Techniques, Hassan II University of Casablanca, Mohammedia 28810, Morocco
* Correspondence: iouzzi.nisrine@gmail.com (N.I.); mouldi.benmeftah@poliba.it (M.B.M.); Tel.: +39-080-5963508 (M.B.M.)

Abstract: The estuary ecosystem's health and ecological integrity are essential for preserving environmental quality, habitats, and economic activity. The main objective of the present study is to comprehend the wave hydrodynamic impact on the Oued Sebou estuary, which is situated in the Kenitra region on Morocco's north Atlantic coast in North Africa. Specifically, it focused on the dredging effect (caused by sand extraction) on the wave motion and its impact on the estuary environment. Different scenarios of wave-propagation simulations were carried out, varying the significant wave height, in deep water (from 1.5 to 4 m), and considering the bathymetry before and after two dredging cases of 2- and 4-m depths. The change of wave height at the Oued Sebou estuary shoreline was simulated by using the third version of the Simulating Waves Nearshore Model (SWAN). The SWAN model formulates the wave evolution in terms of a spectral energy balance on a structured grid. The effect of dredging on the wave spreading in addition to the flow hydrodynamic structures were extensively analyzed. According to the simulated results, the dredging activities in the Oued Sebou estuary mainly affect the river mouth and the southern breakwater area, increasing the potential erosive action. The areas at the northern coastal strip and near the northern breakwater are subject to possible accumulation of sediments.

Keywords: Oued Sebou estuary; bathymetry; dredging; SWAN; wave spreading; flow hydrodynamic structures

Citation: Iouzzi, N.; Mouakkir, L.; Ben Meftah, M.; Chagdali, M.; Loudyi, D. SWAN Modeling of Dredging Effect on the Oued Sebou Estuary. *Water* **2022**, *14*, 2633. https://doi.org/10.3390/w14172633

Academic Editor: Giuseppe Oliveto

Received: 14 July 2022
Accepted: 21 August 2022
Published: 26 August 2022

1. Introduction

Estuaries are one of the most important interconnections between land and sea. They are frequently important areas for leisure and economic activities. Estuaries are also ecosystems that are very susceptible to pollution and environmental disturbances and must be adequately preserved. In estuaries, rivers and oceans interact through a number of intricate phenomena: exchange of water of different densities, sediments, pollutants, nutrients, organic matter, and biota. Therefore, it is necessary to have a better understanding of the hydrodynamics of estuaries and coasts in order to effectively assess and monitor the environment quality in these regions and to predict coastal evolution [1]. It takes appropriate methodologies combining theoretical analysis and modeling studies to anticipate the hydrodynamics of estuaries and coasts, which are caused by complicated mechanisms linking mass exchange to heat transfer processes [1].

Recently, several research studies were conducted to investigate the hydrodynamics of estuaries [2–12]. Many driving conditions, such as river flow characteristics, tidal amplitude, sediment properties, surges, waves, currents, and bathymetry, affect the dominant physical processes in an estuary. The primary sources of energy input into estuaries are typically tides, surges, and waves. Wave dispersion in shallow water is significantly impacted by the change in bathymetry, leading to refraction, diffraction, reflection, and shoaling. Natural beaches and manmade coastal structures that reflect waves can have a significant impact on the hydrodynamic structures and, consequently, the transport of sediment in front of the reflector [13]. According to Chang and Hsu [13], the prediction of wave reflection coefficients is still a challenging task. Before marine structures are created, wave reflection at them is typically evaluated with physical models for engineering practice. Incident waves are also reflected by the sloping bathymetry. The drag coefficient estimation is complicated by the superposition of incident and reflected waves (due to the variation in bathymetry and the existence of any natural or man-made structures). In their study, Chang and Hsu [13] compiled many approaches for predicting the reflection coefficient of the waves suggested in the literature and highlighted their limitations. Most of these methods, according to the authors, are ineffective at predicting the reflection coefficients of waves on a bed with arbitrary or natural bathymetry. Based on a linear wave shoaling theory (without considering wave breaking), determining changes in wave height and phase due to bathymetry variations, Chang and Hsu [13] proposed a simple frequency-domain method for separating incident and reflected waves to account for normally incident linear waves propagating on an inclined bed with arbitrary 2D bathymetry. According to the authors, their method is applicable to both laboratory and field conditions predominantly for normal shores on which deep-water waves are propagated.

In intermediate and shallow waters, waves from deep waters begin to interact with the seabed, undergoing various transformations. These transitions are the results of many phenomena, for example, refraction, diffraction, reflection, shoaling, breaking, and swash. Wave prediction is of crucial importance to human activities as it provides useful information for many coastal engineering applications such as the coastal protection, environmental monitoring, navigation, safe port management, and good control of recreational activities in popular coastal areas [14]. Both observational and measurement data, as well as physical and numerical models, are often used to assess extreme marine events around the world [15]. Chi and Rong [15] confirmed that long-term in situ observations can accurately estimate the sea level variation but are usually spatially limited. The authors also indicated that many numerical models have been developed and validated to reproduce the spatial and temporal variation of wave spreading and uncover the underlying dynamics. The sea wave motion is strongly nonlinear and greatly influenced by many factors, i.e., the seabed, wind velocity, current circulations, and induced radiation stress. The interaction of the waves with different factors leads to complex hydrodynamic behaviors, making their numerical simulation very uncertain, which always requires validations using measured data.

The Simulating Waves Nearshore (SWAN) model is one of the most very popular wind–wave models [16–27], used by many government organizations, research institutes and consulting companies worldwide, to predict the size and forces of waves, allowing for changes in wave propagation from deep water to the surf zone [14]. Based on SWAN manuals and as outlined in Lin [14], the model's primary function is to resolve the spectral action balance equations, which represent the effects of spatial propagation, refraction, shoaling, generation, dissipation and nonlinear wave–wave interactions, without imposing any a priori constraints on the spectral evolution of wind waves. Hoque et al. [16] applied the SWAN model to forecast storm wave conditions in the shallow nearshore region off the Mackenzie Delta in the Canadian Beaufort Sea. The standard setup for the SWAN model was implemented by the authors, comparing different methods for quantifying the wave whitecapping dissipation. Hoque et al. [16] found that, after examining the bottom friction effects and triad interactions in predicting shallow-water waves, the simulated results in terms of significant wave heights and peak period are in good agreement with the observed data. Amunugama et al. [17] analyzed wind waves with SWAN on structured mesh and unstructured mesh during the arrival of a typhoon. After comparing the simulated results with measured data, they confirmed that the wind–wave characteristics obtained by both approaches (structured SWAN and unstructured SWAN) were substantially consistent with some advantages of unstructured SWAN, especially in complex geometries. The authors recommended the combination of both approaches where necessary. Due to a lack of time-series wave data, Gorman et al. [19] have used SWAN and wind data to hindcast the generation and propagation of deep-water waves incident on the New Zealand coast over a 20-year period (1979–1998). The SWAN model was also used to analyze spatial and temporal variations in cold front events [20], predict waves generated by cyclones [21], simulate wave characteristics at shorelines [22,24], and estimate wave energy potential [23], etc.

This study aims at investigating the effect of extreme dredging depths (due to sand extraction) on the wave characteristics in the Oued Sebou estuary, located in the Kenitra region on the north Atlantic coast of Morocco by applying the SWAN model (Cycle III version 41.20). The wave hydrodynamic characteristics around the river outlet were examined. The dredging impact on wave dispersion, bottom friction velocity field, flow agitation in the river, and stress on the structures housing the estuary and wave energy budget (before breaking) were extensively analyzed.

2. Study Area

This study covers the Oued Sebou estuary area. The Sebou estuary is located on the Atlantic coast in the Kenitra region of Morocco in North Africa. It is a region where numerous coastal developments are located. Two concentrically spaced longitudinal breakwaters channel the Oued Sebou over a distance of about 800 m, extending to a bathymetric line of 7 m. The sandbar that prevents shipping in the canal was reduced by the construction of the breakwaters in 1931 (Figure 1). During low tide, these structures give the tidal current the ability (velocities and forces) to drive the sand along the shoreline in either direction. Additionally, this river is controlled by water agencies that guarantee its stability. The Mehdia beach (of the city side in the southern part of the river mouth) is surrounded by a corniche composed of a wall with an average height of 1.50 m and 2 m for the northern and southern halves, respectively. In the southern and northern beach zones, respectively, the wall's height above sea water level (SWL) is nearly 11 and 13 m. It is also important to note that the Mehdia shoreline dune is occupied by various manmade structures.

Dredging operations were carried out every year with an average volume of almost 460,000 m^3/year. This makes it possible to ensure adequate depth, which is necessary for the navigation of the fishing vessels and ships in direction of the Mehdia ports, located inside the channel (Figure 1).

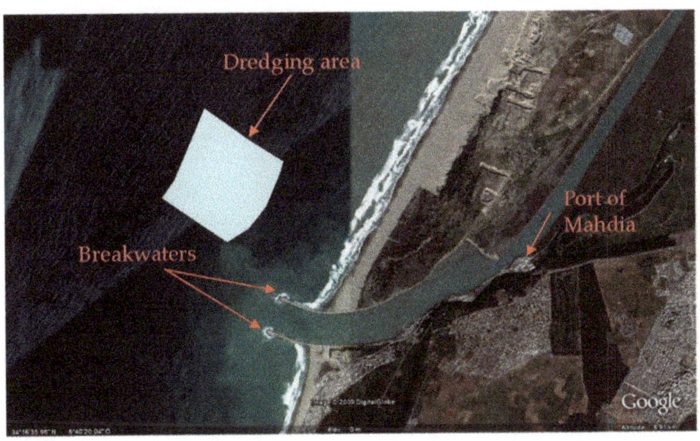

Figure 1. *Cont.*

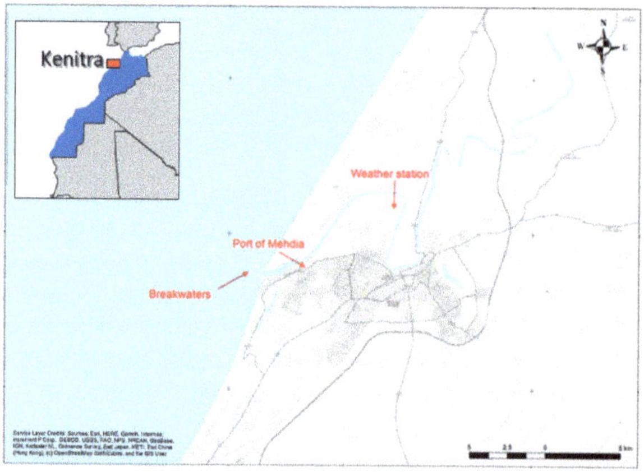

Figure 1. Outlet of the Oued Sebou into the Atlantic Ocean (above: Google Maps Platform) and a general overview of the study region (below).

3. Model Theoretical Formulation

The selected SWAN model is a spectral numerical model designed to simulate waves evolving in coastal regions, lakes, and estuaries under defined wind, bathymetry, and current conditions. It is based on the Energy Density Balance equation linking the advection term to the source and sink terms. Therefore, the wave energy evolves in both geographic and spectral space and changes its aspect due to the presence of wind at the surface, friction with the bottom, or during the breaking of the waves. The SWAN model is a stable model based on unconditionally stable numerical schemes (implicit schemes).

SWAN, in its third version, is in stationary mode (optionally non-stationary) and is formulated in Cartesian or spherical coordinates. The unconditional numerical stability of the SWAN model makes its application more effective in shallow water.

In SWAN, the waves are described with the two-dimensional spectrum of the wave action density N,

$$N(x, y, \sigma, \theta) = \frac{E(x, y, \sigma, \theta)}{\sigma} \quad (1)$$

where x and y are the horizontal components of geographic space, σ is the relative frequency, θ is the wave direction, and E is the energy density.

The spectrum considered in the SWAN model is that of the wave action density N (σ, θ) rather than the spectrum of the energy density E (σ, θ). This is because, in the presence of currents, the wave action density is conserved while the energy density is not [27].

Because wave action propagates in both geographic and spectral space under the influence of genesis and dissipation terms, wave characteristics are described in terms of two-dimensional wave action density. The action density spectrum balance equation relating the propagation term to the effects of the source and sink terms, in Cartesian coordinates, is [28]

$$\frac{\partial N}{\partial t} + \frac{\partial(C_x N)}{\partial x} + \frac{\partial(C_y N)}{\partial y} + \frac{\partial(C_\sigma N)}{\partial \sigma} + \frac{\partial(C_\theta N)}{\partial \theta} = \frac{S}{\sigma}. \quad (2)$$

On the left-hand side of Equation (2), the first term represents the local temporal variation of the wave action density, the second and third terms represent the propagation of wave action in the geographical space of velocities C_x and C_y, the fourth term represents the shifting of the relative frequency due to variations in bathymetry (with propagation velocity C_σ) and currents (with propagation velocity C_θ), and the fifth term represents the refraction induced by the combined effects of depth and currents. It is worth mentioning that C_x, C_y, C_σ, C_θ propagation velocities were obtained from linear wave theory. The term in the right-hand side of Equation (2) represents processes that generate, dissipate, or redistribute wave energy, and S can be expressed as [24]

$$S = S_{in} + S_{wc} + S_{brk} + S_{bot} + S_{n14} + S_{n13}\,, \tag{3}$$

where S_{in} is the wind energy input. The dissipation term of wave energy is represented by the contribution of three terms: the whitecapping S_{wc}, bottom friction S_{bot}, and depth-induced breaking S_{brk}. S_{nl4} and S_{nl3} represent quadruplet interaction and triad interactions, respectively.

Adopting a finite difference scheme for each of the five dimensions: time, geographic space, and spectral space made the numerical implementation in SWAN effective. The following guidelines must be followed in order to obtain the discretization adopted at the SWAN model level for numerical computation: (i) time of a constant and identical time step Δt for the propagation term and the source term, (ii) geographical space of a rectangular grid with constant spatial steps Δx and Δy, (iii) spectral space of a constant directional step $\Delta\theta$ and a constant relative frequency step $\Delta\sigma/\sigma$; (iv) frequencies between a fixed minimum maximum values of 0.04 Hz and 1 Hz, respectively, unlike the WAM and WAVEWATCHIII models, which use dynamic values, and (v) as an option, the direction θ can also be delimited by the minimum and maximum values $\theta_{\min}$ and $\theta_{\max}$.

4. Model Forcing Data

4.1. Bathymetry

The bathymetry of the study area is generated by using the SHOM charts, SHOM7702 SHOM Nautical Charts—Morocco. Available online: https://www.nauticalchartsonline.com/charts/SHOM/Morocco (accessed on 1 February 2021).

Figure 2a shows the bathymetry of the coarse grid domain, while Figure 2b illustrates a more detailed bathymetry. The seabed zone is characterized by comparatively regular isobaths parallel to the coastline. According to the bathymetry of the research region, shallow water extends up to 550 m from the coast with a depth range between 2 and 3 m.

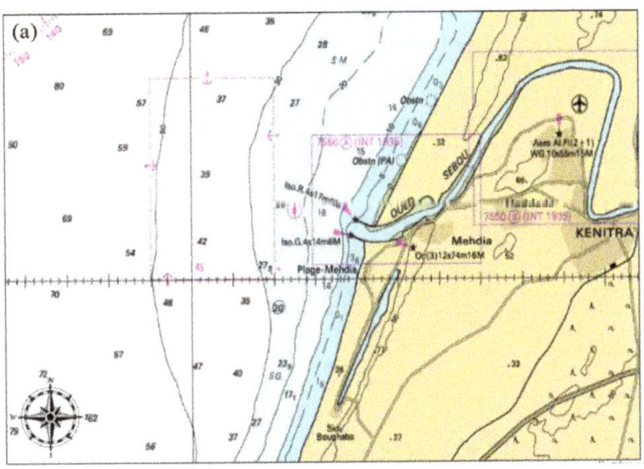

Figure 2. *Cont.*

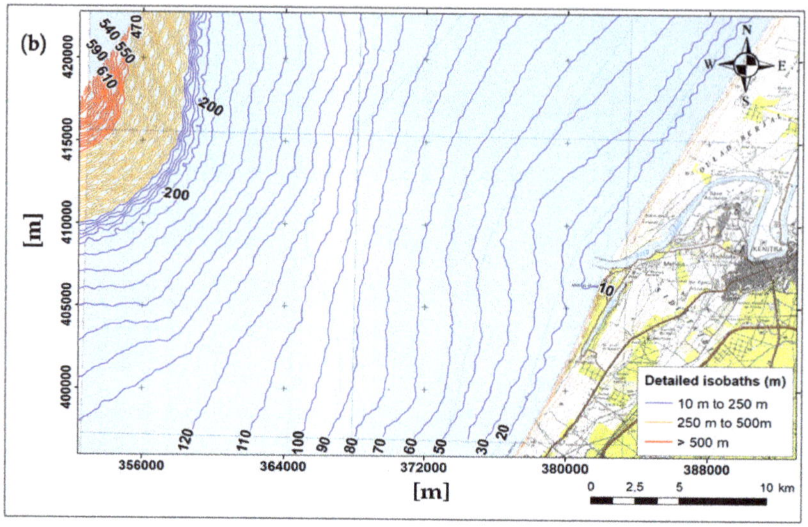

Figure 2. Bathymetry of Sebou estuary. (**a**) Bathymetry of the coarse grid domain. (**b**) Detailed isobaths of the study area.

Generally, for wave simulation, the research and commercial models use flexible grid mesh (structured and unstructured grids). In previous studies [17,21], it was observed that simulations with structured or unstructured grids were substantially consistent. In this study, we use a nested regular grid with a resolution of approximately 25 m, as shown in Figure 3. The bathymetry meshing was generated by using the BlueKenue software.

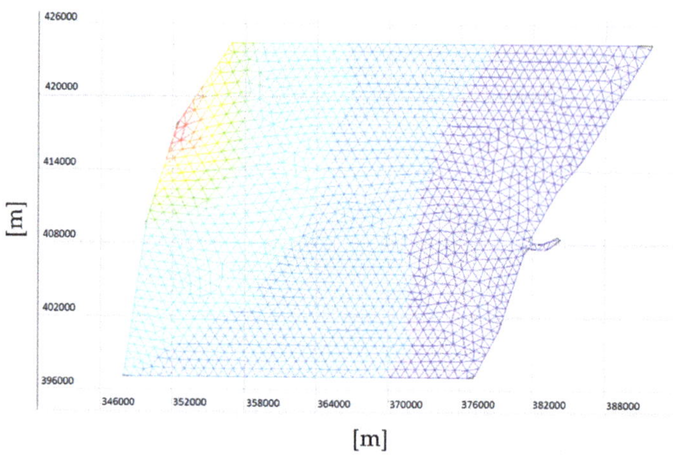

Figure 3. Mesh used of the study area: regular grid of approximately 25 m resolution.

4.2. Wind and Wave Fields

Because the accuracy of the wind field has a large impact on the predicted wave fields [16], in this study, the main meteorological parameters were well analyzed. The average annual values of the wind properties indicate that the study area is characterized by: (i) a winter regime for which the dominant wind (or most frequent) mainly comes from the eastern sector (onshore wind). During this period, the strongest wind speed (>9 m/s) comes from the directions ranging between the southwest and west sectors, with an occurrence of almost 3%, (ii) a summer regime (from March to October), for which the dominant wind comes mainly from the western sector (sea wind). In this period, the strongest wind speed (>9 m/s) comes from the west and north sectors, with an occurrence of almost 3%.

The meteorological data (Figure 4) were collected from the nearest weather station to the study area. It is situated at the Kenitra airport, which is 8 km from the study location. The collected measured data correspond to the period from 1990 to 2009. The annual predominant wind direction is from the west and its speed range from 4 to 9 m/s (Figure 4c).

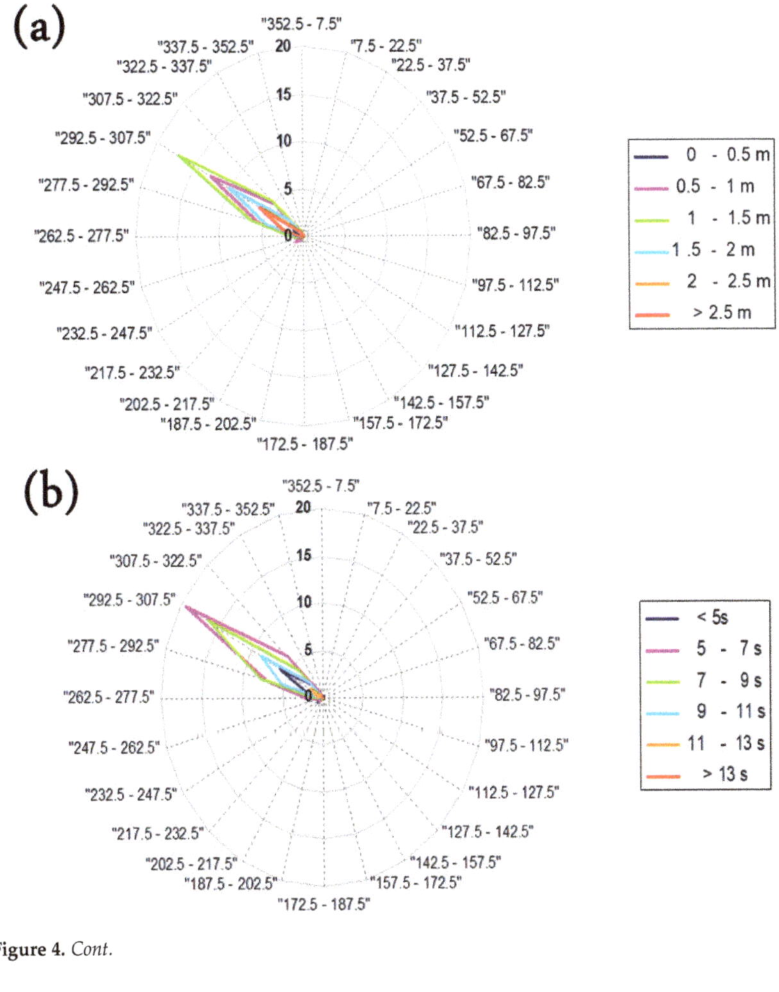

Figure 4. *Cont.*

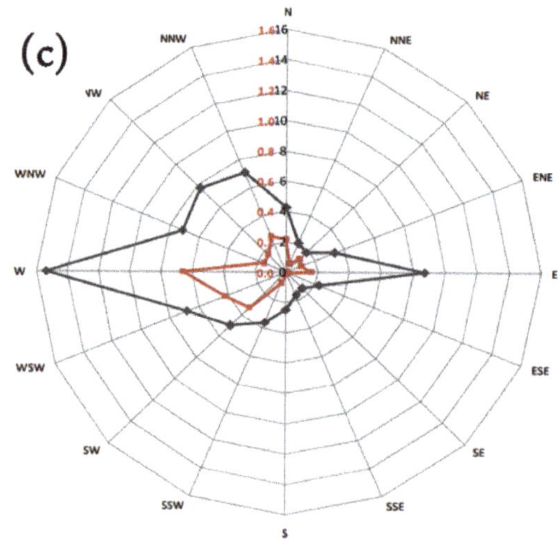

Figure 4. Wave rose (**a**) of the significant wave heights, (**b**) corresponding periods of the significant wave, (**c**) and average annual wind rose.

Figure 4 shows the annual wave rose of the significant wave heights (Figure 4a) and their corresponding periods (Figure 4b). The dominant wave (the most frequent) comes from the northwest (300°N to 320°N). Storm swells are more westerly and come from the west–northwest sector (280°N to 300°N). Five significant wave heights were considered in this study (results of 19-year data analysis), H_{s0} = 1.5, 2, 2.5, 3, 3.5, and 4 m. Here, we denote by H_{s0} the significant wave height in deep water.

4.3. Tidal Current and Water Level

The average sea level in the region of Kenitra is almost equal to 2.17 m. The tide is of semi-diurnal type with a period of 12 h 25 min. Previous studies confirmed that the tidal current and water level significantly affect wave behaviors [24]. As a result, in this study, the tidal forcing was also considered as SWAN mode input data l. Table 1 provides a summary of the astronomical tide level values for the study area.

Table 1. Tidal levels in the study area.

Table	Low Tide (m)	High Tide (m)	Tidal Range (m)
Exceptional high water	0.50	3.90	3.40
Medium high water	0.80	3.50	2.70
M.M. (medium)	1.40	3.00	1.60
Medium still waters	1.50	2.70	1.20

4.4. Model Setup

The site of the Oued Sebou estuary is exposed to the west–northwest direction, which is also the direction of the main dominant wave. Based on the layout of the coastline and the regularity of the hydrodynamic solicitations, the main sediment transport occurs along the channel profile. Monitoring of the estuary area shows that there is no significant littoral transit.

Regarding the primary swell direction (N300), the most exposed areas to wave ac-tion are located downstream of the river. These locations have been given the labels A, B, C, and D, as shown in Figure 5.

Figure 5. The different zones subject to analysis: stress on structures (A, C), B-agitation in the river outlet (B), littoral erosion (D).

After considering the wave and tidal conditions, we chose to proceed with a selection of eight (8) modeling cases (see Table 2), combining the considered five significant wave heights, indicated above, and different tide levels ranging between 0 and 3.4 m. The selection of wave characteristics covers the most representative samples. The current bathymetry (without dredging), the bathymetry with a dredged area of uniform extraction depth of 2 m (with respect to the current bathymetry), and the dredged area of uniform extraction depth of 4 m were all taken into consideration. Moreover, the SWAN model was used with the following additional hydrodynamic parameters: energy spectral density distribution (JONSWAP) of a spectrum width parameter at 3.3 and a low angular spread, spectrum discretization of 36 angular sectors and 32 frequency intervals (between 4 and 20 s), the refraction and diffraction phenomena were considered, and the wind turbulence effects were not considered.

Table 2. Specification of the hydrodynamic cases considered.

	Wave		Tide
Case Ref.	Significant Height Hs_0 (s)	Period T_p (m)	Level (m)
1	4.0	12	0.0 (low tide)
2	2.0	10	0.0 (low tide)
3	1.5	8	0.0 (low tide)
4	1.5	6	0.0 (low tide)
5	4.0	12	2.2 (intermediate level)
6	2.0	10	2.2 (intermediate level)
7	1.5	8	2.2 (intermediate level)
8	4.0	12	3.8 (high tide)

5. Results and Discussion

As an illustration, Figure 6 depicts the simulated outcomes of the local significant wave height (H_s) distribution, before (with current bathymetry), as shown in Figure 6a, and after dredging of 2 m, as shown in Figure 6b. The data illustrated in Figure 6 refer to case 3 (Table 2) of significant wave height in deep water H_{s0} = 1.5 m and a period T_p = 8 s.

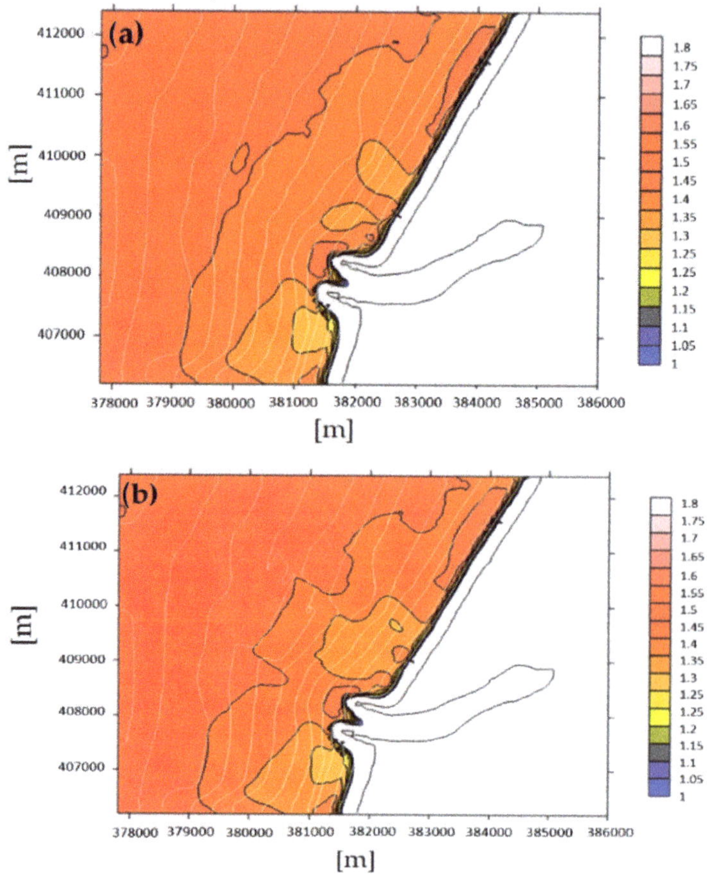

Figure 6. Map of local significant wave height distribution for (H_{s0} = 1.5 m, T_p = 8 s), (**a**) before dredging (with current bathymetry), (**b**) after dredging of 2 m depth. The legend indicates the values of H_s in meters.

The results show that the presence of the dredging affects the wave dispersion. There was an increase in H_s in and at the edges of the dredging area, which can be explained by the sudden increase in the water depth. In general, the simulations show a decrease in the transfer of wave energy out of the extraction zone, causing a lateral energy flow with respect to the excavation. The waves propagating across the excavation area tend to refract toward the areas of shallow water along the edges, increasing the wave heights (known as wave energy focusing). In terms of wave amplitude, there is a modest decrease downstream (in the wave propagation direction) of the excavation and a slight increase inside it and on its sides. This behavior is also confirmed with extreme waves, as shown in Figure 7. In Figure 7, the data of H_s refer to case 8 (Table 2) of H_{s0} = 4 m and a period T_p = 12 s, simulated with and without the dredging of 2 m depth.

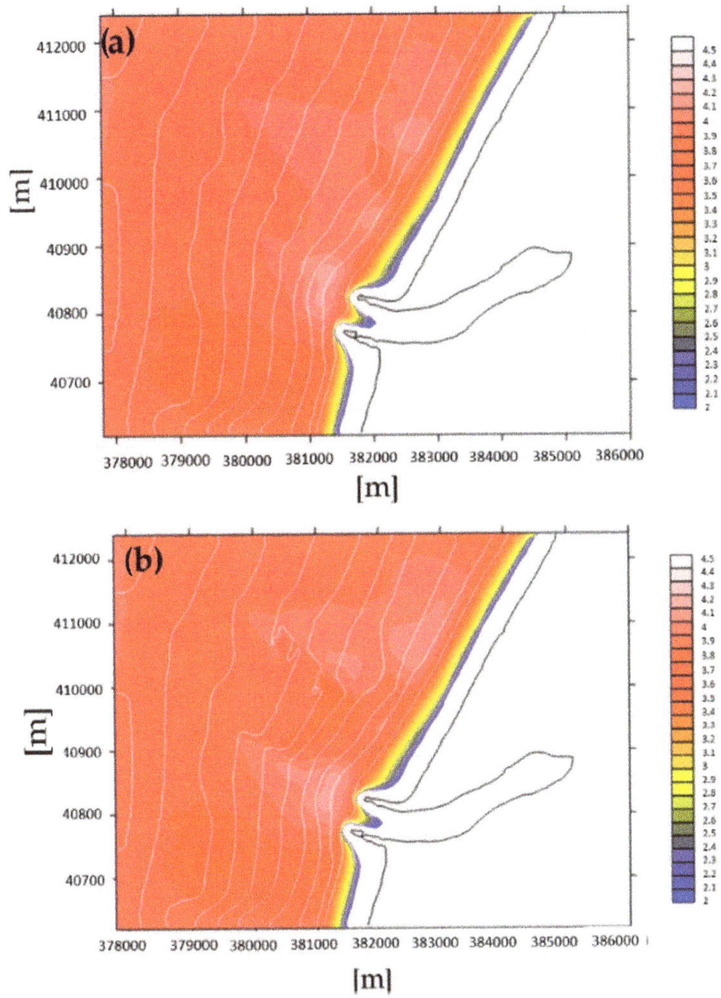

Figure 7. Map of local significant wave height distribution for (H_{s0} = 4 m, T_p = 12 s), (**a**) before dredging, (**b**) after dredging of 2 m depth. The legend indicates the values of H_s in meters.

To evaluate the impact of dredging on the estuary environment, two indicators have been considered: (i) impact on sediment transport processes, where the analysis focuses on the evolution of the bottom friction velocities, responsible for the movement of sediment particles and proportional (velocity squared) to the erosion rate, and (ii) impact on structures, where the analysis focuses on the variation of the energy before the wave breaks, at each concerned zone (Figure 5). For this final point, the impact of the significant heights (squared), measured at the bathymetry of 10 m, and on each of the four selected areas was integrated. In practice, the values of $H_s{}^2$ determined at different locations S_A, S_B, S_C, and S_D (Figure 8), along the isobath of 10 m, were performed to indicate the wave energy impact on the selected zones, as shown in Figure 5. A comparison was made between relative configurations with and without dredging.

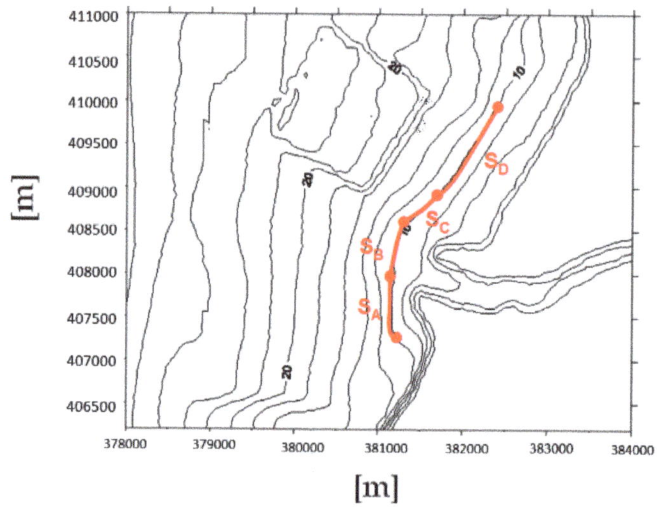

Figure 8. Segmentation for the analysis of energy flows in the concerned zone. The contour lines indicate the bathymetry values.

Figure 9 depicts examples of the relative bottom friction velocity (reflecting the bottom resistance) distribution in the target area. The relative friction velocity is defined as the difference between the simulated friction velocity with bathymetry with the dredged area and that without dredging (current bathymetry). The data shown in Figure 9 refer to the dredging of 2 m depth. Figure 9a shows the results of the relative friction velocity obtained with the most frequent case (H_{s0} = 1.5 m, T_p = 8 s) (see Table 2), whereas Figure 9b shows that of the highest wave condition (H_{s0} = 4 m, T_p = 12 s). Figure 9 shows the clear effects of the dredging on the bottom friction velocity distribution.

The bottom friction velocity changes significantly as the wave gets closer to the excavation site. Locally, around the borders of the excavation, a minor increase of bottom resistance is noted, as demonstrated by the positive relative friction velocity values in Figure 9. The bottom friction velocity significantly decreases inside the dredging area, reaching maximum magnitudes (of negative values). The effect of the dredging area on the friction velocities extends downstream of it, reaching zones C and D (Figure 5).

The variation of the bottom friction velocity due to the excavation certainly affects the sediment transport potential [29–31] and the wave energy flux linked to the redirected waves, particularly in zones C and D. Figure 9 indicates a clear increase in the bottom friction velocity at the lateral sides (in the wave direction) of the excavation, which is more pronounced with the high tide condition (H_{s0} = 1.5 m, T_p = 8 s). The possibility for erosion action increases as bottom shear stress increase. The friction velocity at the sea bottom slightly decreases in zones C and D, which are located downstream of the extraction site. This suggests that zones C and D are likely subject to sediment accumulation following dredging activities. An excessive sediment buildup can cause several environmental problems. It can reduce the seawater depth, preventing the passage of ships. It can also lead to contamination that poses a threat to aquatic plants (*Posidonia oceanica*) and wildlife. Zones A and B of the coastal area in the southern part (from breakwaters) are almost not affected by the excavation site. Additionally, findings indicate that dredging up to 2 m depth has no significant effect along the channel profile.

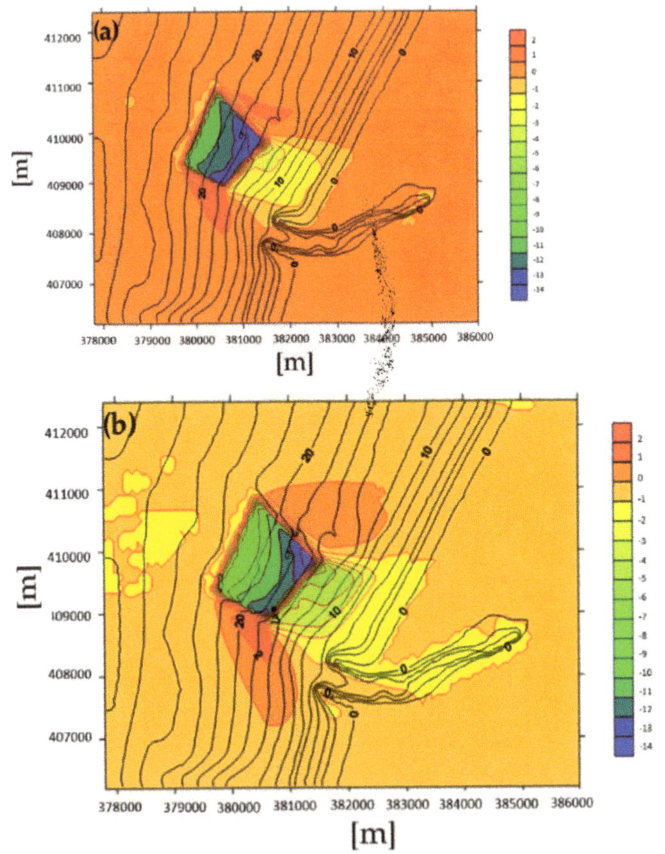

Figure 9. Relative bottom friction velocity, (**a**) (H_{s0} = 1.5 m, T_p = 8 s), (**b**) (H_{s0} = 4 m, T_p = 12 s) after dredging of 2 m depth. The legend indicates the values of relative friction velocity in [cm/s]. The contour lines indicate the bathymetry values.

Figure 10 shows the wave height, H_s (Figure 10a) and its relative velocity (Figure 10b) distribution maps, with additional excavations up to 4 m deep. Figure 10a demonstrates how waves propagating across the excavation area tend to refract toward the areas of shallow water along the edges, increasing the wave heights (known as wave energy focusing). This causes the amplification of H_s at the level of the river mouth, affecting zone B (Figure 5). The increase in wave height at the mouth of the river continues to propagate upstream along the channel. As a kind of obstruction, the excavation causes the incident waves to converge and creates a wake region downstream of it where the wave height decreases.

Similar behavior to the wave height distribution is shown in Figure 10b, where the relative friction velocity oscillates near the bottom. The friction velocities increase along the excavation's lateral edges and decrease downstream of it. Compared to Figure 9a, dredging up to 4 m deep has a greater impact on the estuary environment than dredging up to 2 m deep. With 4 m of dredging, the area is subject to increased erosion. Zones C and D, however, are subject to sediment accumulation.

Estuary ecosystems are significantly impacted by wave energy. The bathymetry in some coastal regions causes a concentration of wave energy, which raises wave height [32]. Figure 11 displays the fraction of wave relative energy percentage that corresponds to zones A, B, C, and D (Figure 5). The energy was estimated at 10 m isobath, as shown in Figure 8, and the relative percentage of energy was determined as the difference between the energy of the simulated wave with the excavation present and without it. The percentage is in relation to the simulated wave energy without excavation. The data shown in Figure 11 relates to a 2 m-deep excavation.

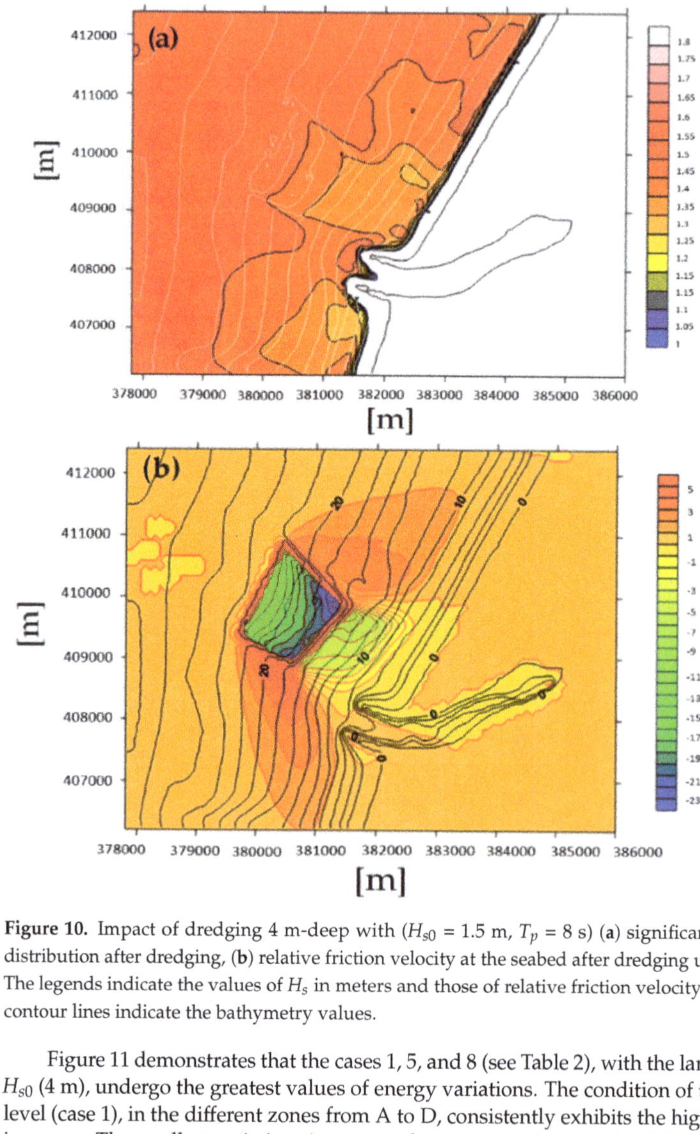

Figure 10. Impact of dredging 4 m-deep with (H_{s0} = 1.5 m, T_p = 8 s) (**a**) significant wave height distribution after dredging, (**b**) relative friction velocity at the seabed after dredging up to 2 m depth. The legends indicate the values of H_s in meters and those of relative friction velocity in [cm/s]. The contour lines indicate the bathymetry values.

Figure 11 demonstrates that the cases 1, 5, and 8 (see Table 2), with the largest values of H_{s0} (4 m), undergo the greatest values of energy variations. The condition of the lowest sea level (case 1), in the different zones from A to D, consistently exhibits the highest variation in energy. The smallest variations in energy always appear in cases 4 and 7, with the lowest values of H_{s0} (1.5 m). Figure 11 further demonstrates that zone A, which has a maximum variation value of order 4%, is the most affected zone by wave energy. The energy impact gradually decreases in the order from zone A to zone D, which is especially prominent in zones C and D. In general, it can be concluded that the dredging activities have a minimal but noticeable impact on the estuary environment of Oued Sebou that is not significant.

The simulated results of wave dispersions and their hydrodynamic structures are useful for estimating the sediment transport rates that will be an extension of this work.

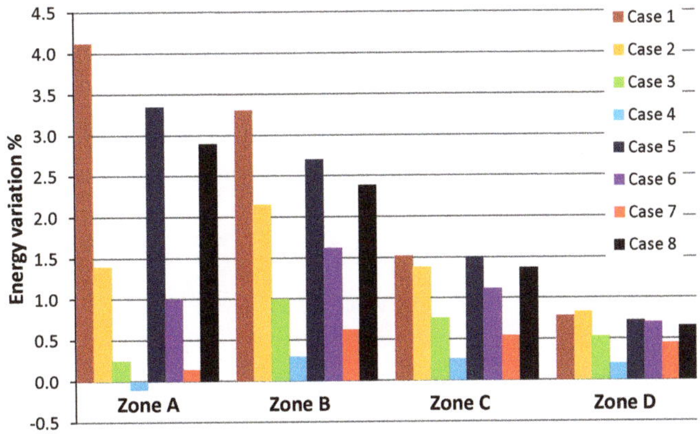

Figure 11. Energy percentage change in zones A, B, C and D.

6. Conclusions

The SWAN model was used to simulate a large number of wave motions at the Kenitra site in order to better understand the effects of dredging activities (due to sand extraction) in the Oued Sebou estuary. In various meteorological conditions, comparisons between the simulated results with and without dredging bathymetries were made. Eight configurations that accurately reflected the actual hydrodynamic circumstances that characterize the study area were tested.

The effects of the bathymetric changes, due to the sand extraction, on wave dispersion, bottom friction velocity field, and energy budget (before breaking) were extensively analyzed.

The sharp variation of the bathymetry due to the dredging that increases water depth causes an increase of the local significant wave height in and at the edges of the area susceptible to sand extraction. According to the simulation results, the waves propagating across the excavation area tend to refract toward the areas of shallow water along the edges, increasing the wave heights (known as wave energy focusing). This causes the amplification of H_s at the river mouth level and along the southern breakwater structure, which is more pronounced with deeper dredging. The excavation plays the role of a kind of obstacle, causing the incident waves to converge and creates a sort of wake region downstream of it where the wave height decreases.

The results show that zones C (at the northern breakwater) and D (northern coastal area) are subject to possible accumulation of sediments, whereas zones A (at the southern breakwater) and B (river mouth) are subject to an increased potential for erosive action and a risk of scouring processes at the southern breakwater.

Dredging activities in the Oued Sebou estuary mainly affect the river mouth (zone B) and the southern breakwater area (zone A), which is very noticeable with dredging up to 4 m deep.

In general, it can be concluded that the dredging activities show a certain level of impact on the estuary environment of Oued Sebou that is not very significant.

Despite the crucial role dredging plays in the nation's economy and maritime engineering management (i.e., it helps make the water navigable, removes contaminants from seabeds and recreates damaged areas, maintains many marine infrastructures, and many other advantages), dredging could have serious and long-lasting negative impacts on the environment, leading to contamination that poses a threat to aquatic plants (*Posidonia oceanica*) and wildlife

The simulation results, which will be validated by some measured wave characteristic data, are useful for examining sediment variations along the estuary coastal area, a subject we are currently working on.

Author Contributions: N.I. and M.C., performed the numerical simulations and methodology; M.B.M. and N.I. formal analysis, study design, writing–original draft preparation; L.M., M.C. and D.L. contributed suggestions, discussions, and reviewed the manuscript. All authors have read and agreed to the published version of the manuscript.

Funding: This research received no external funding.

Data Availability Statement: The data presented in this study are available on request from the first author (Nisrine Iouzzi).

Acknowledgments: This study was carried out at Hassan II University of Casablanca, Faculty of Sciences Ben M'Sik, (Morocco).

Conflicts of Interest: The authors declare that they have no conflict of interest.

References

1. Guo, Y. Hydrodynamics in estuaries and coast: Analysis and modeling. *Water* **2022**, *14*, 1478. [CrossRef]
2. Sohrt, V.; Hein, S.S.V.; Nehlsen, E.; Strotmann, T.; Fröhle, P. Model based assessment of the reflection behavior of tidal waves at bathymetric changes in estuaries. *Water* **2021**, *13*, 489. [CrossRef]
3. Prandle, D.; Rahman, M. Tidal response in estuaries. *J. Phys. Oceanogr.* **1980**, *10*, 1552–1573. [CrossRef]
4. Sutherland, J.; O'Donoghue, T. Wave phase shift at coastal structures. *J. Waterw. Port Coast. Ocean Eng.* **1998**, *124*, 90–98. [CrossRef]
5. Khalil, U.; Yang, S.Q.; Sivakumar, M.; Enever, K.; Sajid, M.; Bin Riaz, M.Z. Investigating an innovative sea-based strategy to mitigate coastal city flood disasters and its feasibility study for Brisbane, Australia. *Water* **2020**, *12*, 2744. [CrossRef]
6. Ralston, D.K.; Talke, S.; Geyer, W.R.; Al-Zubaidi, H.A.; Sommerfield, C.K. Bigger tides, less flooding: Effects of dredging on barotropic dynamics in a highly modified estuary. *J. Geophys. Res. Ocean.* **2019**, *124*, 196–211. [CrossRef]
7. Eyre, B.; Hossain, S.; McKee, L. A suspended sediment budget for the modified subtropical Brisbane River estuary, Australia. *Estuar. Coast. Shelf Sci.* **1998**, *47*, 513–522. [CrossRef]
8. Cai, H.; Yang, Q.; Zhang, Z.; Guo, X.; Liu, F.; Ou, S. Impact of river-tide dynamics on the temporal-spatial distribution of residual water level in the Pearl River channel networks. *Estuaries Coasts* **2018**, *41*, 1885–1903. [CrossRef]
9. Jay, D.A.; Borde, A.B.; Diefenderfer, H.L. Tidal-fluvial and estuarine processes in the lower Columbia River: II. water level models, floodplain wetland inundation, and system zones. *Estuaries Coasts* **2016**, *39*, 1299–1324. [CrossRef]
10. Liu, W.C.; Ke, M.H.; Liu, H.M. Response of salt transport and residence time to geomorphologic changes in an estuarine system. *Water* **2020**, *12*, 1091. [CrossRef]
11. Chen, X. A sensitivity analysis of low salinity habitats simulated by a hydrodynamic model in the Manatee River estuary in Florida, USA. Estuar. *Coast. Shelf Sci.* **2012**, *104–105*, 80–90. [CrossRef]
12. Bale, A.J.; Uncles, R.J.; Villena-Lincoin, A.; Widdows, J. An assessment of the potential impact of dredging activity on the Tamar estuary over the last century: I. Bathymetric and hydrodynamic changes. *Hydrobiologia* **2007**, *588*, 83–95. [CrossRef]
13. Chang, H.K.; Hsu, T.W. A two-Point method for estimating wave reflection over a sloping beach. *Ocean Eng.* **2003**, *30*, 1833–1847. [CrossRef]
14. Lin, J.G. An improvement of wave refraction-diffraction effect in SWAN. *J. Mar. Sci. Technol.* **2013**, *21*, 12. [CrossRef]
15. Chi, Y.; Rong, Z. Assessment of extreme storm surges over the changjiang river estuary from a wave-current coupled model. *J. Mar. Sci. Eng.* **2021**, *9*, 1222. [CrossRef]
16. Hoque, M.A.; Perrie, W.; Solomon, S.M. Application of SWAN model for storm generated wave simulation in the Canadian Beaufort Sea. *J. Ocean Eng. Sci.* **2020**, *5*, 19–34. [CrossRef]
17. Munugama, M.; Suzuyama, K.; Manawasekara, C.; Tanaka, Y.; Xia, Y. Analysis of wind-waves with SWAN on structured mesh and unstructured mesh during the arrival of typhoon. *J. Jpn. Soc. Civ. Eng. Ser. B3 (Ocean Eng.)* **2019**, *75*, 283–288. [CrossRef]
18. Muscarella, P.; Brunner, K.; Walker, D. Estimating coastal winds by assimilating high-frequency radar spectrum data in SWAN. *Sensors* **2021**, *21*, 7811. [CrossRef]
19. Gorman, R.M.; Bryan, K.R.; Laing, A.K. Wave hindcast for the New Zealand region: Nearshore validation and coastal wave climate. *N. Z. J. Mar. Freshw. Res.* **2003**, *37*, 567–588. [CrossRef]
20. Xu, P.; Du, Y.; Zheng, Q.; Che, Z.; Zhang, J. Numerical study on spatio-temporal distribution of cold front-induced waves along the southeastern coast of China. *J. Mar. Sci. Eng.* **2021**, *9*, 1452. [CrossRef]
21. Panigrahi, J.K.; Umesh, P.A.; Padhy, C.P.; Swain, J. Nearshore propagation of cyclonic waves. *Nat. Hazards* **2012**, *60*, 605–622. [CrossRef]
22. Ekphisutsuntorn, P.; Wongwises, P.; Chinnarasri, C.; Vongvisessomjai, S.; Zhu, J. The application of simulating Waves Nearshore Model for wave height simulation at Bangkhuntien shoreline. *Am. J. Environ. Sci.* **2010**, *6*, 299–307. [CrossRef]
23. Lo Re, C.; Manno, G.; Ciraolo, G.; Besio, G. Wave energy assessment around the Aegadian Islands (Sicily). *Energies* **2019**, *12*, 333. [CrossRef]
24. Lv, X.; Yuan, D.; Ma, X.; Tao, J. Wave characteristics analysis in Bohai sea based on ECMWF wind field. *Ocean. Eng.* **2014**, *91*, 159–171. [CrossRef]

25. Hakkou, M.; Castelle, B.; Benmohammadi, A.; Zourarah, B. Wave climate and morphosedimentary characteristics of the Kénitra-Bouknadel sandy coast, Morocco. *Environ. Earth. Sci.* **2011**, *64*, 1729–1739. [CrossRef]
26. Ayad, H.; Mouhid, M.; Loudyi, D.; Mouakkir, L.; Chagdali, M. Numerical modelling of the impact of dredging on stability of Oum Rabiâ estuary (Morocco) using SWAN model. In *Artificial Intelligence in Intelligent Systems*; Lecture Notes in Networks and Systems 229; Springer: Cham, Switzerland, 2021; p. 2. [CrossRef]
27. Whitham, G.B. *Linear and Nonlinear Waves*; Wiley: New York, NY, USA, 1974.
28. Hasselmann, K.; Olbers, D. Measurements of wind-wave growth and swell decay during the joint North Sea project (JONSWAP). *Ergänzung Deut. Hydrogr. Z. Reihe* **1973**, *12*, 1–95.
29. Ben Meftah, M.; De Padova, D.; De Serio, F.; Mossa, M. Secondary Currents with Scour Hole at Grade Control Structures. *Water* **2021**, *13*, 319. [CrossRef]
30. Ben Meftah, M.; De Serio, F.; De Padova, D.; Mossa, M. Hydrodynamic Structure with scour hole downstream of bed sills. *Water* **2020**, *12*, 186. [CrossRef]
31. Ben Meftah, M.; Mossa, M. New Approach to Predicting Local Scour Downstream of Grade-Control Structure. *J. Hydraul. Eng.* **2020**, *146*, 04019058. [CrossRef]
32. Mouakkir, L.; El Hou, M.; Mordane, S.; Chagdali, M. Wave Energy Potential Analysis in the Casablanca-Mohammedia Coastal Area (Morocco). *J. Mar. Sci. Appl.* **2022**, *21*, 92–101. [CrossRef]

water MDPI

Article

Real-Time Properties of Hydraulic Jump off a Tidal Bore, Its Generation and Transport Mechanisms: A Case Study of the Kampar River Estuary, Indonesia

Ulung Jantama Wisha [1,2,3], Yusuf Jati Wijaya [1,4] and Yukiharu Hisaki [1,*]

1 Physical Oceanography Laboratory, Department of Physics and Earth Sciences, Graduate School of Engineering and Science, University of the Ryukyus, Nishihara-cho 903-0213, Japan
2 Research Institute for Coastal Resources and Vulnerability, Ministry of Marine Affairs and Fisheries, Padang 5245, Indonesia
3 Research Center for Oceanography, National Research and Innovation Agency, Jakarta 14430, Indonesia
4 Department of Oceanography, Faculty of Fisheries and Marine Science, Diponegoro University, Semarang 50275, Indonesia
* Correspondence: hisaki@sci.u-ryukyu.ac.jp

Abstract: Since the hydraulic jump off a tidal bore in the Kampar Estuary has never been well-described, real-time measurements of hydraulic jump properties are crucial to understanding the tidal bore characteristics. This study aims to determine the real-time properties of a tidal bore generation, hydraulic jump, and transport mechanism in the Kampar River estuary. Tidal harmonic and range are analyzed using least-square-based tidal modeling. The tidal bore height and turbulent velocity records based on ADCP surveys in the estuary and upstream area are used to determine the hydraulic jump properties. Furthermore, an acoustic-based approach is also employed to quantify the suspended sediment concentration and flux during the passage of the bore. Kampar Estuary is predominated by semidiurnal co-tidal components (M2 and S2), where, based on the phase lag magnitude, it is categorized as an ebb-dominant estuary. This finding is proven by the more intense and prolonged ebb phases, especially during spring tidal conditions where the tidal range reaches 4 m. Of particular concern, the tidal bore height declines by 1.5 m every 20 km upstream with an erratic turbulent velocity. A sudden increase in transverse and vertical velocity during the passage of bore (ranging from −0.9 to 0.2 m/s) reflects the potency of sediment resuspension in the surrounding river edge marked by the significant increase in suspended sediment flux of about 3.7 times larger than at the end of the ebb tide. However, long-term measurement and regular bathymetry surveys are crucial to monitor the tidal bore behavior and morpho-dynamics in the Kampar River estuary.

Keywords: turbulent velocity; transport mechanism; tidal range; suspended sediment flux

Citation: Wisha, U.J.; Wijaya, Y.J.; Hisaki, Y. Real-Time Properties of Hydraulic Jump off a Tidal Bore, Its Generation and Transport Mechanisms: A Case Study of the Kampar River Estuary, Indonesia. *Water* **2022**, *14*, 2561. https://doi.org/10.3390/w14162561

Academic Editor: Mouldi Ben Meftah

Received: 6 July 2022
Accepted: 18 August 2022
Published: 19 August 2022

1. Introduction

An estuary is a significant and unique water area where oceanographic and anthropogenic factors shape its characteristics [1], impacting the river and estuary's physical, biological, geological, and even chemical states [2]. As the place of sea-river water confluence, tidal influence is considerable in estuaries, playing a significant role in controlling transport mechanisms in the form of oscillations in the estuary [3]. These tidal oscillations are associated with large mixing within estuarine water [4], transporting and scouring bed sediment along the river [5]. These large transports can result in tidal bore propagation often seen in estuarine waters.

A tidal bore is a tidal wave commonly observed in estuarine waters generated by a relatively high tidal range and large river streams, surging upstream because of different hydraulic pressure [1,2,6,7]. According to [8], a tidal bore is an unsteady water movement induced by a rapid surface water-level rise at the mouth of the estuary during the early high tidal phase. With time, the first wave crest becomes steeper and steeper until it forms a wall of water propagating upstream. The first wave crest is characterized by the train of a secondary wave following the non-breaking surge front [1,9]. According to [10], the hydraulic jump off the tidal bore induces scouring and turbulence beneath the first wave crest. Therefore, revealing the real-time hydraulic jump properties is essential in determining the process and evolution of bore-affected estuaries.

It is estimated that there are more than 400 bore-affected estuaries worldwide [1,11]. There are about five tidal bore symptoms scattered throughout the Indonesia archipelago. However, tidal bore records and studies are relatively limited due to the difficult access to sites. By contrast, a tidal bore in the Kampar Estuary is the most well known and frequently studied, precisely located in the Pelalawan Regency, Riau Province, Indonesia (Figure 1). This tidal bore, called Bono, was scientifically reported for the first time by [12], who conducted a qualitative survey and modeled the symptoms of abrasion and accretion in the estuary of the Kampar River. The Kampar River estuary is a macro-tidal estuary with a funnel shape (V-shape) whereby the river width and depth gradually decline upstream [4]. In subsequent studies, [13] surveyed one-line parallel bathymetry data along the Kampar River, reporting that the tidal bore propagates upstream, reaching 60 km from the estuary. On the other hand, local researchers and scholars have conducted several environmental-based studies in the Kampar Estuary [14–18].

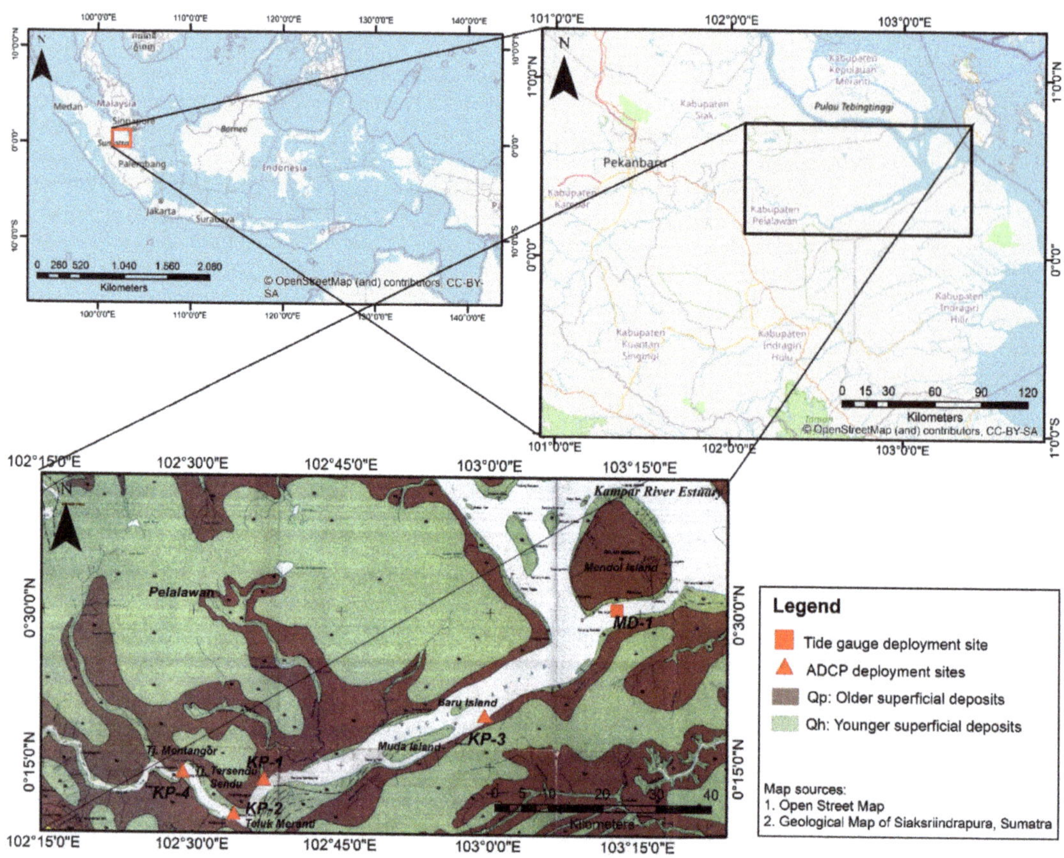

Figure 1. Details of the Kampar River estuary and observation sites.

Numerous numerical hydraulic jump models in the Kampar River have also been developed [3,19–21]. Overall, numerical model approaches employed by previous studies have never been well validated using field data measurements or calibrated using local bathymetry. Even though some reports explicitly mentioned the hydraulic jump records in the Kampar Estuary [4,17], these data did not describe the properties of a tidal bore well. Since the established data of tidal bore motions in the Kampar Estuary are limited, the hydraulic jump properties of the tidal bore are crucial to determine. Moreover, to date, the turbulent velocity regimes during the passage of bores in the Kampar Estuary are not yet clearly explained. As previously reported by [5,10,22–25], a rapid increase in water level (hydraulic jump) is characterized by a sharp rise in turbulent velocity, playing a role in the transport mechanism throughout the river. Therefore, we report herein the record of hydraulic jump and turbulent velocities off a tidal bore in the Kampar River estuary, based on the ADCP (Acoustic Doppler Current Profiler) survey deployed simultaneously in several significant areas along the river. A few studies have reported the influence of the passage of bores on suspended sediment fluctuation [26,27], and this is one aspect that should be investigated. Therefore, this study aims to determine the real-time tidal bore generation, hydraulic jump properties, and transport mechanism in the Kampar River estuary.

2. Materials and Methods

2.1. Study Site and Field Observation

The study area is situated in the Kampar River estuary, Pelalawan Regency, Riau Province, Indonesia, categorized as a river watershed area geographically positioned at 0°40′0″–0°13′20″ north and 102°40′0″–103°26′40″ east (Figure 1). Geologically, Kampar River lies between two rock formations; younger superficial deposits (Qh) composed of clay, silt, gravel, plant waste, peat swamp, and coral reefs; and older superficial deposits (Qp) composed of clay, silt, sandy clay, plant waste, and sandy granite (Figure 1) [4].

Kampar River is the largest river in Riau Province, with an approximate length of 400 km from the estuary to upstream, sourced from Bukit Barisan mountains, passing through several regencies (Indragiri Hulu, Indragiri Hilir, Kampar, Kuantan Singing, Pelalawan, Siak, and Pekanbaru), and eventually ending up in the Malacca Strait. Moreover, the area of Kampar River reaches 24,548 km^2 with a river discharge of approximately 600 m^3/s upstream and 200 to 400 m^3/s downstream [28].

Kampar River is a shallow downstream river with a funnel (V-shaped) formation. In several areas close to the estuary, enormous abrasions have frequently occurred, and unstable sediment transport is highly controlled by a destructive tidal bore phenomenon called Bono [12].

Field surveys were conducted twice (April and August 2016) by the Research Institute for Coastal Resources and Vulnerability, Ministry of Marine Affairs and Fisheries, Indonesia. A simultaneous ADCP (Acoustic Doppler Current Profiler) deployment was carried out throughout the Kampar River estuary (Nortek™ Aquadopp Profiler 600 kHz and Teledyne RDI Workhorse ADCP). Due to safety reasons, the instruments were deployed no longer than 48 h. The instrument setup and specifications are shown in Table 1. Moreover, except for station KP-4, the data yielded from the ADCP surveys consisted of surface elevation changes, turbulent velocity profiles, and water temperature.

On the other hand, to understand the tidal-generated bores in the estuary, a Tide Master Valeport vented strain gauge was deployed for 38 days from 23 April to 31 May 2016, mounted close to Mendol Island (Figure 1). The filtered tidal data were then analyzed to gain harmonic tidal constituents and tidal range profiles, which will be elucidated in Section 2.3.

Table 1. Details of deployment setup and instrument specifications.

Observation Station	Area Location	Deployment Coordinates		Instrument Specifications	Deployment Setup	Survey Period (Western Indonesian Time UTC+07:00)
		Longitude	Latitude			
MD-1	Mendol Island	103.212° E	0.503° N	**Tide Master Valeport** Range: min 0.8 m and max 20 m Beam angle: ± 6° Frequency: 25 GHz Accuracy: ± 10 mm Precision: 1 mm	• Mode: B4 • Pressure units: meter • User pressure cal: Gain 1.026967, Offset −0.011763 • Vale pressure cal: P0 = 2.49 × 10^{-9}E-09, P1 = 0.003178, P3 = −1.880864 • Calibrated: 12 November 2015	23 April 2016 15:45 to 31 May 2016 16:00
KP-1	Tanjung Tersendu-sendu	102.624° E	0.225° N	**ADCP Nortek Aquadopp** Acoustic frequency: 0.6 MHz Max profile range: 30–40 m Cell-size: 1–4 m Minimum blanking: 0.5 m Max cell: 128 Velocity range: ± 10 m/s Accuracy: 1% of measured value ± 0.5 m/s Max sampling range: 1 Hz	• Profile interval: 300 s • Number of cell sizes: 8 • Cell size: 1 m • Blanking distance: 0.5 m • Measurement load: 100% • Average interval: 60 s • Compass update range: 300 s	24 April 2016 13:30 to 25 April 2016 14:55
KP-2	Teluk Meranti	102.568° E	0.165° N			25 April 2016 17:00 to 26 April 2016 17:25
KP-3	Baru Island	103.005° E	0.336° N			20 August 2016 11:00 to 22 August 2016 19:55
KP-4	Tanjung Mentangor	102.480° E	0.241° N	**Teledyne RD Instrument Workhorse Sentinel** Center working frequency: 614 kHz Max. typical profiling range: 70 m Nadir angle: 20 deg Max. number of cells per beam: 256 Min. Blanking distance: 0.5 m Cell size: 0.25–4 m Max. ping range: 3 Hz Max. velocity: 10 m/s Cell overlap: 25%	• Ensemble interval: 300 s • Frequency: 614.4 kHz • Beam angle: 20 deg • Deployment hours: 120 • Pings/Ens: 50 • Time/ping: 6 min • First cell range: 2.11 m • Cell size: 1 m	21 August 2016 11:30 to 23 August 2016 15:00

2.2. Mounting Scheme and Geometry of Kampar River

Based on national bathymetric data calibrated using a single beam echosounder measurement, the bathymetry profile of the Kampar River estuary varies considerably. It ranges from 2.3 to 6.2 m in the surrounding mouth of the estuary. It becomes significantly shallow in the surrounding Muda Island (wave energy mixing zone) [4] with a water depth ranging from 0.1 to 1 m, and it gradually gets a little bit deeper after passing Tanjung Tersendu-sendu area (station KP-1), ranging from 0.6 to 2.5 m (Figure 2A). Several sandbank formations are observed in the Tanjung Mentangor area, where the tidal bore is reported to be decaying at this point.

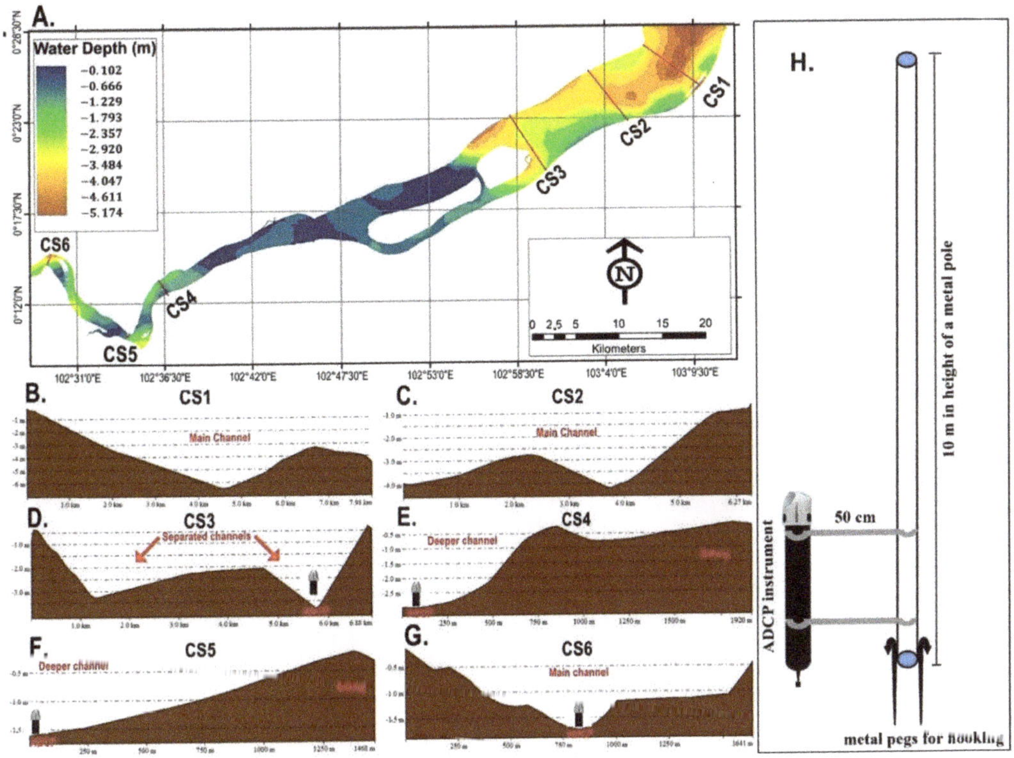

Figure 2. Bathymetry and geometry profiles of the Kampar River estuary calibrated using a field survey in 2016 (**A**); cross-section in the mouth of the estuary (**B**); cross-section in the main channel shifting (**C**); cross-section in separated channels in front of Baru Island with an ADCP deployment scheme (Station KP-3) (**D**); cross-section in the remnant observation stations (**E–G**); mounting scheme of ADCP instruments (**H**). Dotted lines denote the bathymetry (depth level) in every cross-section.

Based on the cross-section of the bottom morphology (Figure 2B–G), the Kampar River estuary has a funnel-shaped formation (V-shape) where the river width declines upstream, ranging from 1 to 8 km from upstream to downstream. Because of the unstable bore-induced sedimentation, the bathymetry of the river (and mouth) often changes, resulting in complex hydrodynamic systems that regularly alter the position of the Kampar River's main channel. Passing through Baru and Muda Island, located right in the middle of the river, the channel is separated, and the depth decreases considerably, with a magnitude of approximately 4 m. Shoaling effects in this super shallow area are possible because of the significant bathymetry changes, inducing a greater propagation of bores and local erosion [4,17,29]. The bores coalesce after passing Muda Island, forming higher waves [12].

The processes that occur during the passage of bores are presumed to induce a varied bottom sediment transport and resuspension.

Referring to [4], mixed-sediment types have been observed in the mouth of the estuary (clayey silt, silt, and sandy silt) and the area of tidal bore decay (sand, silty sand, sandy silt, and silt). In contrast, sand sediment is predominated in the middle of the river, where mixed wave energy occurs. Therefore, sand sediment settling in the area around Muda Island creates a super rigid bottom substrate. This state also becomes a reason why a mounting scheme of ADCP could not be performed throughout this area.

Concerning the ADCPs, they were deployed in the area with a depth of more than one meter, attached to a metal pole to maintain the position of the instruments. The first deployment was situated at stations KP-1 and KP-2, representing the middle river zone. The second measurement was positioned at stations KP-3 and KP-4, representing the downstream and upstream areas (Figure 2B–G), respectively. The mounting scheme of ADCPs applied in the field is shown in Figure 2H, where using a 10 m metal pole was possible to maintain the position of the ADCP. The pole was hooked using several metal pegs in the bottom part. The instrument's distance from the pole was 50 cm, attached by other more minor metals connected to the central pole.

2.3. Tidal Data Analyses

Since tidal data measurement in August 2016 could not be performed due to technical reasons, we employed a month's tidal prediction data provided by the Geospatial Information Agency of Indonesia (BIG), retrieved from a webpage: https://srgi.big.go.id/tides, accessed on 12 May 2022. We collected the data at a coordinate of 103.2075° E and 0.5143° N, right at station MD-1 (Figure 1).

The collected tidal data were modeled for 18.6 years using a NAO.99b program developed by National Astronomical Observatory, Tokyo, Japan [30]. All the data were then filtered to be analyzed using the ERG program, a program developed by Bandung Institute of Technology, Bandung, Indonesia, consisting of three primary sub-programs: ERGRAM, ERGELV, and ERGTIDE [31]. These programs are developed based on a least-square method [32]. The significant elevation-dependent mean sea level and harmonic constituents were obtained from this simulation.

As previously reported by several scholars [3,33–35], in the study area, the semidiurnal components (M2 and S2) are predominant. Therefore, because the influence of diurnal components is extremely low on shaping the tidal asymmetry [36], we focused on the two most significant semidiurnal constituents to be analyzed in more detail. The tidal model was equipped with an online Fourier analysis, quantitatively estimating the co-tidal constituent distribution across the study area. The co-tidal charts were then interpolated using spline interpolation to minimize the total surface curvature [37].

Aside from harmonic analysis, the field-measured and predicted tidal data were also filtered according to the moon phase. As an astronomically influenced oceanographic parameter, the surface elevation and the tidal forcing vary depending on the position of the moon, the earth, and the sun [38,39]. The following stage involved quantifying the tidal range and the displacement period of sinusoidal tides by subtracting the top and trough elevation data starting from the slack after flood tides up to the ebb slack point during neap and spring phases [40]. Furthermore, the displacement period for one cycle of tides was calculated using the record of tidal modulations as follows:

$$TR = \left|\zeta_{high\ tide} - \zeta_{low\ tide}\right| \tag{1}$$

$$DP = \left|t_{high\ tide} - t_{low\ tide}\right| \tag{2}$$

where:

TR = tidal range

ζ = tidal elevation (m)

DP = displacement period
t = time (hours)

2.4. Calculating the Approximate Tidal Bore Height

A previous study conducted by the authors in another site (Rokan Estuary) elucidated the same workflow concerning tidal data analyses and tidal bore height calculation [25]. The tidal bore front features could be approached from a quasi-steady flow analogy with tidal bore celerity U (Figure 3), considering mass and momentum conservation [1,26] as follows:

$$(V_1 + U)wl_1 = (V_2 + U)wl_2 \tag{3}$$

$$\frac{1}{2}\rho g\left(wl_2^2 - wl_1^2\right) = \rho(V_1 + U)wl_1(V_1 - V_2) \tag{4}$$

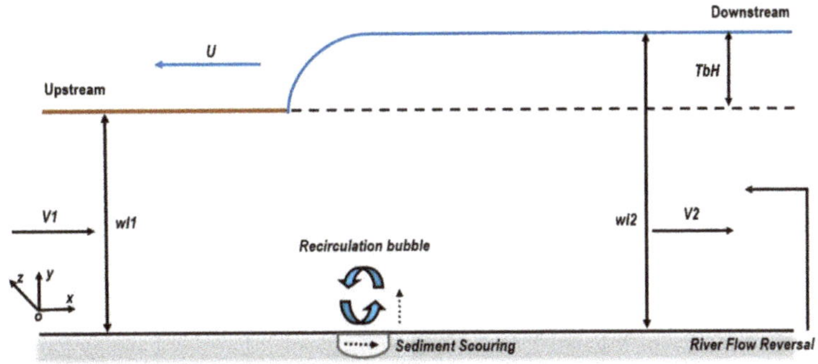

Figure 3. Definition sketch of a quasi-steady flow of tidal bores, modified from [22]. The blue line shows the tidal bore surface flow, and the brown line shows the initial water level before the tidal bore passage. Horizontal arrows denote the main water flow direction.

According to [22], Equations (3) and (4) are valid for stationary hydraulic jumps ($U = 0$), the tidal bore passage ($U > 0$), and positive surges flowing downstream ($U < 0$). Then, the solution to Equations (3) and (4) results in the height of the tidal bore Tb_H and Froude number Fr as follows:

$$Tb_H = \frac{wl_1}{2}\left(\sqrt{1 + 8Fr^2} - 3\right) \tag{5}$$

$$Fr = \frac{V_1 + C}{\sqrt{gwl_1}} \tag{6}$$

$$Tb_H = wl_2 - wl_1 \tag{7}$$

where:

U = tidal bore celerity measured in the field positive upstream (m/s)
V_1 = initial cross-sectional average flow velocity (m/s)
V_2 = cross-sectional average flow velocity after the bores propagate (m/s)
wl_1 = initial water level (m)
wl_2 = water level after the bores propagate (m)
ρ = water density (kg/m^3)
g = acceleration of gravity (9.81 m/s^2)
Tb_H = tidal bore height (m)
Fr = Froude number

2.5. Acoustic-Based Suspended Sediment Estimation

Aside from tidal bore and velocity measurements, the strength of reverberated sound was analyzed to determine the relationship between the relative acoustic backscatter (RB) and estimated SSC (suspended sediment concentration) [41–43] as follows:

$$SSC = 10^{(A+B.RB)} \tag{8}$$

where A and B are the intercept and slope, respectively, which can be calculated using an empirical formula of simple linear regression, and RB (relative backscatter) is the measured backscatter corrected from transmission losses (units are dB) [42].

After gaining the value of SSC during the passage of bores, the instantaneous suspended sediment flux per unit area was quantified based on the value of SSC and the longitudinal turbulent velocity (V_x) [27] as follows:

$$Q_s = SSC \times V_x \tag{9}$$

where:

Q_s = suspended sediment flux per unit area (kg/m^2/s)
SSC = suspended sediment concentration (kg/m^3)
V_x = longitudinal turbulent velocity (m/s).

3. Results and Discussion

3.1. Tidal Harmonic Analysis in the Kampar River Estuary

Based on tidal harmonic analysis, semidiurnal components are predominant in the Kampar River Estuary with a Formzahl of 0.26 (mixed tide with prevailing semidiurnal). Several scholars also reported that the semidiurnal component significantly controls water mass transfer within estuaries throughout the Malacca Strait [3,25,34]. In the Malacca Strait, the tidal distribution is transformed from semidiurnal to a mixed tide with prevailing semidiurnal from Aceh to Riau Province [33]. Even though the semidiurnal components are predominant, the diurnal constituents could not be neglected, influencing the water level fluctuation within the Malacca Strait.

The tidal harmonics results recorded at station MD-1 were compared with the result of the tidal model (Table 2). Overall, both simulated and observed data showed an agreement in constituent amplitude and phase lag with a deviation (RMSE) of less than 5 cm. This indicates that the tidal simulation result could represent the natural conditions of the study area. Because of the gap in time series between the field and model data, a digression is possible [25]. The amplitude of semidiurnal main components was 3 cm higher than the simulated data. The main diurnal component showed a slight difference (<2 cm).

Table 2. Tidal harmonic analysis in the Kampar River estuary.

Tidal Constituents	Simulated Data		Observed Data		Angular Frequency (Degree/Hour)	RMSE (cm)	Species
	Amplitude (cm)	Phase Lag (°)	Amplitude (cm)	Phase Lag (°)			
M2	113.02	199.25	115.74	164.77	28.98	1.92	Principal lunar semidiurnal
S2	52.78	35.09	55.80	49.92	30.00	2.21	Principal solar semidiurnal
N2	24.87	188.15	29.98	157.23	28.44	3.61	Larger lunar elliptic semidiurnal
K2	11.26	104.60	11.96	191.47	30.08	0.49	Lunisolar semidiurnal
K1	36.61	89.55	34.19	50.56	15.04	1.71	Lunar diurnal
O1	23.36	42.29	27.57	69.69	13.94	2.98	Lunar diurnal
P1	14.40	−54.15	11.92	−45.38	14.96	1.75	Solar diurnal
M4	15.14	−4.08	15.02	−2.41	57.97	0.08	Shallow water over-tides of principal lunar
MS4	13.96	187.41	12.01	170.24	58.98	1.38	Shallow water quarter diurnal

Compared to all analyzed tidal constituents, the principal lunar and solar semidiurnal components (M2 and S2) showed a strong amplitude in the Kampar Estuary (Table 2). According to [40], the co-tidal amplitude could determine the distribution of tidal type, and the phase lag differences could reflect the tidal range cycles. In addition to phase lag, despite a large gap between model and field data with <60° resulting from different data frequencies, the phase lag resulting from tidal harmonic analysis could determine interactions among constituents and reflect the variability of the tidal current [44]. In semidiurnal tidal regimes, the flood or ebb dominance is quantified from the relationship between M2 and M4 phase lags [40]. From the harmonic analysis, we calculated that the value of $2g_{M2} - g_{M4}$ ranged between −90° and 90°. Therefore, the Kampar Estuary is characterized by a more prolonged and intense ebb than flood current (ebb dominant). The flood–ebb phase and duration in the Kampar Estuary will be elucidated in the next section.

Concerning the diurnal constituents, the Kampar River mainstream extent from station KP-3 to KP-4 (approximately 70.16 km) was selected for the tidal analysis with the specified observation points right at the ADCP deployment coordinates (Figure 4). During the time frame of field measurement, the study area was subjected to neither the extreme effect of winds nor large freshwater flows. The extent of semidiurnal constituent amplitude distribution upstream was observed and its influence evoking the free surface elevation extended up to approximately 104 km upstream. Overall, the spatial magnitude of the M2 and S2 components gradually increased upstream. The peak magnitude was observed at station KP-2 (Teluk Meranti), with an amplitude of 1.63 m and 1.05 m for M2 and S2, respectively (Figure 4C,D). The magnitude decreased significantly toward station KP-4 (Tanjung Tersendu-sendu), indicating less tidal influence evoking flood discharge throughout the river. These results show that the tidal bore commenced decaying after passing Teluk Meranti station (KP-2). However, aside from tidal regimes, tidal bore propagation relies on bottom morphology, channel formation, and river discharge [4]. These primary co-tidal constituents (M2 and S2) play a significant role in evoking the tidal range-induced tidal bore characteristics in the estuary of Kampar River.

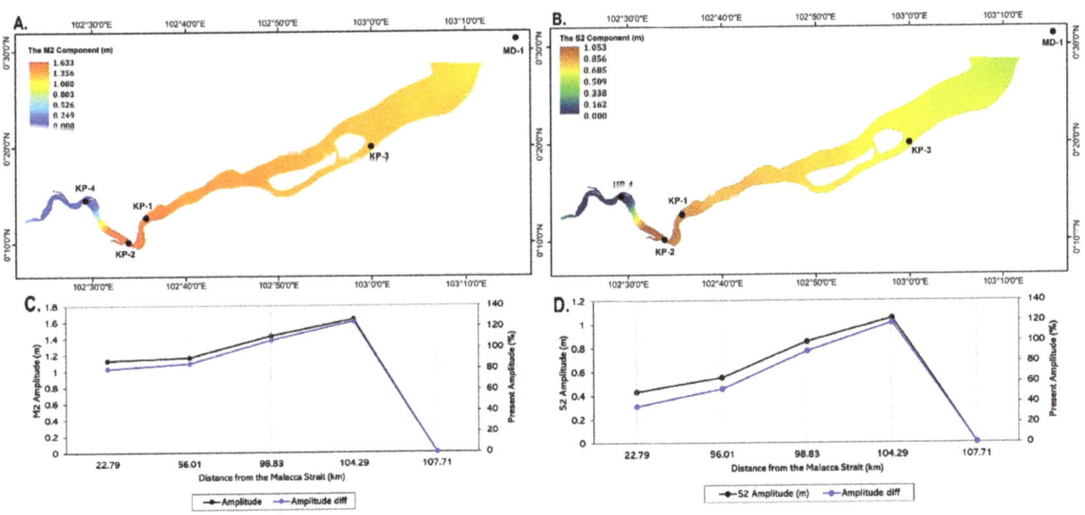

Figure 4. Spatial variability of the dominant tidal component distribution throughout the study area. M2 component distribution (**A**); S2 component distribution (**B**); amplitude and percentage of amplitude difference of M2 (**C**) and S2 (**D**). The amplitude difference percentage computes the amplitude fraction at each station to the amplitude at the mouth of the river (M1).

3.2. *Tidal Range Profiles and Tidal Bore Generation*

Due to the importance of tidal fluctuation characteristics in the estuary, the minimum range to evoke tidal bore passage has been previously explained (exceeding 4–6 m) [10,26]. The analysis of tidal range and flood–ebb duration is essential to understanding the variability of tidal bore propagation. The process in the estuarine system relies on tidal dynamics related to mixing and sediment transport. Therefore, since the tidal variability depends on the moon phase, the propagation of tidal bores generated in the Kampar Estuary could be distinguished by various astronomical forces. Generally, the tidal range peaked during the full moon phase (spring tidal state) and declined considerably during the third quarter of the moon phase (neap tidal state). However, the flood–ebb displacement duration was longer during neap phases (Figures 5 and 6).

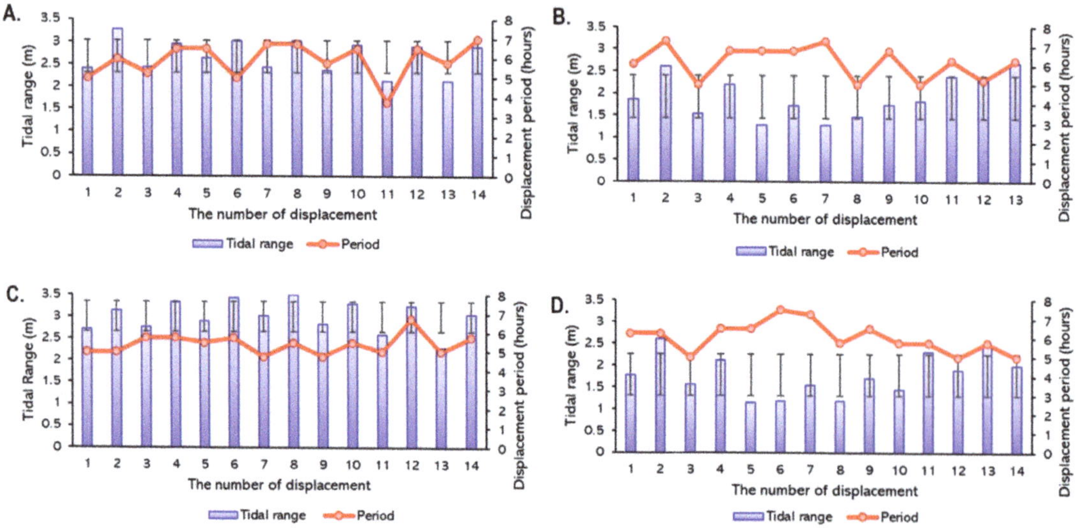

Figure 5. Flood tidal range at the Kampar Estuary during the new moon phase (**A**); during the first quarter of the moon phase (**B**); during the full moon phase (**C**); and during the third quarter of the moon phase (**D**). The grey lines denote the standard deviation.

During the new moon phase, the tidal range during flood tides ranged from 2.3 to 3.3 m, with a flood duration ranging from 4 to 7 h (Figure 5A). The magnitude decreased by around one meter, with a longer flood duration (on average 6 h) during the first quarter of the moon phase (Figure 5B). The highest elevation was observed during the full moon phase, ranging from 2.2 to 3.5 m, with a mean flood duration of 5.5 h (Figure 5C). The remnant phase showed the lowest range in tidal elevation, ranging from 1.3 to 2.6 m, with a flood duration of approximately 6 h (Figure 5D).

A more significant range was observed during ebb tides compared to the flood tides (Figure 6). The tidal range ranged from 2 to 3.7 m with an average ebb duration of 6.5 h during the new moon phase. The tidal range declined significantly during the first quarter of the moon phase, where the ebb-tidal range ranged from 0.9 to 3 m, and the ebb displacement duration ranged from 5 to 8 h. During the full moon phase, the more extreme ebb tides peaked at more than 4 m in the tidal range, with a longer ebb duration of approximately 7 h. In contrast, the lowest range was observed during the third quarter of the moon phase, with a tidal range of 0.5–3 m and an ebb duration of 6.5 h.

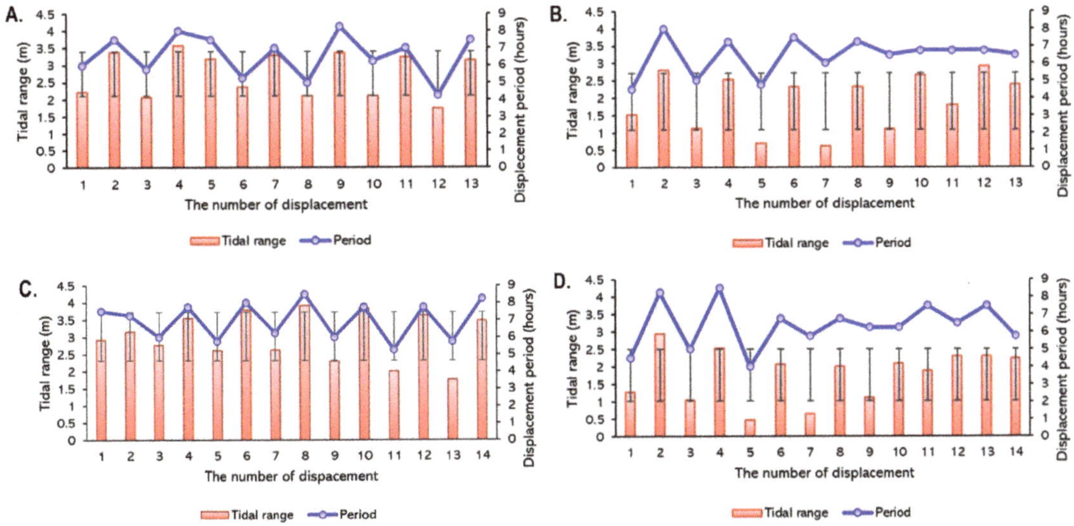

Figure 6. Ebb tidal range at the Kampar Estuary during the new moon phase (**A**); during the first quarter of the moon phase (**B**); during the full moon phase (**C**); and during the third quarter of the moon phase (**D**). The grey lines denote the standard deviation.

Generally, spring tidal states for both the new moon and full moon phases (spring tidal condition) showed a higher tidal range than the moon quarter phases (neap tidal condition). By contrast, the displacement duration was longer during the moon's quarter phases, although the deviation was not too significant. According to [45], the full moon phase evokes the highest sinusoidal tidal level, where the tidal range would be significant during the spring tidal phase. Concerning the tidal range-induced tidal bore propagation, the higher magnitude of the tidal range parameter synchronized with a higher river discharge would induce a higher tidal bore height with a more robust turbulent velocity [7,26]. This aspect is addressed in Section 3.3.

In addition to the ebb dominance of Kampar Estuary, the previously discussed results are in accordance with the tidal phase lag analysis. This condition may vary and be different in the other estuaries throughout the eastern coast of Sumatra due to the difference in latitude-induced co-tidal characteristics [44]. Therefore, examining the characteristics of tidal currents in all estuaries throughout the east coast of Sumatra for further studies is necessary to understand the variation in tidal-induced water motion in the estuarine area.

3.3. Hydraulic Jump Properties of Tidal Bores

A hydraulic jump off a tidal bore is marked by a sudden transition in velocity in rivers or canals where the robust velocity deforms to be a slower motion followed by a sudden rise in free-surface elevation [2,11]. Moreover, the hydraulic jump can be observed from the tidal bore front when rapid changes in water level and velocity take place. Based on the ADCP measurement, we displayed the record of hydraulic jumps at the four observation stations (stations KP-1 to KP-4). Of particular concern, the tidal bore height declined by approximately 1.5 m every 20 km upstream with a non-uniform pattern of turbulent velocity in every observation station (Figure 7). It is expected that the morphology and channel formation play a significant role in determining the variability of the turbulent velocity throughout the Kampar Estuary.

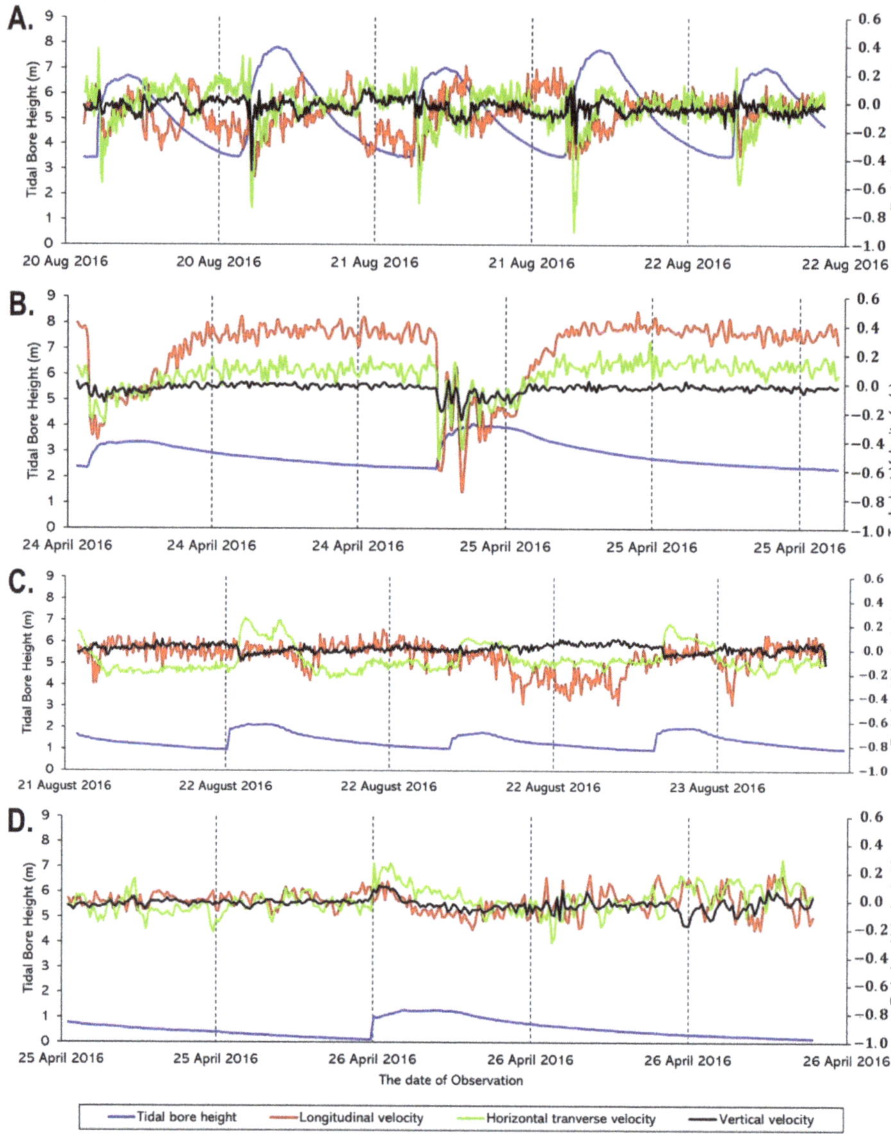

Figure 7. Hydraulic jump properties of tidal bores at station KP-3 (Baru Island) (**A**), KP-1 (Tanjung Tersendu-sendu) (**B**), KP-2 (Teluk Meranti (**C**), and KP-4 (Tanjung Mentangor) (**D**).

At station KP-3 (Baru Island), the water elevation altered from about 3.5 m to 6.8 m and 8 m during tidal bore passage at midday and night, respectively (Figure 7a). Therefore, the tidal bore height in this station was 3.3 m during midday propagation and 4.5 m during night propagation. Before the train of bores passed through the station, a rapid transitional change in turbulent velocity was observed. The longitudinal velocity V_x was negatively upstream, ranging from 0 to −0.4 m/s and positively downstream, ranging from 0 to 0.2 m/s. Its sharp increase in magnitude occurred at the end of the ebb tide, with approximately 0.6 m/s upstream. The longitudinal velocity profile was not significantly different between midday and night passages of bores even though the free surface level was 1.5 m different in magnitude. Moreover, the negative value of longitudinal velocity detected while the bores were passing by indicates the flow reversal within the channel [26,46].

Unlike the longitudinal velocity, a sudden increase in transverse velocity was observed after the train of bores passed the station. A sharp increase was detected simultaneously with the peak phase of the tidal bore (reaching −0.9 m/s during night propagation). Since the V_y was negative to the river shoreline and positive to the nearby island [5], the sufficiently high negative value indicates the potential of shoreline erosion in the Baru Island. This consideration was proven by [29] based on Landsat records from 1990 to 2007, where significant erosion had occurred in Baru Island with an average change of 2.48 m and a rate of 0.15 m/year. On the other hand, we found that the upward vertical velocity V_z, ranging from −0.4 to 0.1 m/s, significantly evoked a downward movement at the time of the peak surface level during night passages with respect to the theory of a recirculation bubble (turbulence) beneath the first wave crest, as previously explained [1] (see Figure 3).

At station KP-1 (Tanjung Tersendu-sendu), the tidal bore height ranged from 1 to 1.5 m with 2.5 m water depth prior to the passage of bores (Figure 7B). Unlike the KP-3 station, the positive longitudinal velocity was higher by about 0.2 m/s, indicating the ebb stream flow was significant in this area. The reversal flow of the tidal bore was marked by a sudden decrease in longitudinal velocity of about 0.6 m/s negative upstream. The pattern of horizontal transverse velocity tended to be as stable as the previous station, ranging from −0.4 to 0.2 m/s. In this station, an indication of erosion was not expected because no sharp decrease in transverse velocity was detected. Moreover, the variability of turbulent velocity was significant during the night passage of bores. The vertical velocity was not too significant, ranging from 0 to −0.2 m/s at the end of the ebb tides (Figure 7B).

The following observation station (station KP-2) is located 6.26 km from station KP-1. The surface-level change was not too significant, approximately 1 m at both night and midday propagation (Figure 7C). This area is super shallow, with a water depth at the end of the ebb tide of about 1 m, and while the bores were passing this station, the water depth increased by 1 m. The feature of turbulent velocity did not reflect the characteristics of hydraulic jump, where the transverse horizontal velocity showed a particular positive pattern during the passage of bores, indicating that the main flow rotates toward the Teluk Meranti shoreline. The longitudinal velocity was shallow, below 0.2 m/s at the end of the ebb tide, and it slightly decreased by −0.4 m/s while the bores propagated, although this pattern was not applicable for all flood flows. In contrast, the opposite pattern of the horizontal transverse velocity feature was identified where its value was positive during the passage of bores, indicating the water motion flowed toward the northern river shoreline, whereas the vertical velocity showed a small magnitude, ranging from 0.1 to −0.1 m/s (Figure 7C).

The remnant station is KP-4 (Tanjung Mentangor). This station was shallower than KP-2, where the water depth at the end of the ebb tide was close to 0 m. This state was also proven by the presence of sandbanks surrounding this station. The tidal bore height was not too significant (<1 m), indicating a decayed area of tidal bore propagation [20]. No more sharp increases in turbulent velocity were observed in this area ranging from −0.1 to 0.1 m/s.

Overall, the turbulent velocity and water level change properties are consistent with the earlier field study in other bore-affected estuaries [2,7,25,26,47]. As elucidated beforehand, the turbulent velocity profiles, particularly the horizontal transverse circulation, are possibly linked with the irregular channel cross-section and the evolution of the estuary. In this study, the record of hydraulic jump properties is very limited. Therefore, a long-term measurement is recommended to monitor the variability of tidal bore features, which is imperative for future decision-making and development.

The turbulent velocity regime may control sediment transport throughout the river. A previous study determined that the variability of transverse and vertical velocity affecting the secondary currents behind the bore front plays a significant role in evoking sediment resuspension advected upstream [26]. This transport mechanism is linked to very-strong turbulent mixing determining the erosion and accretion in the upper estuarine zone. Therefore, the fluctuation of suspended sediment beneath the passage of bores is essential to understand.

3.4. Suspended Sediment Concentration and Flux during the Passage of Bores

As the less dynamic tidal bore flow upstream, we displayed the fluctuation of suspended sediment recorded at station KP-3 (Baru Island), whereby the result depicted was measured in the surface bottom layer. The period shown in Figure 8 was sampled during the night passage of the bore, when the tidal bore height was more significant. At the end of the ebb tide, the suspended sediment concentration (SSC) ranged from 0.4 to 0.9 kg/m^3. Compared to Landsat OLI imagery surface layer detection [17], a lower SSC in Kampar River was observed ranging from 0.042 to 0.241 kg/m^3 at the end of the ebb tide.

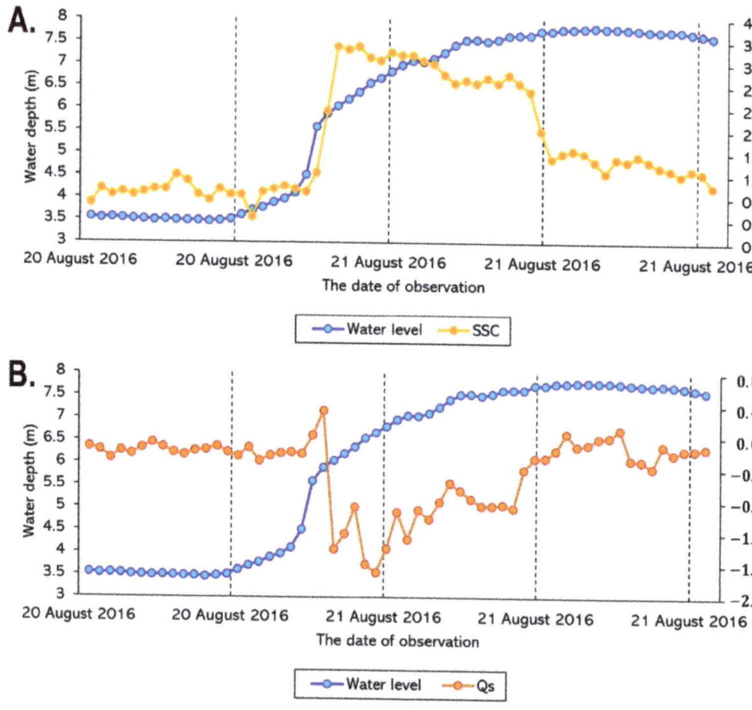

Figure 8. Time series of bore propagation-induced sediment transport at station KP-3. The SSC fluctuation beneath the first wave crest (**A**); the mean sediment flux per unit area (**B**).

When the sudden change in water level occurred, the SSC significantly increased, reaching approximately 3.62 kg/m^3, and it gradually decreased as the bore propagated on average about 2.9 kg/m^3. The peak of SCC occurred beneath the second trough of the tidal bore passage. According to [26], the eroded materials placed in suspension behind the bore front were advected by the "whelps" and secondary current movements. When the water level reached its peak magnitude, it gradually declined, followed by a gradual decrease of SSC in the surface bottom ranging from 0.9 to 1.34 kg/m^3 (Figure 8A).

The result of SSC estimation was consistent with the turbulent velocity regimes where the horizontal transverse and vertical velocity showed a sharp increase in magnitude behind the first wave crest, reflecting an overwhelming erosion that occurred on the surface bottom and near the river edge. The suspended materials are brought by the train of bores upstream, and they are deposited in the area of tidal bore decay (station KP-4), proven by the formation of sandbanks in the middle of the river composed of coarse sediments [4].

The instantaneous suspended sediment flux per unit area Q_S data are displayed in Figure 8B. The sediment flux data revealed a downstream positive mass flux at the end of the ebb tide. Before the tidal bore passage, the suspended sediment flux per unit area was an average of -0.2 kg/m^2/s. When the sharp increase in longitudinal velocity magnitude occurred, it induced a sudden increase in the flux of suspended materials, with a deviation of 0.5 kg/m^2/s. As previously defined by [27], an abrupt reversal flow occurred during the flood tide, marking the passage of bores, and the suspended sediment flux was negative. Shortly after the bore passage, the sediment flux per unit area reached its peak magnitude, with negative values ranging from –0.8 to –1.8 kg/m^2/s, indicating a significant fluctuation of sediment flux. About 2 h after the first wave passed by the station, the sediment flux was approximately -0.3 kg/m^2/s.

The sediment flux data were integrated over time to provide the net sediment mass transfer per unit area over a given period [26,27]. Prior to the passage of the tidal bore (23:50 > t > 24:00), the net sediment mass per unit area was transferred approximately +321.6 kg/m^2 within 10 min of the data. Following the tidal bore passage, the net sediment mass transfer equaled -3028.5 kg/m^2 within 25 min (24:00 > t > 00:25). In contrast, at the end of the bore passage before ebb, the net sediment mass transfer equaled -4068.9 kg/m^2 within 90 min (01:30 > t > 03.00) (Figure 8B). Overall, during the passage of bores, the net sediment mass transfer was estimated to be 3.7 times larger in magnitude than at the early flood tide and 2.6 times larger than the net sediment mass transfer magnitude toward the ebb tide. These trends observed in the present study are consistent with several earlier studies that highlighted that the intense sediment mixing, and upstream advection of suspended materials occurred during the passage of bores [26,27,48].

In addition to sedimentary movement, the surface bottom of the river is subjected to significant stress reversals when it receives cyclic loading during the tidal bore resuspending the cohesive sediment underneath the second trough of the tidal bore train. Because the riverbed is saturated, pore pressure changes occur during "rapid cycling," i.e., the pace of cycling is such that pore pressure variations are not dissipated [49]. As a result, the riverbed may liquefy, causing bed materials to be suspended and transported upstream by the bores.

3.5. Implication of the Tidal Bore Passage to the Surrounding Environment

As mentioned previously, the very high magnitude of horizontal transverse and vertical velocity potentially induces local erosion of the surface bottom and the river edge. The indication of river shoreline erosion at station KP-3 is shown in Figure 9A. Due to the dynamic tidal bore passage, the river shoreline gradually changes over time. According to [29], over 16 years (1990–2016), the area prone to abrasion was detected in the surrounding Baru and Muda Island, with the eroded area reaching 2.36 ha/year. The remarkable celerity of the tidal bore was recorded in several river edges (Figure 9B), where the line track of the tidal bore showed the high erosion induced by the significant transverse velocity. According to [26], the tidal bore can cause significant damage to riverbanks, and navigational hazards in tidal bore have affected estuaries.

On the other hand, due to the long wave propagation surging upstream (tidal bore), the Kampar River is well known for its long-distance surfing. Local government and other third parties promote this attraction for enhancing the local tourism [50], where it is reported that the annual festival "Bakudo Bono," meaning riding the tidal bore through surfing techniques, is held in Teluk Meranti, attracting not only domestic but also foreign tourists [51].

Significant river shoreline changes over time threaten local society where tidal flooding occurs at certain times and houses are devastated (Figure 9C). This tragedy also happens in the Qiantang River, where its banks are overtopped by tidal bore, and dozens of local settlements are drowned yearly [26]. Other tragic examples of drownings in tidal bores and "whelps" include multiple human deaths in the Colorado River (Mexico), Bamu and Fly River (PNG), and Seine River (France) [2]. On the other hand, indigenous boats are often sunk in the middle of the river due to the intense morphological alteration [6,17]. Therefore, a sustainable water level and regular bathymetry survey are crucial to understanding tidal bore behavior better. Moreover, a model-based estimation can also predict extreme phases of the tidal bore in the Kampar River estuary. The local and central governments must initiate these efforts to construct future development and management.

Figure 9. The impact of tidal bore propagation on the surrounding environmental conditions. River-line abrasion (**A**); coastal erosion at the edge of the river (**B**); local settlement destruction due to tidal bore (**C**); non-woven geotextile fabrics as coastal protection applied in several areas of interest (**D**).

The unstable river shoreline changes in the Kampar Estuary previously reported by [29], where a significant river shoreline change was detected from 2010 to 2016 at a rate of 3.56 m/year, may be worsened by some man-made structures constructed in recent years. In several vital areas, particularly the shoreline near the wood-cutting factory, temporary coastal structures were built around the 2000s [52], such as non-woven geotextile container technology (Figure 9D). This kind of structure could significantly protect the river shoreline from destructive waves [53], but on the other hand, the possibility of unstable erosion–sedimentation in the nearby areas due to coastal structures cannot be neglected [4]. As reported by [54], in a coastal area with rapid erosion, declivous slope, and significant waves regimes, a geotextile container could effectively dampen the waves and induce sedimentation behind the structure by around 50 cm within a year. However, other erosion symptoms have also been reported in some nearby areas [54].

Another anthropogenic activity that causes significant river shoreline changes in the Kampar Estuary is sand mining since the propagation of tidal bores relies on bottom morphology. A significant change in bottom morphology due to sand mining activity throughout Kampar River will alter the feature of the tidal bores [4]. As reported by the Department of Energy and Mineral Resources of Pelalawan Regency, the most substantial area of sand mining was identified in Teluk Meranti village, with a total area of 47,701,000 m^2 and sediment volume of 83,400,000 m^3 (Table 3). These results also correlate with the previous report of river shoreline changes [29]. Even though sand mining activity has declined, the practice is still occurring illegally [28]. Therefore, the potency of shoreline erosion due to tidal bores and anthropogenic activities is still high. Sustainable monitoring of tidal bore properties and sediment dynamics is necessary to control the shoreline instability and minimize other impacts on local society.

Table 3. Record of sand-mining area and volume in 2009 in the Kampar River estuary reported by the Department of Energy and Mineral Resources, Pelalawan Regency [55].

Location	Area (m^2)	Volume (m^3)
Teluk Meranti village	41,701,000	83,400,000
P. Muda village	23,270,000	46,540,000

4. Conclusions

As a semidiurnal-affected water region, water motion throughout the Kampar Estuary is controlled by semidiurnal components (M2 and S2) and their derivatives, characterized by more prolonged and intense ebb than flood currents (ebb dominant). The tidal range during ebb tides is slightly higher, with a longer displacement time than the flood condition. However, spring tidal conditions produce higher elevation (peaking at 4 m) and shorter tidal duration. The amalgamation between the tidal range–duration and the fluctuation of river discharge determines the tidal bore generation.

The tidal bore height declines by approximately 1.5 m every 20 km upstream with an arbitrarily erratic turbulent velocity profile. The rapid increase in horizontal transverse and vertical velocity detected in the estuary determines a strong turbulent mixing affecting the erosion and accretion in the upper estuarine zone (river edge erosion). In the upstream area, the sign of hydraulic jump is not too significant, reflecting the decay phase of the tidal bore indicated by several sandbanks in the middle of the river body, where the suspended sediment deposition is predominant.

The SSC increases significantly by four times in magnitude during the passage of bores and gradually declines as ebb tides occur. In contrast, an abrupt reversal flow occurring during the flood tide induces a negative suspended sediment flux per unit area, where the net sediment mass transfer is estimated to be 3.7 times larger in magnitude than at the early flood tide and 2.6 times larger than the net sediment mass transfer magnitude toward the ebb tide. The properties of suspended sediment induced by the hydraulic jump off a tidal bore play a significant role in the morphodynamical alteration of Kampar Estuary. Therefore, long-term monitoring of tidal bore properties is crucial to support future studies and research-based decision-making regarding further developments in Pelalawan Regency, Riau Province.

Author Contributions: Conceptualization, U.J.W., Y.J.W. and Y.H.; methodology, U.J.W., Y.J.W. and Y.H.; software, U.J.W.; visualization, U.J.W.; validation, U.J.W. and Y.J.W.; formal analysis, U.J.W.; resources, U.J.W.; writing—original draft preparation, U.J.W.; writing—review and editing, U.J.W. and Y.J.W.; supervision, Y.J.W. and Y.H.; funding acquisition, Y.H. All authors have read and agreed to the published version of the manuscript.

Funding: This study was financially supported by a Grant-Aid for Scientific Research (C-2) from the Ministry of Education, Culture, Sports, Science, and Technology of Japan (20K04708).

Institutional Review Board Statement: Not applicable.

Informed Consent Statement: Not applicable.

Data Availability Statement: The data used during the study appear in the submitted article. All secondary data for this paper are cited and referred to in the reference list. Tidal prediction data can be retrieved from a webpage: https://srgi.big.go.id/tides currently accessed on 12 August 2022.

Acknowledgments: We would like to thank the Research Institute for Coastal Resources and Vulnerability, Ministry of Marine Affairs and Fisheries of Indonesia for supporting the observational data used in this study, Ing. Semeidi Husrin, and Wisnu A. Gemilang for the constructive advice regarding theoretical and conceptual aspects of a tidal bore. Gratitude is also given to Department of Physics and Earth Sciences, University of the Ryukyus and MEXT Scholarship.

Conflicts of Interest: The authors declare no conflict of interest.

References

1. Li, Y.; Pan, D.Z.; Chanson, H.; Pan, C.H. Real-time characteristics of tidal bore propagation in the Qiantang River Estuary, China, recorded by marine radar. *Cont. Shelf Res.* **2019**, *180*, 48–58. [CrossRef]
2. Chanson, H. Environmental, Ecological, and cultural Impacts of Tidal Bores, Burros and Bonos. In Proceeding of the International Workshop on Environmental Hydraulics; Theoretical, Experimental, and Computational Solutions (IWEH), Valencia, Spain, 29–30 October 2009. [CrossRef]
3. Kurniawan, A.; Wisha, U.J.; Husrin, S.; Karjadi, E.A. Tidal Bore Generation at the Estuaries of the East Coast of Sumatra. In Proceedings of the 37th IAHR World Congress, Kuala Lumpur, Malaysia, 13–18 August 2017.
4. Gemilang, W.A.; Wisha, U.J.; Rahmawan, G.A. Particle size characteristics of riverbed sediments transported by tidal bore 'BONO' in Kampar Estuary, Riau-Indonesia. *Mar. Res. Indones.* **2018**, *43*, 25–35. [CrossRef]
5. Chanson, H. Current knowledge in hydraulic jumps and related phenomena: A survey of experimental results. *Eur. J. Mech. B Fluids* **2009**, *28*, 191–210. [CrossRef]
6. Bayu, A.C.; Pudjaprasetya, S.R.; Wisha, U.J.; Husrin, S. Numerical simulation of tidal bore Bono at Kampar River. *J. Appl. Fluid Mech.* **2019**, *12*, 311–318. [CrossRef]
7. Wolanski, E.; Williams, D.; Spagnol, S.; Chanson, H. Undular tidal bore dynamics in the Daly Estuary, Northern Australia. *Estuar. Coast. Shelf Sci.* **2004**, *60*, 629–636. [CrossRef]
8. Docherty, N.J.; Chanson, H. Physical Modeling of Unsteady Turbulence in Breaking Tidal Bores. *J. Hydraul. Eng.* **2012**, *138*, 412–419. [CrossRef]
9. Peregrine, D.H. Calculations of the development of an undular bore. *J. Fluid Mech.* **1966**, *25*, 321–330. [CrossRef]
10. Chanson, H. Mixing and dispersion in tidal bores: A review. In Proceeding of the International Conference on Estuaries and Coasts ICEC, Hangzhou, China, 9–11 November 2003.
11. Chanson, H. *Tidal Bores, Aegir, Eagre, Mascaret, Pororoca: Theory and Observation*; World Scientific: Singapore, 2011; pp. 1–220. [CrossRef]
12. Yulistiyanto, B. The phenomenon of bono rising wave in the Kampar River estuary. *Din. Tek. Sipil* **2009**, *9*, 19–26. (In Indonesian)
13. Rahmawan, G.A.; Wisha, U.J.; Husrin, S.; Ilham, I. Bathymetry and tidal analyses in the estuary of Kampar River: The generation of tidal wave "Undular Bore Bono". *J. Ilm. Geomatika* **2017**, *22*, 57–64. [CrossRef]
14. Asiah, N.; Sukendi, S.; Harjoyudanto, Y.; Junianto, J.; Yustiati, A. Water Quality Analysis Based on Plankton Community Structure in Kampar River, Riau Province. In Proceeding of the IOP Conference Series: Earth and Environmental Science, the 9th International and National Seminar on Fisheries and Marine Science, Pekanbaru, Indonesia, 10–11 September 2020. [CrossRef]
15. Harjoyudanto, Y.; Rifardi, R.; Windarti, W. Water Quality Analysis Around the Floating Net Cage Culture Activities in the Kampar River, Buluhcina Village, Kampar District. In Proceeding of the IOP Conference Series: Earth and Environmental Science, the 8th International and National Seminar on Fisheries and Marine Science, Pekanbaru, Indonesia, 12 September 2019. [CrossRef]
16. Mubarak; Sulaiman, A.; Efriyeldi. Environmental Effect of Tidal Bore Propagation in Kampar River. In Proceeding of the MATEC Web of Conferences, the International Symposium on Civil and Environmental Engineering, Melaka, Malaysia, 5–6 December 2016. [CrossRef]
17. Wisha, U.J.; Dhiauddin, R.; Kusumah, G. Remote estimation of total suspended solid (TSS) transport affected by tidal bore "BONO" of Kampar Big River estuary using Landsat 8 OLI imagery. *Mar. Res. Indones.* **2017**, *42*, 37–45. [CrossRef]
18. Wisha, U.J.; Maslukah, L. Nutrient condition of Kampar Big River estuary: Distribution of N and P concentrations drifted by tidal bore "Bono". *Indones. J. Mar. Sci.* **2017**, *22*, 37–45. [CrossRef]
19. Abdullah, F.A.R.; Ningsih, N.S.; Rachmayani, R. Numerical simulation of tidal bore in Kampar River: A preliminary study. In Proceeding of the IOP Conference Series: Earth and Environmental Science, the first Maluku International Conference on Marine Science and Technology, Ambon, Indonesia, 24–26 October 2018. [CrossRef]
20. Mubarak. Modeling of Kampar River discharge as a solitary wave. *Inter. J. Eng. Technol. UAE* **2018**, *7*, 138. [CrossRef]
21. Putra, Y.S.; Noviani, E.; Nurhasanah; Nurhanisa, M.; Azwar, A. A numerical study of Hydro-Hydraulic energy on Undular Tidal Bore phenomenon. In Proceeding of the Journal of Physics: Conference Series, The 10th International Conference on Theoretical and Applied Physics, Mataram, Indonesia, 20–22 November 2020. [CrossRef]
22. Chanson, H. Momentum considerations in hydraulic jumps and bores. *J. Irrig. Drain. Eng.* **2012**, *138*, 382–385. [CrossRef]
23. Leng, X.; Chanson, H. Coupling between free-surface fluctuations, velocity fluctuations and turbulent Reynolds stresses during the upstream propagation of positive surges, bores and compression waves. *Environ. Fluid Mech.* **2016**, *16*, 695–719. [CrossRef]
24. Madsen, P.A.; Simonsen, H.J.; Pan, C.H. Numerical simulation of tidal bores and hydraulic jumps. *Coast. Eng.* **2005**, *52*, 409–433. [CrossRef]
25. Wisha, U.J.; Wijaya, Y.J.; Hisaki, Y. Tidal bore generation and transport mechanism in the Rokan River Estuary, Indonesia: Hydro-oceanographic perspectives. *Reg. Stud. Mar. Sci.* **2022**, *52*, 102309. [CrossRef]

26. Chanson, H. Current knowledge in tidal bores and their environmental, ecological and cultural impacts. *Environ. Fluid Mech.* **2011**, *11*, 77–98. [CrossRef]
27. Keevil, C.E.; Chanson, H.; Reungoat, D. Fluid flow and sediment entrainment in the Garonne River bore and tidal bore collision. *Earth Surf. Process. Landf.* **2015**, *40*, 1574–1586. [CrossRef]
28. Wisha, U.J.; Rahmawan, G.A.; Ilham, I. *Bono Kuala Kampar, Primadona di Timur Sumatera Yang Terancam Hilang*, 1st ed.; AMAFRAD Press: Jakarta, Indonesia, 2018; pp. 1–89. (In Indonesian)
29. Putra, A.; Wisha, U.J.; Kusumah, G. Spatial analysis of the river line and land cover changes in the Kampar River estuary: The influence of the Bono tidal bore phenomenon. *Forum Geogr.* **2017**, *31*, 220–231. [CrossRef]
30. Matsumoto, K.; Takanezawa, T.; Ooe, M. Ocean tide models developed by assimilating TOPEX/POSEIDON altimeter data into hydrodynamical model: A global model and a regional model around Japan. *J. Oceanogr.* **2000**, *56*, 567–581. [CrossRef]
31. Pratama, M.B. Tidal Flood in Pekalongan: Utilizing and Operating Open Resources for Modelling. In Proceeding of the IOP Conference Series: Materials Science and Engineering, International Conference on Science and Engineering, Gowa, Indonesia, 24 October 2018. [CrossRef]
32. Zainuri, M.; Helmi, M.; Griselda, M.; Novita, A.; Kusumaningrum, H.P.; Koch, M. Improved performance of geospatial model to access the tidal flood impact on land use by evaluating sea level rise and land subsidence Parameters. *J. Ecol. Eng.* **2022**, *23*, 1–11. [CrossRef]
33. Rizal, S.; Damm, P.; Wahid, M.A.; Sündermann, J.; Ilhamsyah, Y.; Iskandar, T.; Muhammad. General circulation in the Malacca Strait and Andaman Sea: A numerical model study. *Am. J. Environ. Sci.* **2012**, *8*, 479–488. [CrossRef]
34. Setiawan, I.; Rizal, S.; Haditiar, Y.; Ilhamsyah, Y.; Purnawan, S.; Irham, M.; Yuni, S.M. Study of current circulation in the Northern Waters of Aceh. In Proceeding of the IOP Conference Series: Earth and Environmental Science, the 2nd International Conference on Marine Science; Better Insight for the Healthy Ocean, Bogor, Indonesia, 6–7 September 2017. [CrossRef]
35. Wyrtki, K. *Physical Oceanography of the Southeast Asian Waters: Scientific Results of Marine Investigations of the South China Sea and the Gulf of Thailand*, 2nd ed.; Scripps Institution of Oceanography, University of California: La Jolla, CA, USA, 1961; pp. 1–195.
36. Chawla, A.; Jay, D.A.; Baptista, A.M.; Wilkin, M.; Seaton, C. Seasonal variability, and estuary-shelf interactions in circulation dynamics of a river-dominated estuary. *Estuaries Coast.* **2008**, *31*, 269–288. [CrossRef]
37. Segeth, K. Some splines produced by smooth interpolation. *Appl. Math. Comput.* **2018**, *319*, 387–394. [CrossRef]
38. Dias, J.M.; Lopes, J.F.; Dekeyser, I. Tidal propagation in Ria de Aveiro lagoon, Portugal. *Phys. Chem. Earth Part B Hydrol. Ocean. Atmos.* **2000**, *25*, 369–374. [CrossRef]
39. Menéndez, M.; Woodworth, P.L. Changes in extreme high water levels based on a quasi-global tide-gauge data set. *J. Geophys. Res. Ocean.* **2010**, *115*, 1–15. [CrossRef]
40. Khojasteh, D.; Chen, S.; Felder, S.; Heimhuber, V.; Glamore, W. Estuarine tidal range dynamics under rising sea levels. *PLoS ONE* **2021**, *16*, e0257538. [CrossRef]
41. Baranya, S.; Józsa, J. Estimation of suspended sediment concentrations with ADCP in danube river. *J. Hydrolog. Hydromech.* **2013**, *61*, 232–240. [CrossRef]
42. Dwinovantyo, A.; Manik, H.M.; Prartono, T.; Susilohadi; Ilahude, D. Estimation of Suspended Sediment Concentration from Acoustic Doppler Current Profiler (ADCP) Instrument: A Case Study of Lembeh Strait, North Sulawesi. In Proceeding of the IOP Conference Series: Earth and Environmental Science, The 3rd International Symposium on LAPAN-IPB Satellite for Food Security and Environmental Monitoring, Bogor, Indonesia, 25–26 October 2016. [CrossRef]
43. Venditti, J.G.; Church, M.; Attard, M.E.; Haught, D. Use of ADCPs for suspended sediment transport monitoring: An empirical approach. *Water Resour. Res.* **2016**, *52*, 2715–2736. [CrossRef]
44. Mao, Q.; Shi, P.; Yin, K.; Gan, J.; Qi, Y. Tides and tidal currents in the Pearl River Estuary. *Continent. Shelf Res.* **2004**, *24*, 1797–1808. [CrossRef]
45. Kvale, E.P. The origin of neap-spring tidal cycles. *Mar. Geol.* **2006**, *235*, 5–18. [CrossRef]
46. Trevethan, M.; Chanson, H.; Takeuchi, M. Continuous high-frequency turbulence and suspended sediment concentration measurements in an upper estuary. *Estuar. Coast. Shelf Sci.* **2007**, *73*, 341–350. [CrossRef]
47. Simpson, J.H.; Fisher, N.R.; Wiles, P. Reynolds stress and TKE production in an estuary with a tidal bore. *Estuar. Coast. Shelf Sci.* **2004**, *60*, 619–627. [CrossRef]
48. Furgerot, L.; Mouazé, D.; Tessier, B.; Perez, L.; Haquin, S.; Weill, P.; Crave, A. Sediment transport induced by tidal bores. An estimation from suspended matter measurements in the Sée River (Mont-Saint-Michel Bay, northwestern France). *Comptes Rendus Géosci.* **2016**, *348*, 432–441. [CrossRef]
49. Donnelly, C.; Chanson, H. Environmental impact of undular tidal bores in tropical rivers. *Environ. Fluid Mech.* **2005**, *5*, 481–494. [CrossRef]
50. Rianto, S.; Santri, S. Obstacles and efforts to develop Bono tourism object in the Kampar River, Teluk Meranti Sub-District, Pelalawan Regency. *J. Spasial.* **2017**, *3*, 71–81. (In Indonesian) [CrossRef]
51. Hidir, A.; Asriwandari, H.; Kartikowati, S.R. Development Strategy for Coastal Society Based on the Development of Bono Tourism (Tidal Bore) in the Pelalawan Regency. In Proceeding of the National Seminar of Politic Bureaucracy, and Social Change in Efforts to Develop National Characters, Pekanbaru, Indonesia, 21 May 2013. (In Indonesian).

52. Persoalan Lahan Masyarakat Rantau Kasih vs. Perusahaan Kayu, Ada Penyelesaian? Available online: https://www.mongabay.co.id/2021/09/15/persoalan-lahan-masyarakat-rantau-kasih-vs-perusahaan-kayu-ada-penyelesaian/ (accessed on 14 August 2022). (In Indonesian).
53. Saathoff, F.; Oumeraci, H.; Restall, S. Australian and German experiences on the use of geotextile containers. *Geotext. Geomembr.* **2007**, *25*, 251–263. [CrossRef]
54. Mahabror, D.; Indriasari, V.Y.; Sofyan, A.; Nugroho, D.; Akhwady, R. Prototype engineering technology of elongated geotextile container as an alternative construction for mitigating abrasion in Pademawu Beach, Pamekasan Regency. *Naturalis* **2021**, *10*, 32–45. (In Indonesian) [CrossRef]
55. Anonymous. *Final Report of Natural Resources Inventory in the Pelalawan Regency 2009*; Department of Energy and Mineral Resources: Pekanbaru, Indonesia, 2009; pp. 221–308. (In Indonesian)

water

Article

Flow Field around a Vertical Cylinder in Presence of Long Waves: An Experimental Study

Rosangela Basile [1] and Francesca De Serio [1,2,*]

1 DICATECH, Department of Civil, Environmental, Land, Building Engineering and Chemistry, Polytechnic University of Bari, 70125 Bari, Italy; rosangela.basile@poliba.it
2 CoNISMa-Interuniversity Consortium for Marine Sciences, 00196 Rome, Italy
* Correspondence: francesca.deserio@poliba.it; Tel.: +39-080-5963557

Abstract: Long waves caused by storm surges or river floods can significantly impact marine and fluvial structures such as bridge piers. Apart from the forces that they generate on the structure, they also contribute to the formation of turbulent eddies downstream of the obstacle. This is relevant, as in this way they can affect both an erodible bottom and the ecosystem. The present study describes a medium-scale experiment, in which the propagation of two different long waves released on a steady current is investigated in the presence of a bottom-mounted rigid emergent cylinder. Velocity measurements were acquired by a Particle Image Velocimetry (PIV) system, providing instantaneous flow velocity vectors on selected 2D planes. For each experimental condition, the time-varying velocity field near the cylinder was examined in selected vertical and horizontal planes. First, we tested which analytical theory or approximated method can best represent the experimental waves. After this, we estimated the horizontal maps of velocity and vorticity downstream of the obstacle and finally processed the velocity signals by means of a wavelet-based technique, to derive the length scales of turbulent eddies. In such a way, we specifically derived how the spreading of coherent turbulent structures downstream of the cylinder depends on the features of the flume, cylinder, and wave.

Keywords: solitary wave; turbulent coherent structures; length scales; wavelet transform

Citation: Basile, R.; De Serio, F. Flow Field around a Vertical Cylinder in Presence of Long Waves: An Experimental Study. *Water* **2022**, *14*, 1945. https://doi.org/10.3390/w14121945

Academic Editor: Giuseppe Pezzinga

Received: 27 May 2022
Accepted: 15 June 2022
Published: 17 June 2022

1. Introduction

In a marine and fluvial environment, civil infrastructures are often exposed to extreme wave actions. This is certainly the case of coastal and offshore facilities during storm surges, as well as of riverine bridge piers when a flood occurs. These events, typical of extreme conditions, are becoming more and more frequent and dangerous with climate change. Therefore, to provide proper structural designs and safety assessments, we must evaluate the impact of such extreme waves on these structures [1]. Generally, many experimental and numerical investigations have focused on the detection of both drag and inertia forces exerted by a flow on a vertical cylinder using the well-known Morison equation [2], requiring the detailed knowledge of the velocity field in which the cylinder is immersed.

Specifically, current–structure interaction and wave–structure interaction have been extensively examined, but separately. Among the many, Zdravkovich [3] defined the regions of disturbed flow field around a circular cylinder; Rockwell [4] characterized the spanwise flow structure around the cylinder in terms of instantaneous velocity and vorticity; Vested et al. [5] analyzed the force distribution on a vertical, circular cylinder exposed to shoaling regular waves; Antolloni et al. [6] investigated the wave-induced vortex generation around a slender, vertical cylinder; Duan et al. [7] described the shedding of vortices and the hydrodynamic forces resulting from a wave passing over a submerged circular cylinder; Chen and Wang [8] investigated with experiments and computations the wave interaction with fixed, partially submerged, vertical cylinders; Mo and Jensen [9] performed laboratory experiments and numerical tests for solitary waves breaking on a constant slope, where a vertical cylinder was installed to detect the horizontal velocity profiles. More complex structures have also been examined in presence of waves, especially to evaluate their effect in reducing the transmitted energy [10,11]. Nevertheless, limited studies have investigated the joint wave–current–structure interaction. Further, when the mutual effects of all these three components have been considered, the waves analyzed were mainly regular ones. On the contrary, based on their great lengths and periods, extreme waves are often assumed as solitary waves. Only more recent research has addressed the interaction of the solitary wave with pile structures in experimental terms and in numerical terms. As an example, Yang et al. [12] carried out a series of experiments on a solitary wave–current interaction in the presence of a vertical obstacle, focusing on the changes in wave height and water velocity. Kim et al. [13] conducted numerical simulations of a wave and current interacting with fixed offshore substructures.

It is thus clear that more investigations are needed to better understand the mechanism of the interaction between wave, current, and structure. For this reason, the present study has a twofold purpose. Firstly, we aim at contributing to this deepening of knowledge by providing a set of high-quality experimental data for solitary waves travelling on a shallow current and interfering with vertical, bottom-mounted, rigid cylindrical structures. Secondly, we aim at analyzing in detail how the flow field downstream of the obstacle is affected by the transit of the wave in terms of vorticity and turbulence. In particular, we used medium-scale laboratory models and focused on the time-varying velocity field (i) upstream of the cylinder due to the wave transit, and (ii) downstream of the cylinder due to both the wave transit and the presence of the obstruction. Specifically, the first analysis is needed to detect how the velocity distribution changes due to the contemporary presence of waves and currents, while the second is dedicated to the observation of vorticity and coherent turbulent eddies that detach and spread from the cylinder. The underlying motivation is to contribute to a thorough knowledge of the distribution of eddies downstream, characterized by different length scales. In fact, it is well-known that the condition of incipient movement of sediment particles is significantly influenced by turbulent eddies given that they can penetrate up to the channel bed [14,15]. Moreover, resembling the transport of turbulent kinetic energy, eddies can contribute to the entrainment and transport of sediments or eventually tracers, thus they can affect a mobile bottom or even the biota and the local ecosystem [16,17].

This study is arranged as follows: in the next section we introduce the experimental setup used for the investigation and the theoretical frame for the wave's description. In Section 3, first the main features of the streamwise velocities are depicted and compared with the available theories to define the best matching one based on a proper index. After this, the vertical profiles of the flow velocity are examined during the wave transit. Further, vorticity and coherent turbulent structures in the horizontal plane around the cylinder are examined. Finally in Section 4 the main conclusions are summarized.

2. Materials and Methods

2.1. Experimental Equipment and Procedure

We performed the experiments at the Hydraulic Laboratory of the DICATECH of the Polytechnic University of Bari (Italy). The used rectangular flume, 25 m long and 0.4 m wide, has sidewalls and bottom fabricated of Plexiglass and is well-suited for optical measurements (Figure 1). The head tank feeding the flume can be served independently by both a low-pressure and a high-pressure water circuit. The low-pressure main circuit provided a stationary flow condition in the flume during the test (base flow). The high-pressure adduction pipe released an additional water discharge in the head tank by means of an electro valve (Figure 1) managed by a process PC, equipped with a LabVIEW software. In this way, by conveniently tuning the added water release, the desired long wave was generated in the channel, superimposed to the uniform base flow. At the downstream end of the channel, a second tank is located, receiving the discharged flow. It is provided with a triangular sharp-crested weir used to estimate the flow rate. The water level is controlled by a sloping gate at the end of the flume. In order to reduce the reflection of the generated waves, a structure with a high degree of porosity, consisting of a 2 m in length metal cage with a 0.01 m mesh filled with d_{50} = 0.015 m gravel, is located in the final part of the flume on the bottom. However, the measurements of the tested waves were acquired in a time period specifically chosen to avoid any reflection.

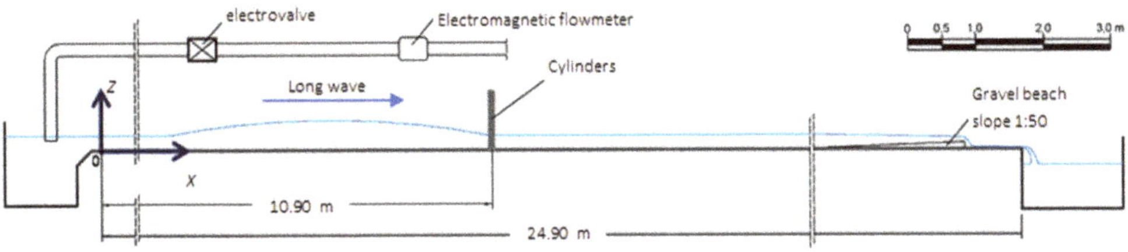

Figure 1. Sketch of the channel with location of the cylinders and (*X*,*Z*) axes. Side view.

The physical model was designed to reproduce a possible bridge pile affected by a river flood, in Froude analogy, thus using a length scale factor equal to 1/10 (model/prototype). Therefore, two rigid cylinders having a diameter d = 0.02 m were located along the Y axis in the same transversal section, at a distance of X = 10.9 m from the head tank O (X being the longitudinal axis of symmetry of the channel as in Figure 1). They were equidistant from the X axis, with Y = 0.10 m and Y = −0.10 m, respectively (Figure 2).

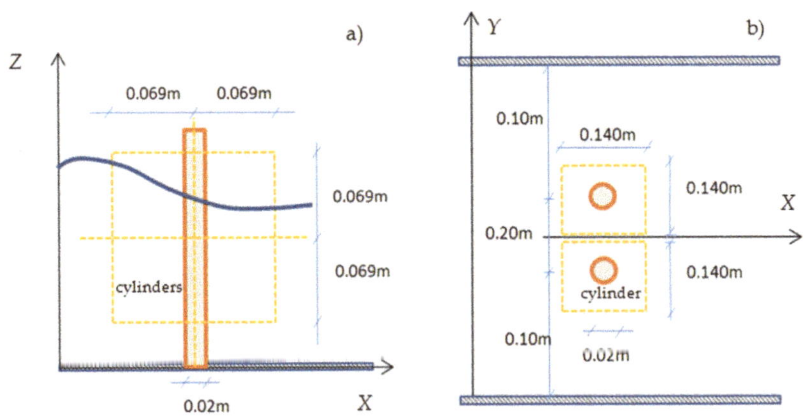

Figure 2. Sketch of the longitudinal (**a**) and plan view (**b**) of the cylinders in the flume with the fixed coordinate system (*X*,*Y*,*Z*). Dashed lines define the field of views (FoV) of the PIV. Measurements not to scale.

In order to obtain the flow velocity vectors on selected 2D planes, the velocity measurements were acquired by a Particle Image Velocimetry (PIV) technique. The 2D PIV system was equipped with a FlowSense EO 4M-32 camera, a laser with pulse energy of 200 mJ at 15 Hz and a synchronizer, controlled and monitored by a PC. The system was handled in double-frame mode; the sampling frequency was settled to 8.13 Hz and the time interval between two frames of the same pair was 150 μs. Considering the importance of seeding particles to obtain good quality measurements [18], we used a neutrally buoyant silver powder as seeding tracer with a mean diameter of O(10 μm).

The data examined in the present work refer to the flow velocity (i) measured in the vertical plane (*X*,*Z*) at Y = 0.10 m (Figure 2a), (ii) and measured in the horizontal plane (*X*,*Y*) located at Z = 0.03 m from the bottom. The velocity components are: u along X, positive towards the gravel beach; v along Y, positive following Y; w along Z, positive upwards, being Z = 0 m at the channel bottom. The field of views (FoVs) were properly selected (see dashed squares in Figure 2). The interrogation area of the images in the adaptive correlation processing was 32 × 32 pixels, thus the velocity vectors were assessed in points regularly spaced at 0.4 mm in the vertical plane and 0.8 mm in the horizontal one, resulting in a high spatial resolution. The number of such grid points for each FoV was 172 × 172, so the calibrated PIV images had the dimensions 69 mm × 69 mm and 138 mm × 138 mm, respectively, in the vertical and horizontal plane. Considering a precision in estimating the displacement within an interrogation window of 0.1 pixel, the accuracy for the detected PIV velocity was close to 1%.

The water depth in the flume was always set to be h = 0.1 m. The base flow rate calculated with the Thomson-type triangular weir provided a reference base flow velocity u_0 equal to 0.16 m/s for all the tests. The base flow Froude number and Reynolds number were $Fr_0 = u_0/\sqrt{gh} = 0.162$ and $\mathrm{Re}_0 = u_0 B/v = 64,000$, respectively, thus representing a turbulent subcritical condition. The water kinematic viscosity ν was set 10^{-6} m^2/s, while B was the channel width.

In order to replicate a flooding wave impacting a bridge pile two different long waves were examined in the experiments named W1 and W2, respectively. Each one was generated by linearly opening and successively closing the electro-valve of the high-pressure circuit (Figure 1) for a specific temporal interval. They were characterized by a wave height H, wave period T, wavelength L, and an Ursell number (index of non-linearity) $Ur = HL/h^2$, as shown in Table 1. The number of images acquired by the PIV was limited by technical reasons related to the storage size of the dedicated PC, thus it was set to 150 for each measurement. Consequently, the total acquisition time for each measurement was close to 18 s, which is lower than the entire wave periods. Nevertheless, it was sufficient to capture most of the ascending and descending part of each examined wave.

Table 1. Characteristics of the two tests.

Test	H [m]	T [s]	L [m]	Ur [–]
W1	0.025	20	32	80
W2	0.05	19	25	125

For both W1 and W2 tests, eight identical waves were generated and successively released in the flume, each of them separated from the previous one by the time-interval necessary to reconfigure the undisturbed condition in the channel. In this way, we could operate a phase averaging technique on the measured velocity signals [17] and obtain the turbulent fluctuations, as written in the following.

Finally, it is worth noting that in our physical model the spacing between the cylinders was 10 times their diameter, and thus each cylinder behaved as isolated, not affecting the flow field around the other one. Rather, a symmetry with respect to the longitudinal X axis in the flow was detected. This motivated our choice to focus our current analysis only on one cylinder, specifically on cylinder 2 (Figure 2).

2.2. Theoretical Background for Waves Description

To describe analytically the generated waves, based on their shape and featuring parameters, we referred to shallow waters following the classical definition [19]. In fact, in our case, the ratio between the water depth and the wavelength is h/L = 0.0008 for W1 and h/L = 0.002 for W2, i.e., less than 1/25 = 0.04.

For this scope, we used the Korteweg–de Vries equation [20] originally derived to describe shallow water waves of great wavelength and small amplitude, thus having $h \ll L$, and $H/h \ll 1$. Its solution provides the time-varying orbital velocity component u along the wave propagation direction x_i:

$$u(x_i, t) = \frac{\omega - 4k^3(2m-1)}{\alpha k} + \frac{12k^2 m}{\alpha} cn^2(kx_i - \omega t + \delta; m) \quad (1)$$

where $k = 2\pi/L$ is the wave number, $\omega = 2\pi/T$ is the angular frequency, δ is the phase and α is a parameter that can be scaled to any real number (but generally posed ± 1 and ± 6) [20]. It is called the cnoidal wave solution for it involves the Jacobi elliptic cosine function, *cn* with modulus m (where $m = 0 \div 1$). Specifically, when m tends to zero, Equation (1) describes the Airy linear (LIN) wave [19]. On the contrary, when $m = 1$, Equation (1) describes a solitary wave with a finite amplitude and propagating entirely above the mean water level with constant speed and without change in shape over a fairly long distance.

As an alternative solitary wave solution, we considered the regularized long-wave equation proposed by Peregrine (PER) to describe the behavior of an undular bore with a smooth wavefront and having the following expression [20]:

$$u(x_i, t) = \frac{\omega - k - 4k^2\omega}{\alpha k} + \frac{12k\omega}{\alpha} \mathrm{sech}^2(kx_i - \omega t + \delta) \quad (2)$$

Moreover, the classical solitary solution of the Boussinesq equation (BOU) was applied in our study [21]:

$$u(x_i, t) = \frac{\omega^2 - c^2k^2 - 4\beta k^2}{\alpha k^2} + \frac{12\beta k^2}{\alpha} \mathrm{sech}^2(kx_i - \omega t + \delta) \quad (3)$$

where typically one sets $\alpha = 3$ and $\beta = 1$, while c is the wave celerity.

Finally, we proposed to test an engineering method not tailored on shallow waters, but widely used in practice due to its simplicity, that is the so-called linear Wheeler stretching (WH) method [22]. It attempts to correct the linear Airy theory, whose straight application is valid from sea bottom to mean water level, by stretching the vertical coordinate Z up to the instantaneous free surface. Thus, a transfer function between wave elevation η and velocity components in the time domain is provided. As for the u component, we have:

$$u(x_i, Z, t) = \frac{\pi H}{T} \frac{\cosh k[Z/(1 + \eta(x,t)/h)]}{\sinh(kh)} \cos(kx_i - \omega t) \quad (4)$$

In the present study, due to some technical limitation during the experiments, wave elevations were not assessed; therefore, we substituted in Equation (4) the exact solitary wave elevation expression computed as [21]:

$$\eta(x_i, t) = H\mathrm{sech}^2\left[\frac{1}{h}\sqrt{\frac{3H}{4h}}(x_i - ct)\right] \quad (5)$$

3. Results and Discussion

The results displayed in the following are based on the vector field maps assessed from pairs of PIV particle images by means of the adaptive correlation method, as implemented in data processing.

3.1. Analytical Characterization of the Experimental Waves

The models described in Section 2.2 were taken into account and their reliability in reproducing the experimental waves generated was tested. To allow a consistent comparison between theoretical and experimental waves, we specifically referred to the streamwise velocity component u, suitably treated. Considering a fixed position (in our case $x_i = X$), we computed $u(t)$ through all the aforementioned analytical models. We used the following procedure for both cases W1 and W2 to extract the velocity measurements for the wave cycle from the experimental time series. We examined the streamwise velocity measured in a large number of points distant and undisturbed by the cylinders, both in the vertical plane upstream (Figure 2a) and in the horizontal plane (Figure 2b). For each measurement, the phase averaging technique was used to process the $u(t)$ signal, so that it was decomposed in a time-averaged value U, a phase-averaged value $u_p(t)$ and a turbulent fluctuation $u'(t)$ [17]. The ensemble-averaged value $<u(t)>$ was also taken into account with the sum $< u(t) >= U + u_p(t)$. This phase averaging operation was applied to the eight waves measured, even if the assumption made of a solitary wave actually implies a non-periodical wave. We decided to follow this procedure because in this way (i) we could account for more robust and consistent measurements, not referring to a single episodic wave event, (ii) and we could operate statistics and extract turbulent components from the signal. Consequently, we were able to compare the trends of the theoretical u varying in the wave cycle with the corresponding phase-averaged experimental values u_p. As an example, for a selected point representative of the undisturbed hydrodynamic condition (far from the cylinder C2 and having coordinates X = 10.95 m, Y = 0.07 m and Z = 0.035 m), Figure 3 shows the comparison among the computed values of u and the measured u_p (normalized by the time-averaged velocity U). The time-averaged velocity in the two experiments is U = 0.28 m/s for W1 and U = 0.40 m/s for W2, respectively.

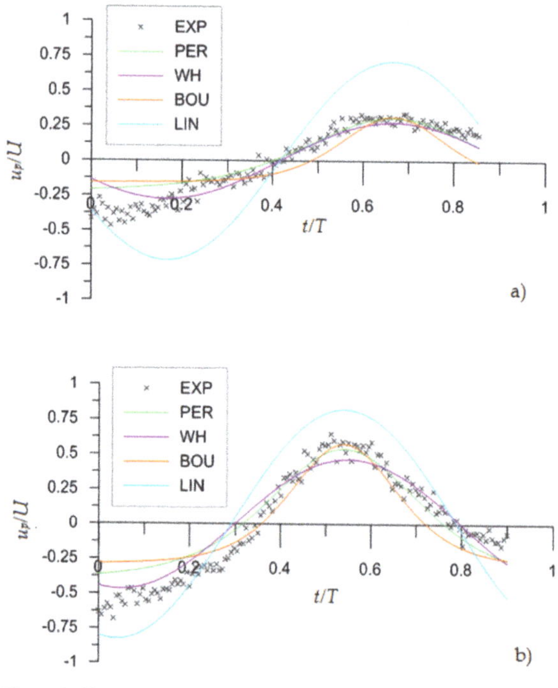

Figure 3. Comparison of computed u and experimental phase-averaged streamwise velocity u_p at point X = 10.95 m, Y = 0.07 m, and Z = 0.035 m for W1 (**a**) and W2 (**b**).

In qualitative terms, for both waves W1 and W2, the WH method seems to show the best matching with the experimental data, while the linear model provides the largest overestimations in absolute terms. To evaluate the agreement also in quantitative terms, the Wilmott index Iw was considered [23]:

$$Iw = 1 - \frac{\sum_{i=1}^{N} (\chi_{ci} - \chi_i)^2}{\sum_{i=1}^{N} (|\chi_{ci} - \overline{\chi}| + |\chi_i - \overline{\chi}|)^2} \tag{6}$$

where χ_c and χ are, respectively, the computed and the experimental values; the overbar stands for the corresponding average value; and N is the total number of data in a wave cycle. The Iw index takes a value of 1 when a perfect agreement is obtained. If calculated for the whole wave cycle, it provides the highest values for the WH method for both W1 and W2 waves, resulting $Iw = 0.99$ and $Iw = 0.97$, respectively.

In any case, at a closer inspection, a worse agreement can be observed between the measured data and the WH data during the wave crest transit. Thus, we also computed Iw for partial data intervals, namely for the positive velocity values (typical of the wave crest transit) and for the negative velocity values (typical of the wave trough transit). The resulting indexes are shown in Table 2.

Table 2. Wilmott index for comparisons of the streamwise velocity trends shown in Figure 3.

Iw	PER	WH	BOU	LIN
W1 trough	0.95	0.97	0.90	0.86
W1 crest	0.94	0.95	0.86	0.70
W2 trough	0.87	0.91	0.80	0.80
W2 crest	0.99	0.98	0.98	0.92

It is evident that for wave W1, the WH method is the best suited one along the whole wave cycle. For wave W2, it remains the best referring to the wave trough; while for the wave crest case, the PER model better reproduces the velocities u_p. It is worth noting that we obtain analogous results (within ±2%, i.e., comparable to PIV accuracy) even when experimental velocities are examined at other undisturbed points. Finally, these results are also consistent with Iwagaki's abacus [15].

3.2. *Upstream Vertical Profiles of the Streamwise Velocity*

For both W1 and W2 waves, we analyzed the FoVs in the vertical plane, upstream the cylinder (Figure 2a). The flow field was observed both in stationary condition (i.e., base flow) and in wave condition, during the transit of the long wave. The vertical profiles of the ensemble-averaged velocity display a quite flat and uniform trend along most Z, during each phase of the wave. Apart from the typical values of the ratio h/L and the indications of Iwagaki's abacus already written in Section 2.2, it was this observation that led us to address theoretically shallow water waves characterized by a condition of homogeneous flow along Z. The rapid release of the added mass into the laboratory flume head tank produced a long flooding wave that promptly affected the entire water depth from the most superficial layer towards the channel bottom. The effect of the bottom on the flow is most visible during the descending wave phase when a reduction in velocity is generally assessed over the entire depth, but is even faster in the range $Z/h < 0.4$. In this range, for both waves, the logarithmic trend typical of uniform flow in open channels is identifiable.

Figure 4 displays, as an example, the vertical distribution of $\langle u \rangle$ normalized by U at four different wave phases, in the profile at $X = 10.842$ m (i.e., at 0.058 m upstream of the cylinder) and $Y = 0.10$ m, for both W1 and W2. For wave W1 the phases addressed are (Figure 4a): $t/T = 0.05$, i.e., the wave trough; $t/T = 0.3$, i.e., the ascending wave phase; $t/T = 0.65$, i.e., the wave crest; and $t/T = 0.8$, i.e., the descending wave phase. Analogously, for wave W2, we have (Figure 4b): $t/T = 0$, $t/T = 0.3$, $t/T = 0.55$, and $t/T = 0.7$, respectively.

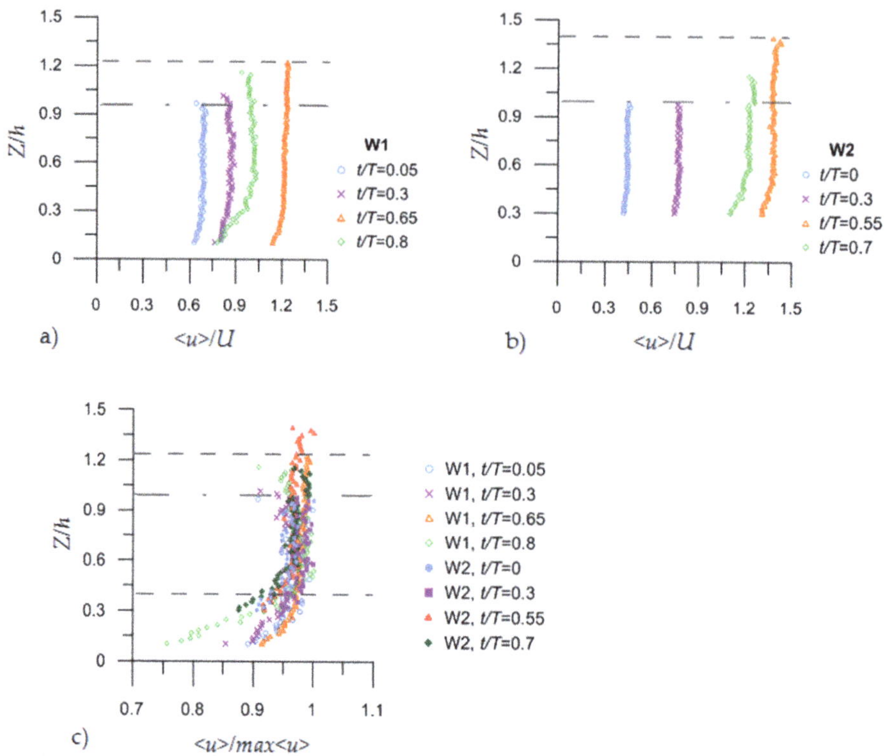

Figure 4. Top row: vertical profiles of the ensemble-averaged streamwise velocity $\langle u\rangle$, normalized by U, measured at wave trough, ascending phase, crest, descending phase. Selected profile at X = 10.842 m and Y = 0.10 m. (**a**) Wave W1 and (**b**) wave W2. Bottom row (**c**): vertical profiles of $\langle u\rangle$ normalized by the maximum value of $\langle u\rangle$ detected for each profile.

The vertical profiles were obtained by stitching the instantaneous velocity maps of the two FoVs upstream of the cylinder. In fact, we originally acquired a top FoV and a bottom FoV, which overlapped 0.005 m along their common edge, as plotted in Figure 2a. In particular, for the W1 wave the lower edge of the bottom FoV was positioned at Z = 0.01 m, while W2 was at Z = 0.03 m, in order to ensure the measurement by the PIV in the upper region during wave crest transit. This transit is evident from the increasing relative heights Z/h where velocities were measured. In fact, in the trough condition the highest point investigated is close to $Z/h \cong 1$ in both waves, while in the crest condition, it is close to $Z/h \cong 1.2$ for W1 and $Z/h \cong 1.4$ for W2 consistently with the wave heights. We observed increasing values of $\langle u\rangle$ due to the wave transit, as expected, reaching maxima values under the wave crest, with $\langle u\rangle = 1.24U$ for W1 and $\langle u\rangle = 1.4U$ for W2. Recalling that $\langle u\rangle$ is the sum of the time-averaged component and the phase-averaged velocity, we can equally write that the initial current flow at most increases by about 24% and 40%, respectively, for W1 and W2 due to the wave crest transit. In addition, approaching from the trough to the crest condition, the $\langle u\rangle$ values double for W1, while increasing 3.5 times for W2, meaning that a more pronounced effect of the wave on the underlying current is evident in the W2 case rather than in the W1 case.

Plotting the same vertical profiles of $\langle u\rangle$ normalized by the maximum value of $\langle u\rangle$ detected at each profile (named $max\langle u\rangle$), we obtain Figure 4c, where all profiles tend to collapse and clearly show a flat trend for $Z/h > 0.4$. For $Z/h < 0.4$ the logarithmic trend of $\langle u\rangle$ is easily recognizable, especially in the W1 case, having measured points closer to the bottom. However, for both cases a bottom boundary layer with a thickness $\delta \cong 0.4h$ is clearly detected, regardless of the wave phase.

3.3. Downstream Horizontal Maps of Streamwise Velocity and Vorticity

The horizontal maps of <*u*>/*U* measured at Z = 0.03 m are displayed in Figure 5. For the sake of brevity, only the maps corresponding to the trough and crest phases already selected and shown in Figure 4 are depicted. For greater convenience, a new local reference system (*x*,*y*) was chosen in the FoV, with the origin in the center of the cylinder.

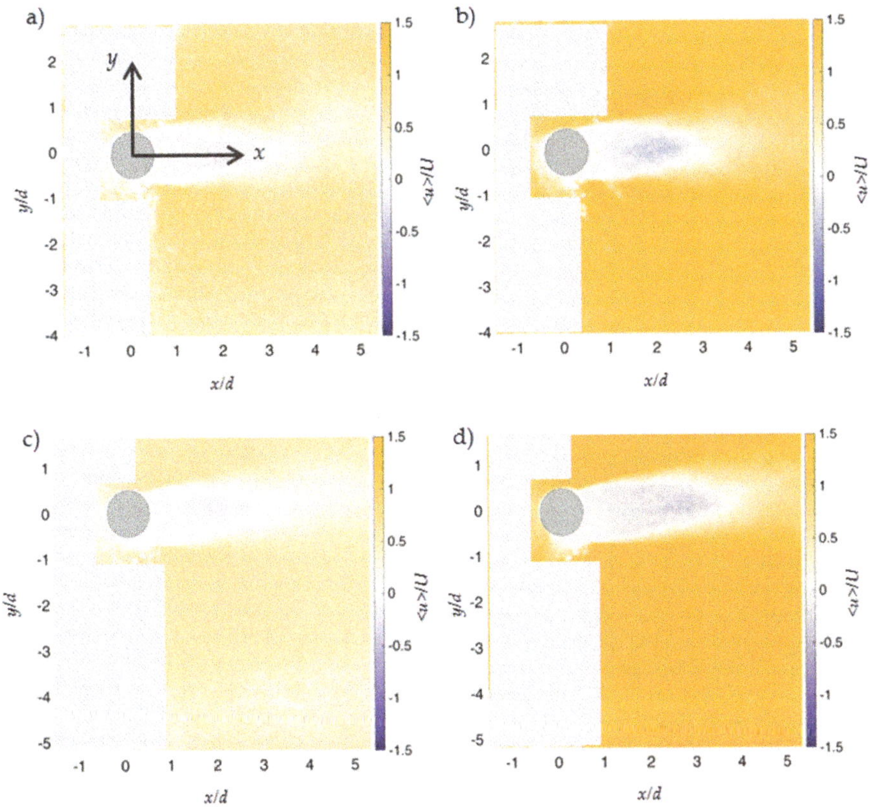

Figure 5. Horizontal maps of the normalized ensemble-averaged streamwise velocity <*u*> at phases (**a**) W1 trough, (**b**) W1 crest, (**c**) W2 trough, (**d**) and W2 crest. Z axis is vertical and positive upward; here Z = 0.035 m.

Some spurious results close to the left edge of the frame are present, due to some technical limitations in the lighting system during PIV acquisition. They do not affect the signal downstream of the cylinder in the area of greatest interest, but we decided to mask them to avoid misinterpretations. For both tests, a shaded area downstream of the cylinder is always present, as expected, showing negative <*u*> values in the near wake. Due to a less intense base flow and a lower wave height in W1, the absolute values of <*u*> are lower than those in W2, both in the trough (Figure 5a) and in the crest (Figure 5b). Further, the higher *U* value for W2 determines a more extended shaded area (Figure 5c,d). In fact we observe negative values of <*u*> for $x/d = 0 \div 3$ in the W1 case and up to $x/d \cong 4$ in the W2 case. Instead, the transversal spreading of such negative velocities is of order O(*d*) for both tests. The inversion of the sign of <*u*> delimits the shaded area, with values close to zero along its edge. As a result, a recirculation arises, with eddies detaching from the cylinder and antisymmetric with respect to the local X axis, as already observed in uniform flows. However, the transit of the solitary wave, and thus the presence of a trough and a crest condition, induces a variation in the vorticity distribution and in the characteristics of coherent turbulent eddies, as explained below.

The vorticity <Wz> in the horizontal plane at $Z = 0.03$ m was computed starting from the ensemble-averaged streamwise <u> and spanwise <v> velocity components measured by the PIV. Figure 6 displays the corresponding horizontal maps for trough and crest phases, for W1 and W2, respectively.

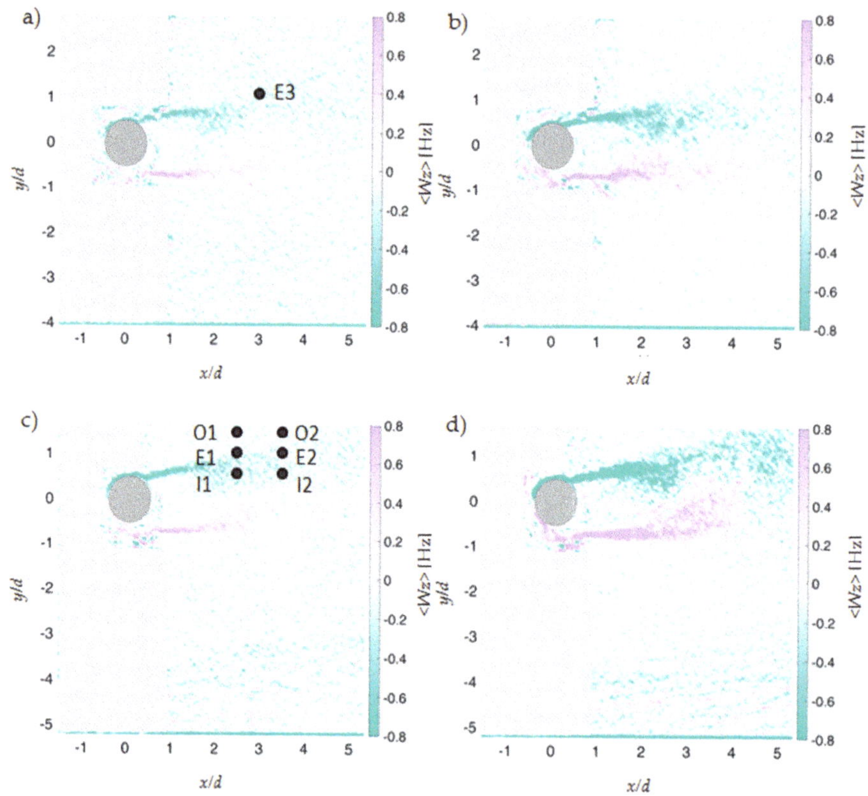

Figure 6. Horizontal maps of the ensemble-averaged vorticity <Wz>, at phases (**a**) W1 trough, (**b**) W1 crest, (**c**) W2 trough, (**d**) and W2 crest. Marked points identify the locations used for the successive wavelet analysis. Z axis is vertical and positive upward; here $Z = 0.035$ m.

Even in these plots the areas with rough data are masked, showing unreliable results. Opposite values of vorticity are observed downstream of the cylinder, mainly clockwise (negative) for $y/d > 0$ and anticlockwise (positive) for $y/d < 0$ consistently with the stationary case of a flow investing a cylinder. During the wave cycle, from trough to crest condition, the cylinder Reynolds number $\mathrm{Re}_c = < u > d/v$ increases, due to the increase in <u>, from 3400 to 6800 for W1 and from 3600 to 11,200 for W2. Thus, a more intense fully turbulent vortex street appears, as evident by comparing Figure 6a,b as well as by comparing Figure 6c,d. The antisymmetry of vorticity along the X axis is generally kept, with the highest values of Wz located at the trailing edge of the shaded area, while a spreading of vorticity with lower intensity is noted outside. In crest conditions, the values of Wz increase, especially for the W2 case where they are in the range $(-0.8 \div 0.8)$ Hz, while for the W1 case they remain in the range $(-0.6 \div 0.6)$ Hz.

3.4. Turbulent Coherent Structures

Methods based on vorticity magnitude were used fairly extensively to identify vortices in the flow. Nevertheless, it should be considered that shear layers also exhibit high vorticity magnitudes. Turbulence structures essentially evolve from the interactions between vorticity and strain rate, so in the present study, to better characterize coherent turbulent vortices in the flow, we applied the continuous wavelet transform (CWT), a power tool capable of pinpointing regions with different spatial and temporal scales in turbulent flows [24], thus providing continuous time-frequency identification of eddy structures.

Classical methods to decompose the velocity signal into frequency bands have invariably been based on the Fourier transform [25]. That is, the underlying assumption is that the same spectral components are always present in a signal. However, this approach is not particularly suited for the treatment of non-stationary signals, such as in the present case, where we distinguish and establish the length and time scales of the coherent structures residing downstream of the cylinder.

The CWT approximates a complex function as a weighted sum of simpler functions, which themselves are obtained from one simple prototype function Ψ, called the mother wavelet. Several functions can be used as the mother wavelet and we assumed the Morlet one [24,26]. In CWT, the temporal distribution of the frequency components of the signal is found by successively passing stretched and compressed versions of the function Ψ, throughout the signal. By decomposing a time series into time-frequency space, one is able to determine both the dominant modes of variability and how those modes vary in time [22].

For both waves we operated in the same way, but for convenience, we firstly focused on the results of the W2 wave. We selected six reference points in the horizontal plane (at Z = 0.03 m), along two different transversal sections at relative distances x/d = 2.5 and x/d = 3.5, and specifically: O1(x/d = 2.5, y/d = 1.5), O2(x/d = 3.5, y/d = 1.5), E1(x/d = 2.5, y/d = 1), E2(x/d = 3.5, y/d = 1) I1(x/d = 2.5, y/d = 0.5), and I2(x/d = 3.5, y/d = 0.5). Points O1 and O2 are outside the cylinder wake; points E1 and E2 are located at the trailing edge; points I1 and I2 are inside the wake (refer to Figure 6 for points location). For each point, the time series of the turbulent streamwise velocity $u'(t)$ to be processed by CWT was obtained by stitching the $u'(t)$ signals of the eight consecutive waves generated in the channel and cutting out the time interval between two succeeding waves. For the six target points, Figure 7a–f display the $u'(t)$ time series, the corresponding scalogram obtained with CWT, and the global wavelet spectrum. The scalogram is the wavelet power spectrum where the highest values in the contour plot correspond to the most energetic frequencies of the signal. The black line represents the cone of influence for the wavelet, meaning that values external to such cone are untrusted because they are too close to the extremes of the time series. The global wavelet spectrum is obtained for each frequency scale from the integral of the wavelet power calculated over the period of investigation.

Considering points O1 and O2 (Figure 7a,d) in the wavelet power spectrum low intensities are generally observed, except than in the range of frequencies ($0.0625 \div 0.25\ s^{-1}$) where more energetic spots are evident during most of the examined time interval. They are related to the presence of large coherent structures which retain the maximum turbulent power, as can easily be seen by observing the global wavelet spectrum. According to [24,25], we can rely on the well-known Taylor's approximation to convert the time scale $1/f$ into the length scale λ. That is $\lambda = 1/fu_s$, where u_s is a proper velocity scale, greater than the turbulent velocity. In the present case, following [26], we assumed the local ensemble-averaged velocity $<u>$ as u_s, resting on the assumption that it conveys downstream of the coherent structures. As a result, the above written large coherent structures outside the wake have corresponding length scales with an order of magnitude in the range $O(10^0 \div 10^{-1}\ m)$, in particular varying between one-tenth of the wavelength $L/10$ and the flume width B, thus controlled by the flume geometry.

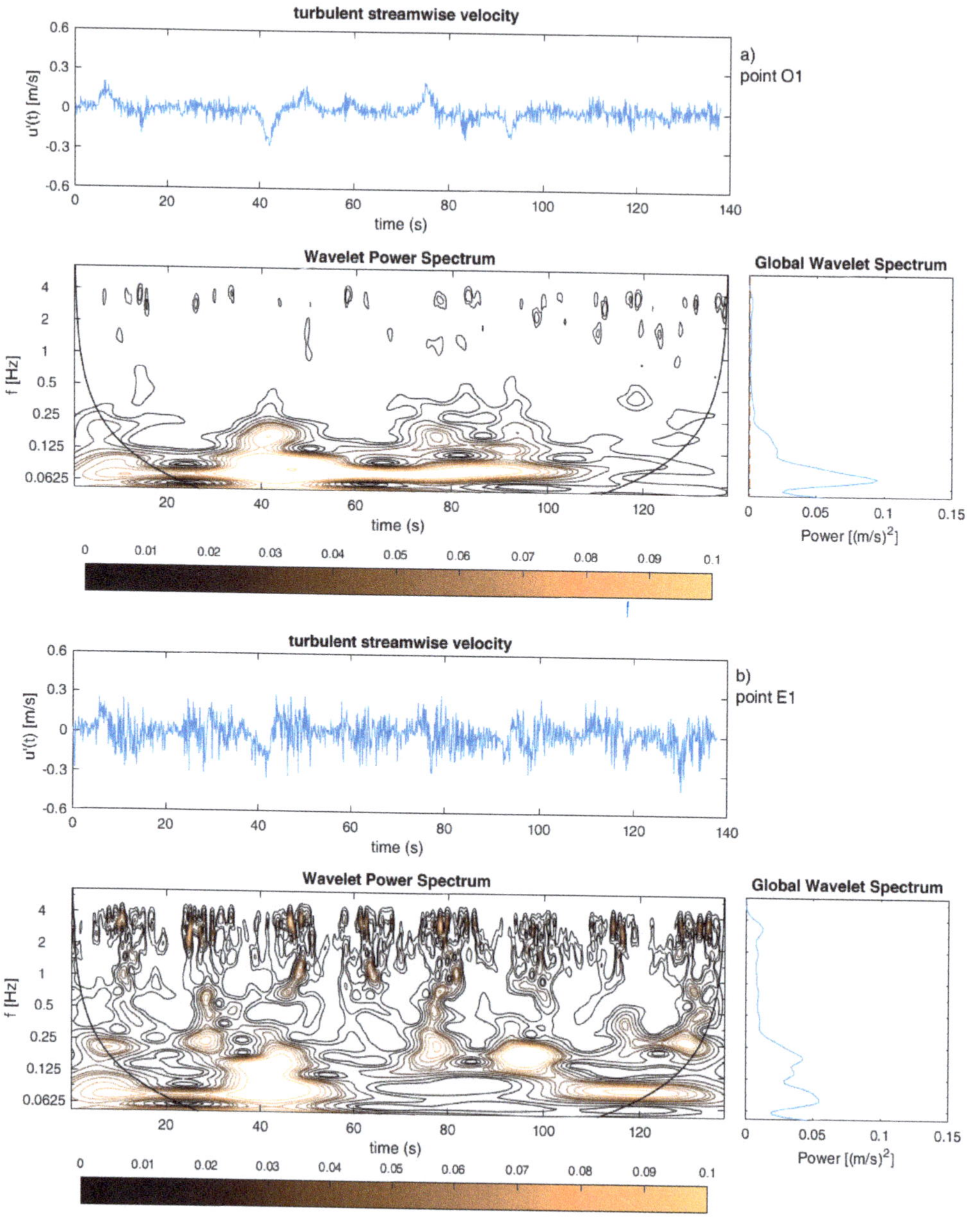

Figure 7. *Cont.*

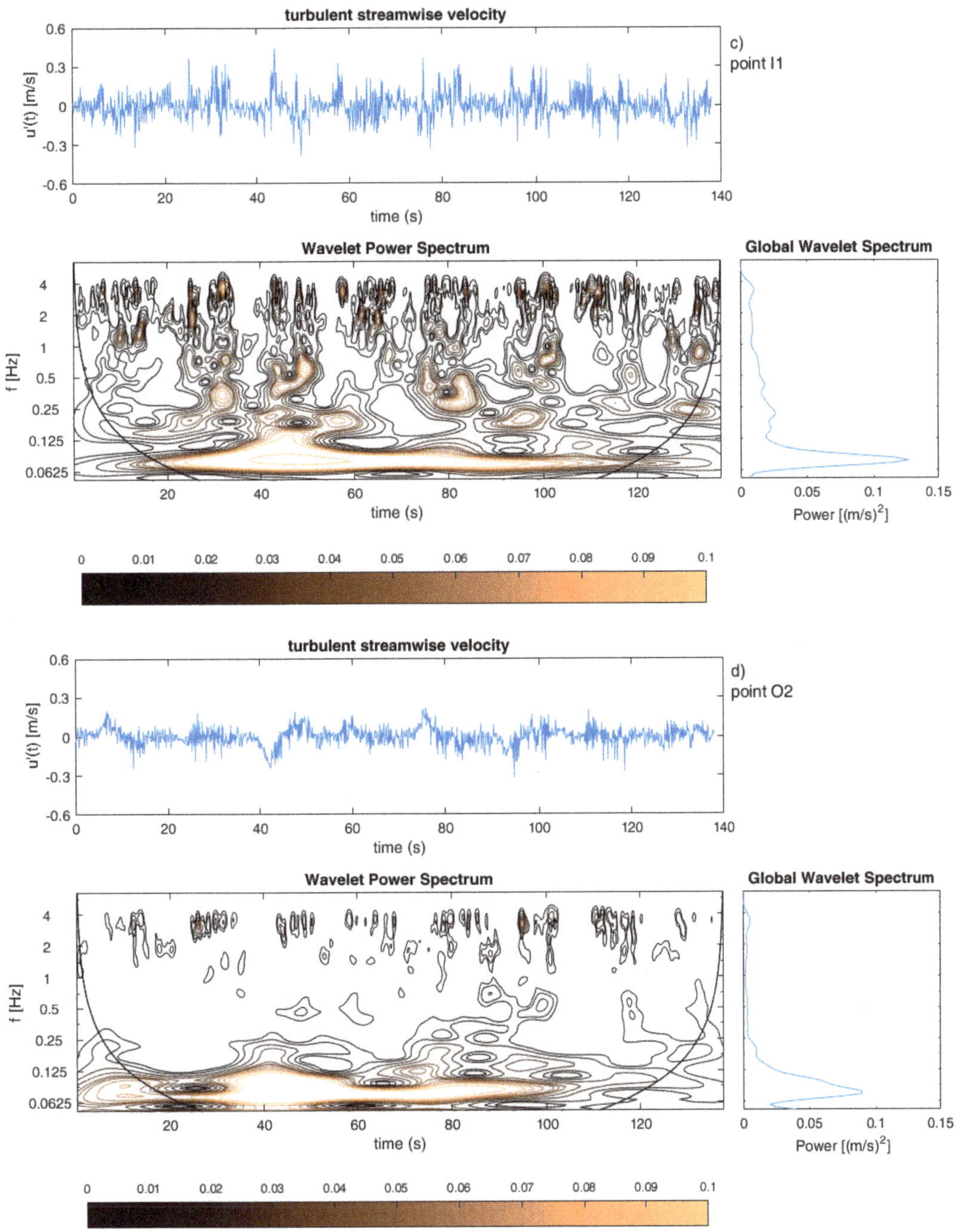

Figure 7. *Cont.*

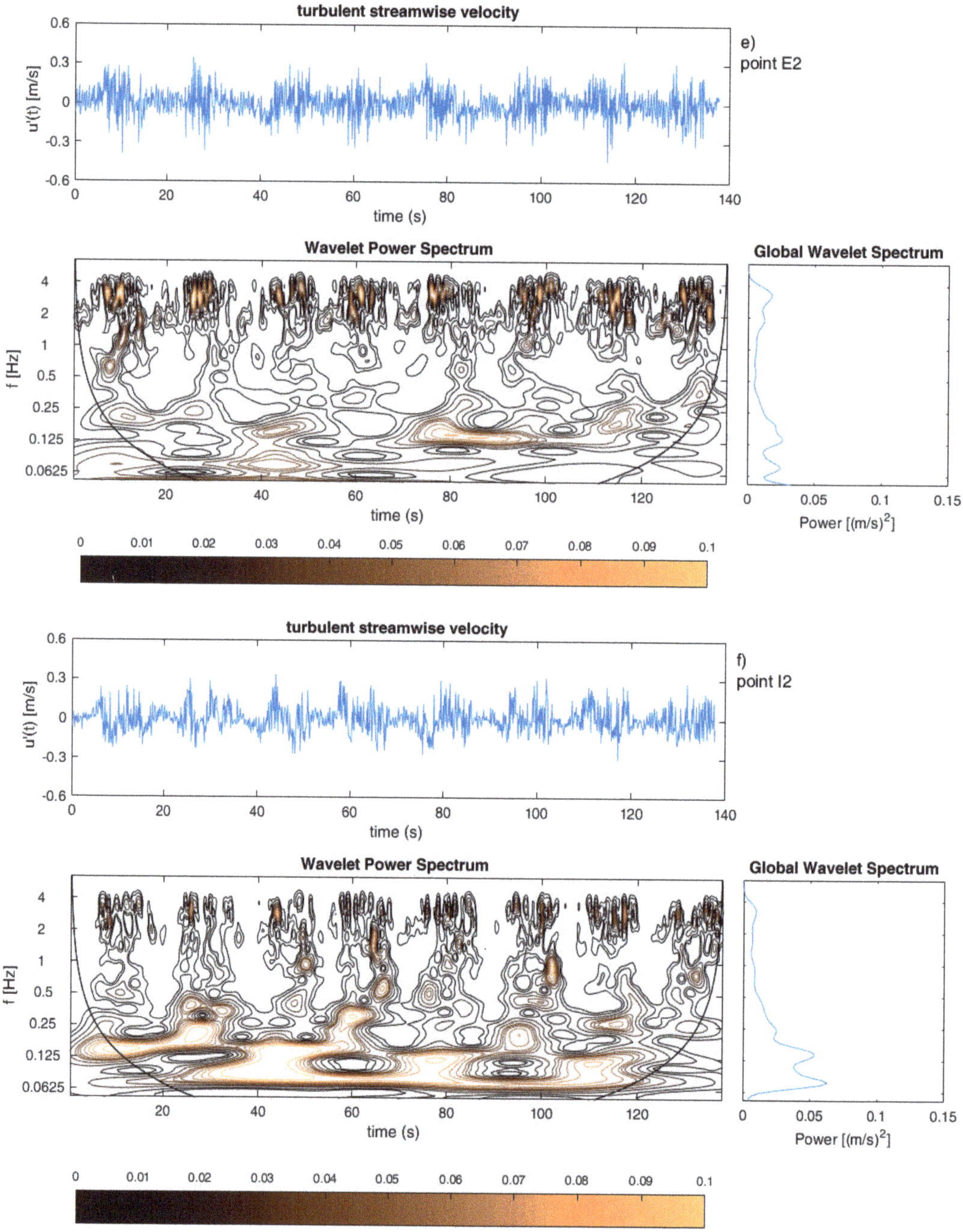

Figure 7. W2 wave: signal of the turbulent streamwise velocity $u'(t)$, spectrogram of $u'(t)$, and global wavelet spectrum for the selected points: O1 (**a**), E1 (**b**), I1 (**c**), O2 (**d**), E2 (**e**), and I2 (**f**).

The results of the wavelet power spectrum for points E1 and E2 along the edge (Figure 7b,e) are also remarkable. Moreover, in this case, high intensities in the wavelet spectrum occur at frequencies ($0.0625 \div 0.25\ \mathrm{s}^{-1}$), manifesting again the presence of large coherent structures similar to what was already noted for O1 and O2 points, even if characterized by lower power values. More interestingly, in E1 and E2, high power spots are visible even at higher frequencies ($2 \div 3.5\ s^{-1}$), thus, we would expect them to represent small eddies. Considering that they appear only during the peak phase of the wave (as evident by moving on the time axis in Figure 7b,e), in the computation of the associated length scales we used the $<u>$ velocity assessed at the wave crest; therefore, their length scales result in the range ($10^{-2} \div 10^{-1}$ m). More precisely, the spectrograms of Figure 7b,e show that, when the wave crest is approaching, coherent structures with the size of order of the cylinder diameter O(d) are present along the edge; during the transit of the crest, such structures become larger and are affected by the wave height with λ values around O($4H$).

Figure 7c,f display the CWT results for points I1 and I2 inside the wake but close to the edge, where a behavior analogous to what was already observed for points E1 and E2 is detected. In this case, the structures at frequencies ($0.0625 \div 1\ s^{-1}$), present during most of the recorded time period, show again a higher turbulent power, comparable to O1 (Figure 7a) and O2 (Figure 7d) cases. The power intensity of the smaller coherent structures arising at higher frequencies ($2 \div 3.5\ s^{-1}$) is quite the same as for points E1 (Figure 7b) and E2 (Figure 7e).

Referring to the coherent structures detected in the W1 case, when investigated with CWT, the turbulent power values in the spectrograms are generally halved compared with the W2 case. Nevertheless, results are similar to the W2 case in terms of frequency of occurrence of the detected coherent structures. Moreover, the order of magnitude of the length scales of the eddies is generally kept, even if their size is reduced compared with the W2 case. The reduced size of vortices behind the cylinder in the case of lower wave amplitudes was also observed by [27] for harmonic waves. For brevity, only one result for the W1 case is shown in Figure 8, referring to a representative point along the trailing edge, namely E3, at $x/d = 3$, and $y/d = 1$.

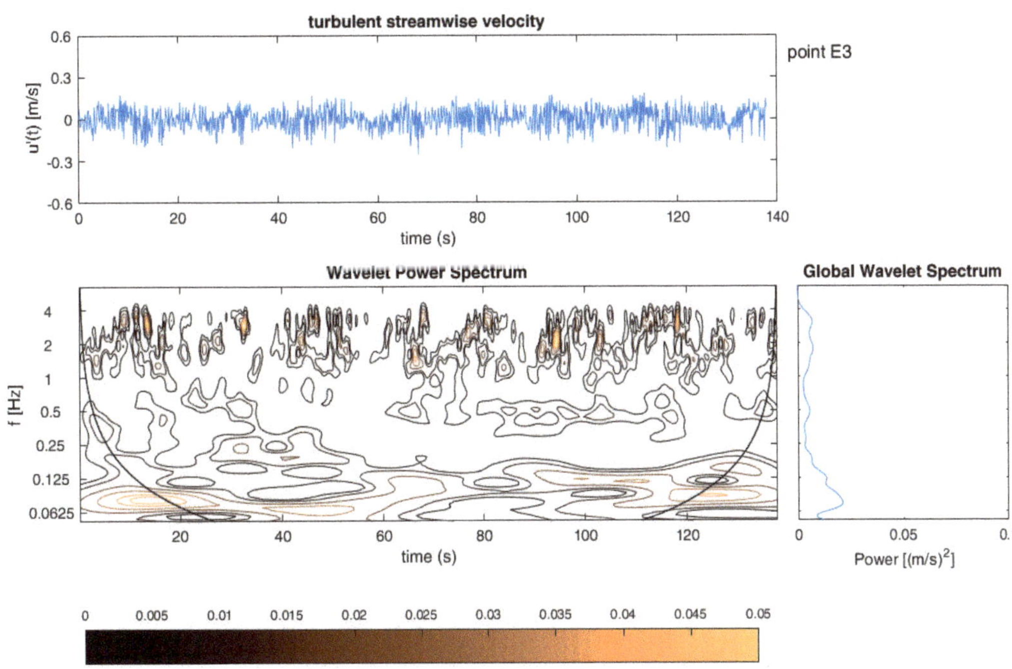

Figure 8. W1 wave: signal of the turbulent streamwise velocity $u'(t)$, spectrogram of $u'(t)$, and global wavelet spectrum for the selected point E3.

We can also consider the dimensionless Strouhal number St to compute the frequency of vortex shedding from the cylinder fs. Using the local <u> as the reference velocity scale and the diameter d as the characteristic length of the obstacle in St and posing $St = 0.2$, we obtain a shedding frequency varying from $fs \sim 2\ s^{-1}$ in absence of the wave (for both W2 and W1 cases) and $fs \sim 3\ s^{-1}$ for the W1 wave crest transit or $fs \sim 4\ s^{-1}$ for the W2 wave crest transit. Thus, such values are consistent with those of the eddies detected with the CWT method along the edge and inside the wake of the cylinder.

The global reading of the analysis conducted with the CWT reveals that the most energetic turbulent eddies in the horizontal plane, downstream of the cylinder, are large and have size varying in the range $O(10^{-1} \div 10^{0}$ m), regardless of the position considered in the FoV. Along the edge and inside the wake other coherent structures appear at higher frequencies, consistent with the Strouhal frequency. In absence of the wave, they have the size of order of the cylinder diameter, while during the transit of the wave crest their size are of order $O(4H)$, i.e., they depend on the wave height. This event, occurring during the transit of the long wave crest, is in fact due to the augmented <u> velocity produced by the wave itself.

4. Conclusions

We experimentally investigated how the flow field can be modified, both upstream and downstream of a rigid, bottom-mounted, emergent cylinder, under the action of a travelling long wave. Two waves were tested, on a constant depth $h = 0.1$ m; that is W1 having a wave height of 0.025 m and W2 having a wave height of 0.05 m. Firstly, they were processed by means of the phase averaging technique and identified in analytical terms. In fact, referring to shallow water waves of great wavelength and small amplitude, we selected some solitary wave solutions that can describe the experimental waves, such as the Peregrine (PER) and the Boussinesq (BOU) solution of the KdV equation. We also compared both laboratory waves with the linear solution by Airy (LIN) and an approximate engineering method by Wheeler (WH). To evaluate the agreement in quantitative terms, the Wilmott index was considered. It resulted that for the W1 wave, the WH method is the best suiting one along the whole wave cycle. For the W2 wave, it provides the best matching in the wave trough description, while the PER model provides a better performance in the wave crest one.

After this characterization, the vertical profiles of the streamwise velocity were analyzed upstream of the cylinder, at selected and representative phases, i.e., wave trough, ascending wave, wave crest, and descending wave. In all cases a quite flat and homogeneous trend is observed, meaning that the long flooding waves promptly affected the entire water depth, from the most superficial layer towards the channel bottom. Regardless of the wave phase, a logarithmic trend with a thickness of $0.4h$ and identifying a bottom boundary layer is recognized. For both waves, the crest transit induces an increase in the initial current flow (by about 24% for W1 and 40% for W2).

The horizontal maps of the velocity acquired at 3 cm form the bottom were then examined, evidencing a recirculation in the near wake of the cylinder, longer for W2 than for W1, due to the higher values of the streamwise velocity. The vorticity maps in the same plane highlight the presence of opposite values downstream of the cylinder, consistently with the stationary case of a flow investing a cylinder. During the wave cycle, from trough to crest condition, a more intense fully turbulent vortex street appears. The antisymmetry of vorticity along the longitudinal axis of symmetry of the cylinder is generally kept, with the highest values of vorticity located at the trailing edge of the shaded area.

Finally, to detect possible coherent structures in the same horizontal plane, the continuous wavelet transform (CWT) was applied. It shows that large coherent structures with size varying in the range $O(10^{-1} \div 10^{0}$ m) arise. They are controlled by the flume dimensions, regardless of the position downstream of the cylinder, and contain the highest turbulent energy. Along the edge and inside the wake, other smaller coherent structures appear with higher frequencies, consistent with the Strouhal frequency. In the absence of the wave, they have length scales with order O(10^{-2} m), i.e., comparable to the cylinder diameter, while during the transit of the wave crest their size increases, depending on the wave height, reaching order O($4H$). It is therefore evident that the long wave impacting the vertical cylinder modifies the turbulence production, spreading and dissipating downstream of the cylinder. This effect must be taken into account, as it can influence connected physical phenomena, such as entrainment and transport of sediments or tracers.

Author Contributions: Conceptualization and methodology, F.D.S.; supervision and funding acquisition, F.D.S.; formal analysis and investigation, R.B. and F.D.S.; writing F.D.S. All authors have read and agreed to the published version of the manuscript.

Funding: The research activity of RB was funded by the Italian Ministry of University and Research (Program PON RI Dottorati Innovativi con caratterizzazione industriale XXXIV ciclo, Decreto MIUR 2983/2018).

Data Availability Statement: The data presented in this study are available on request from the corresponding author.

Acknowledgments: The technical staff of the Hydraulic Laboratory of the DICATECh is gratefully acknowledged for supporting during the research activity.

Conflicts of Interest: The authors declare no conflict of interest. The funders had no role in the design of the study; in the collection, analyses, or interpretation of data; in the writing of the manuscript, or in the decision to publish the results.

References

1. Li, J.; Wang, Z.; Liu, S. Experimental study of interaction between multi-directional focused wave and vertical circular cylinder, part II: Wave force. *Coast. Eng.* **2014**, *83*, 233–242. [CrossRef]
2. Morison, J.R.; Johnson, J.W.; Schaaf, S.A. The force exerted by surface waves on piles. *J. Pet. Technol.* **1950**, *2*, 149–154. [CrossRef]
3. Zdravkovich, M.M. *Flow around Circular Cylinders. Fundamentals*; Oxford University Press: Oxford, UK, 1997; Volume 1, Chapter 6.
4. Rockwell, D. Wave interaction with a vertical cylinder: Spanwise flow patterns and loading. *J. Fluid Mech.* **2002**, *460*, 93–129.
5. Vested, M.H.; Carstensen, S.; Christensen, E.D. Experimental study of wave kinematics and wave load distribution on a vertical circular cylinder. *Coast. Eng.* **2020**, *157*, 103660. [CrossRef]
6. Antolloni, G.; Jensen, A.; Grue, J.; Riise, B.H.; Brocchini, M. Wave-induced vortex generation around a slender vertical cylinder. *Phys. Fluids* **2020**, *32*, 042105. [CrossRef]
7. Duan, W.J.; Wang, Z.; Zhao, B.B.; Ertekin, R.C.; Yang, W.Q. Steady solution of solitary wave and linear shear current interaction. *Appl. Math. Model.* **2018**, *60*, 354–369. [CrossRef]
8. Chen, Y.H.; Wang, K.H. Experiments and computations of solitary wave interaction with fixed, partially submerged, vertical cylinders. *J. Ocean Eng. Mar. Energy* **2019**, *5*, 189–204. [CrossRef]
9. Mo, W.; Jensen, A.; Liu, P. Plunging solitary wave and its interaction with a slender cylinder on a sloping beach. *Ocean Eng.* **2013**, *74*, 48–60. [CrossRef]
10. Huang, Z.; Yuan, Z. Transmission of solitary waves through slotted barriers: A laboratory study with analysis by a long wave approximation. *J. Hydro-Environ. Res.* **2010**, *3*, 179–185. [CrossRef]
11. Maza, M.; Lara, J.M.; Losada, I.J. Solitary wave attenuation by vegetation patches. *Adv. Water Res.* **2016**, *98*, 159–172. [CrossRef]
12. Yang, Z.; Huang, B.; Kang, A.; Zhu, B.; Han, J.; Yin, R.; Li, X. Experimental study on the solitary wave-current interaction and the combined forces on a vertical cylinder. *Ocean Eng.* **2021**, *236*, 109569. [CrossRef]
13. Kim, S.Y.; Kim, K.M.; Park, J.C.; Jeon, G.M.; Chun, H.H. Numerical simulation of wave and current interaction with a fixed offshore substructure. *Int. J. Nav. Arch. Ocean Eng.* **2016**, *8*, 188–197. [CrossRef]
14. Ben Meftah, M.; De Serio, F.; De Padova, D.; Mossa, M. Hydrodynamic structure with scour hole downstream of bed sills. *Water* **2020**, *12*, 186. [CrossRef]
15. Olsen, N.R.B.; Melaaen, M.C. Three-dimensional calculation of scour around cylinders. *J. Hydr. Eng.* **1993**, *119*, 1048–1054. [CrossRef]
16. Nepf, H.M. Flow and transport in regions with aquatic vegetation. *Annu. Rev. Fluid Mech.* **2012**, *44*, 123–142. [CrossRef]
17. De Serio, F.; Mossa, M. A laboratory study of irregular shoaling waves. *Exp. Fluids* **2013**, *54*, 1536. [CrossRef]
18. Bi, C.W.; Zhao, Y.P.; Dong, G.H.; Xu, T.J.; Gui, F.K. Experimental investigation of the reduction in flow velocity downstream from a fishing net. *Aquac. Eng.* **2013**, *57*, 71–81. [CrossRef]
19. CERC. *Shore Protection Manual*; U.S. Army Corps of Engineers: Washington, DC, USA, 1984; Volume 1.
20. Hereman, W. Shallow Water Waves and Solitary Waves. In *Encyclopedia of Complexity and Systems Science*; Meyers, R.A., Ed.; Springer: Berlin, Germany, 2018; pp. 1–18.

21. Madsen, P.A.; Fuhrman, D.R.; Schäffer, H.A. On the solitary wave paradigm for tsunamis. *J. Geophys. Res. Oceans* **2008**, *113*, 012. [CrossRef]
22. De Serio, F.; Mossa, M. Assessment of classical and approximated models estimating regular waves kinematics. *Ocean Eng.* **2016**, *126*, 176–186. [CrossRef]
23. De Padova, D.; Brocchini, M.; Buriani, F.; Corvaro, S.; De Serio, F.; Mossa, M.; Sibilla, S. Experimental and Numerical Investigation of Pre-Breaking and Breaking Vorticity within a Plunging Breaker. *Water* **2018**, *10*, 38722. [CrossRef]
24. Torrence, C.; Compo, G.P. A practical guide to wavelet analysis. *Bull. Am. Meteorol. Soc.* **1998**, *79*, 61–78. [CrossRef]
25. Kanani, A.; da Silva, A.M.F. Application of continuous wavelet transform to the study of large-scale coherent structures. *Environ. Fluid Mech.* **2015**, *15*, 1293–1319. [CrossRef]
26. Longo, S. Turbulence under spilling breakers using discrete wavelets. *Exp. Fluids* **2003**, *34*, 181–191. [CrossRef]
27. Muddada, S.; Hariharan, K.; Sanapala, V.S.; Patnaik, B.S.V. Circular cylinder wakes and their control under the influence of oscillatory flows: A numerical study. *J. Ocean Eng. Sci.* **2021**, *6*, 389–399. [CrossRef]

water

MDPI

Article

Stochastic Assessment of Scour Hazard

David Flores-Vidriales [1], Roberto Gómez [1] and Dante Tolentino [2,*]

[1] Instituto de Ingeniería, Universidad Nacional Autónoma de Mexico, Mexico City 04510, Mexico; ing.dfv.1991@gmail.com (D.F.-V.); RGomezM@iingen.unam.mx (R.G.)

[2] Departamento de Materiales, Universidad Autónoma Metropolitana, Mexico City 02200, Mexico

* Correspondence: dantetl@azc.uam.mx

Abstract: Scour is the most frequent cause of bridge collapses in Mexico. Bridges located along the Mexican Pacific coast are exposed to extreme rainfall originating from tropical storms and hurricanes. Such environmental phenomena trigger sediment loss, which is known as scour. If maintenance actions are not taken after scouring events, the scour depth increases over time until the bridge collapses. A methodology to estimate the scour hazard considering both the scour–fill interaction and the Monte Carlo simulation method is proposed. The general extreme value probability distribution is used to characterize the intensity of the scouring events, the lognormal distribution is used to characterize the sedimentation process (fill), and a homogeneous Poisson process is used to forecast the occurrence of both types of events. Based on the above, several histories of scour–fill depths are made; such simulations are then used to develop time-dependent scour hazard curves. Different hazard curves associated with different time intervals are estimated for a bridge located in Oaxaca, Mexico.

Keywords: scour; stochastic analysis; forecasting; extreme events

Citation: Flores-Vidriales, D.; Gómez, R.; Tolentino, D. Stochastic Assessment of Scour Hazard. *Water* **2022**, *14*, 273. https://doi.org/10.3390/w14030273

Academic Editor: Mouldi Ben Meftah

Received: 31 December 2021
Accepted: 11 January 2022
Published: 18 January 2022

Publisher's Note: MDPI stays neutral with regard to jurisdictional claims in published maps and institutional affiliations.

1. Introduction

A bridge built over a river creates an obstruction to the water flow, changing the local flow field. The erosion around the obstruction is accelerated due to the following three variables: (a) a velocity surge, (b) a difference in pressure between upstream and downstream areas produced by the perpendicular action of the flow, and (c) the generation of vortices. Scour begins when a certain level of shear stress, known as critical Shields stress, is reached. The sediment at the river bed lifts due to the action of vortices and travels downstream; such vortices produce shear stress on the sediment particles lying on the river bed and contribute to the scour process. The critical Shields stress is a function of size, shape, and material of the sediment particles.

There are two current trends that study sediment transport: trend one uses computer fluid dynamics (CFD) to solve the Navier–Stokes equations; such an approach is useful for deterministic analyses and is the best option to precisely simulate the phenomenon under study, but it demands both a great amount of computer time and a high level of capabilities. Trend two consists of a stochastic approach that is not useful to accurately forecast the outcome of a single event, but it is efficient to forecast events in probabilistic terms, and it can be a good tool to study time-dependent problems such as the case of the thickness of the sediment layer on a river bed that changes over time. The sediment transport at any given point in a time lapse, t, is assumed to have two possible outcomes: erosion, which is also known as scour, and accretion, which is also known as fill. Regarding the latter, a mathematical model that consists of both a data generation process (DGP) and a set of suitable simplified sediment transport equations can be used to forecast the fluctuations of the sediment layer thickness within a reasonable error.

The mechanics of sediment transport is a complex issue to study from a stochastic point of view. Thus, assumptions need to be made in order to develop a simplified mathematical approach that is useful to a probabilistic approach. Research dealing with semi-empirical equations for the sediment deposit is scarce. The rate at which sediment is deposited in the river bed is dependent on the availability of sediment and, thus, from the erosion of the bed and walls of the river from upstream. Variables of hydraulics, geometry, and geotechnical are fundamental for sediment transport; they also govern the behavior of secondary variables such as turbulence, flow velocity, and shear stress in the sediment particles [1]. Many studies related to the probabilistic assessment of scour hazard have been developed [2–13]. A simulation that allows forecasting the outcome of an event is a mathematical representation of events that have already happened; such simulation cannot add any new knowledge about the behavior of previous events, but it can provide useful information about possible future events.

In the last 20 years, Mexico has experienced an average of 2.5 bridge collapses per year due to scour effects [14], making scour the leading cause of bridge collapses in Mexico and most of the world [15]. The Mexican coastal highway system comprises geographic zones that are prone to extreme rainfall. For example, in 2005, hurricane Stan severely damaged highway infrastructure and caused the collapse of several bridges in the state of Chiapas. Thus, it is important to develop tools and methodologies to reduce bridge collapses due to scour.

This paper aims to propose a methodology to estimate scour hazard curves considering the following: (a) a preprocessing stage to fix missing data issues using data augmentation, (b) an information criterion for the probability distribution selection, and (c) the interaction of scour and fill events through time. The proposed methodology is illustrated in a bridge with potential scouring problems located in Mexico. Such an approach can deal with missing data, or problems such as outliers, and also has the novelty feature of addressing the accretion process that lowers maximum scour depth, making the mathematical model more complete. Moreover, the methodology is compared with the approach proposed by [5] in which the scour survival function is obtained by the SRICOS EFA method.

2. Data Processing

Most databases in Mexico are somewhat flawed: they either have missing data or have outliers. Such issues introduce an additional source of uncertainty that needs to be reduced in a preprocessing stage. First, outliers are defined as any value above or below 1.5 of the interquartile range, IQR. Such values are removed from the data set and then treated as missing at random, as per Rubin [16]. This mechanism considers data that are not systematically related to hypothetical values of known data. Thus, it is possible to use statistical inference to obtain probable values for the missing data without having to include any information about the missing data. Hence, a methodology that produces the least amount of bias is needed, such as data augmentation, K nearest neighbors, multiple imputation, etc. [17–20]. In this paper, the data augmentation technique [21] is used to address the missing data issue. Such a technique has two advantages: (a) it produces a low amount of bias, and (b) it is relatively simple when compared with complex optimization techniques used for missing data analysis.

2.1. Preprocessing

The available data that are recorded from the observation of a natural phenomenon such as discharge , Q, for a time interval $t \in Z$ can be used to form an observation set, F, in which its members are termed events, and denoted by $\omega \in F$. Let Ω be a set representing all the possible outcomes of the phenomenon. Then, the probability space of the phenomenon is $(\Omega, F, P_\theta \,|\theta \in I)$, where the parameter space, I, has two sources of statistical uncertainty: one is called stochastic uncertainty, which deals with the uncertainty about a fixed parameter θ (Figure 1); the other is called inductive uncertainty.

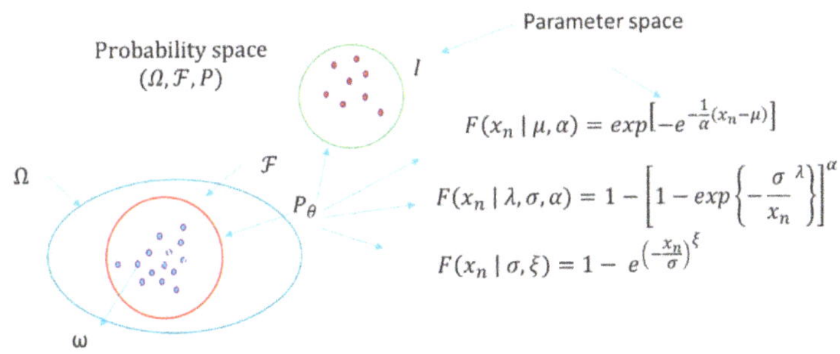

Figure 1. Probability space, parameter space, and probability distribution.

The uncertainty associated with each event of the subset F is represented by a probability function, P, which contains all possible outcomes of ω (if the sample is large enough). If the uncertainty of P is contained in the parameter space, then, for each I there is a possibility of having a different probability distribution, P_θ, $\theta \in I$, where θ is a set of parameters that govern the distribution function. Hence, stochastic problems are referred to as inductive, meaning that they provide predictions about events that have not yet happened using prior observations.

2.2. Statistical Inference of θ

The statistical estimation of the conditional distribution parameters of an observed data set $\theta \in I$ is known as inference. In order to infer the parameters for a given distribution $F = P_\theta[X < x \mid \theta \in I]$, an inference approach must be chosen first.

Frequentist inference is based on the idea of a limiting frequency. One of the issues of using this approach is the intrinsic epistemic uncertainty of the parameters θ. Such approach does not provide statements about θ, which is a fixed value. Thus, the uncertainty lies only in the observations ω (aleatory uncertainty). A frequentist approach usually has a predetermined sample size because p values are calculated over a sample space Ω [22]. In Bayesian inference, θ is considered a random variable. Some knowledge about the distribution of θ is assumed to be available, which is called a prior distribution. Then, after observing some data ω, the distribution of θ must be updated to match the knowledge obtained during the observation process. The Fisherian approach uses the likelihood function, L, to compare different values of θ with the probability of the observed data so that $L(\theta) = P_\theta$ [17]. Such likelihood provides a relative measure of the set of parameters θ, whose exact value is impossible to know. Hence, the parameter values that maximize the likelihood function $L(\theta)$ must be found. Based on the above, the Fisherian approach is used because maximizing the likelihood function $L(\theta)$ does not require previous knowledge of the phenomena, and θ does not have an underlying distribution to propose. The combined likelihood for a given distribution of n parameters can be expressed as

$$L(\theta|data) = \prod_{n=1} P_\theta(data\,|\theta) \tag{1}$$

The maximum of the likelihood function is obtained by two steps. The log-likelihood function is obtained, and then the score function, S, is calculated as follows:

$$S(\theta) \equiv \nabla loglog\, L(\theta) \tag{2}$$

Hence, we obtain the maximum likelihood estimators, MLE, and the solution of θ, $S(\theta) = 0$. MLE is not flawless, and for small data samples, the Simpson's paradox could occur. Moreover, most of the real score functions do not have a closed solution. Thus, a numerical method must be used to approach the solution. Maximum likelihood estimators

(MLE) are often the best way to deal with stochastic uncertainty because they require no prior knowledge about the phenomena.

2.3. Missing Data Inference

The second source of uncertainty is known as inductive uncertainty and deals with the lack of information of the observations available. Rubin [23] uses multiple imputation techniques to reduce the inductive uncertainty. When a small data set is used, stochastic uncertainty becomes relevant, and inductive uncertainty becomes critical [17]. Hence, it is important to rely on data sets with reasonable sizes. In order to reduce the inductive uncertainty, the missing at random mechanism (MAR) is used as follows [21]:

$$f(Q, M \mid \theta, \psi) = f(Q \mid \theta) \times f(M \mid Q, \psi) \quad (3)$$

where M is the missing data indicator with values between 0 and 1, which represent an observed value and no observed value, respectively; $f(\omega, M \mid \theta, \psi)$ is the joint probability distribution of (ω, M). The set of parameters that governs the distribution of the observed data in the subset F_{obs} is θ, whereas for the missing data F_{miss} is ψ. Hence, under the premise of MAR, given a set of observed data ω that is augmented by a quantity Z, the augmented posterior distribution $P(\theta|\omega, Z)$ can be calculated as

$$f(M \mid F_{obs}, F_{miss}, \psi) = f(M \mid F_{obs}, \psi) \times f(F_{obs} \mid \theta) \quad (4)$$

which means that the parameters from the data model θ are independent of the parameters in the missing data. Authors in [24,25] developed the following data augmentation algorithm:

$$F_{miss}{}^{(t+1)} \approx P_r(F_{miss}|F_{obs}, \theta^t) \quad (5)$$

$$\theta^{(t+1)} \approx P_r(F|Q_{obs}, F_{miss}{}^{(t+1)}) \quad (6)$$

where P_r is the probability associated to F_{miss}. Equation (5) is the imputation step, which generates imputed values for F_{miss} using the missing data probability distribution, observed data F_{obs}, and the set of parameters θ at the given iteration t. Equation (6) generates parameter values for the posterior distribution, using the observed data F_{obs} and the imputed values of F_{miss} at the iteration $t+1$. This procedure is performed iteratively until convergence is reached, and a new "complete" data set F_{full} is generated.

2.4. Suitable Probability Distribution

The suitable distribution function of the data $F(\omega,\ t \in T)$ is the backbone of the DGP. In order to select the best fitting distribution to the data set, the Akaike information criterion [26], AIC, is used, which selects the best probability distribution based on the parsimony principle. AIC is obtained using the following equation:

$$AIC = -2n\ \underline{L}(\theta|F) + 2k \quad (7)$$

where n is the number of observations in F, $\underline{L}(\theta|F)$ is the log-likelihood function, and k is the number of parameters in the mathematical model.

2.5. Simulation of Events for the Proposed Approach

The block maxima approach (BMA) is used to estimate the intensity and occurrence of extreme river discharges associated with extreme rain falls. Such an approach divides the observations of the "complete" data set, F_{full}, into non-overlapping blocks that restrict the observation of the maximum on such periods. One-year blocks are used in this research since there is no seasonality in the maximum annual discharge [27]. The generalized extreme value distribution (GEV) is used to describe the annual maximum discharge, whose CDF is

$$F(x \mid \sigma, \mu, \xi \neq 0) = exp\left\{-\left[1+\xi\left(\frac{x-\mu}{\sigma}\right)\right]^{-\frac{1}{\xi}}\right\} \tag{8}$$

$$F(x \mid \sigma, \mu, \xi = 0) = exp\left\{-\frac{x-\mu}{\sigma}\right\} \tag{9}$$

where μ, σ, ξ are the location, scale, and shape parameters, respectively. The law of small numbers provided the basis for the selection of the Poisson stochastic process for the time of occurrence of events that trigger scour and fill. Using this law, the stochastic process of events that exceed a given threshold is

$$P(k \geq K \mid t = T) = e^{-\lambda T}\frac{(\lambda T)^k}{k!} \tag{10}$$

where

$$\lambda_e = \frac{\sum_{i=1}^{n} m \mid \forall\, m \geq \psi}{n} \tag{11}$$

and

$$\psi \cong 0.85IQR \tag{12}$$

Since there is only interest in the probability of occurrence of a single event [28], the cumulative distribution function (CDF) of the arrivals $F_a(t)$ in the Homogeneous Poisson Process (HPP) is obtained as follows:

$$F_a(t) = \int_0^{\infty} 1 - e^{-\lambda t}\, d_t \tag{13}$$

Having characterized the intensity of the events and its occurrence, the joint probability distribution (JPD) is

$$f_{xy}(x,y) = \sum_{y=1}^{n} \frac{\lambda^y e^{-\lambda}}{y!}\left(\int_{-\infty}^{\infty} \frac{1}{\sigma} exp\left\{-\left[1+\xi\left(\frac{x-\mu}{\sigma}\right)\right]^{-\frac{1}{\xi}}\right\}\left[1+\xi_1\left(\frac{x-\mu}{\sigma}\right)\right]^{-1-\frac{1}{\xi}} dx\right) \tag{14}$$

where y is the number of events in a one-year block maxima.

Events that trigger fill have a random nature and have a somewhat fixed occurrence in a year. Such events occur both at the beginning and at the end of the rainy season, when the discharge is at its weakest intensity, but it is still sufficiently strong to transport sediment from the upstream river flow. Therefore, a stochastic model is needed to account for these observations that will trigger the accretion process. The intensity of the discharge, q_t, is considered to follow a lognormal distribution function whose CDF is

$$F(x \mid \mu, \sigma) = \frac{1}{\sigma\sqrt{2\pi}}\int_0^x \frac{1}{t} exp\left\{\frac{-\log(t-\mu)^2}{2\sigma^2}\right\} dt \tag{15}$$

where μ and σ are the shape and location parameters, respectively. The occurrence is obtained in a similar way as extreme events, with the difference in the threshold ψ_2 as follows:

$$\lambda_r = \frac{\sum_{i=1}^{n} m \mid \forall\, m \leq \psi_2}{n} \tag{16}$$

where ψ_2

$$\psi_2 \cong 1.2IQR \tag{17}$$

Thus, the JPD for the fill is

$$f_{xy}(y,t) = \sum_{y=1}^{n} \frac{\lambda^y e^{-\lambda}}{y!} (\frac{1}{\sigma\sqrt{2\pi}} \int_0^x \frac{1}{t} exp\left\{ \frac{-\log(t-\mu)^2}{2\sigma^2} \right\} dt \quad (18)$$

2.6. Simulation of Time Series for SRICOS-EFA Method

A time series process is a set of observations that are well defined and have regular intervals; thus, it is a set of random variables $\{X_t\}$ indexed by integers t (often associated by dates). A stochastic process is a parametrized collection of random variables $\{X_t\}_{t\in T}$ defined on a probability space $(\Omega,\ F,\ P_\theta\ |\theta \in I)$. Then, a time series can be understood as a realization of a stochastic process. Time series are often studied on the time domain using an autoregressive approach. According with the above, an autoregressive moving integrated average model (ARIMA) is used as follows:

$$y\prime_t = c + \phi_1 y\prime_{t-1} + \ldots \phi_p y'_{t-p} + \theta_p \varepsilon_{t-1} + \ldots \theta_q \varepsilon_{t-q} + \varepsilon_t \quad (19)$$

where $y\prime_t$ is the difference series, p is the order of the autoregressive, d is the degree of the first differencing, and q is the order of the moving average. The ARIMA model that best fits the data is selected by the AIC criterion [26].

Once the model is selected, simulations are performed under a random walk approach for the extreme events data set (maximum annual discharge) and then used to estimate the scour depth using the SRICOS-EFA methodology [5]. These results are then compared to the proposed approach explained in Section 2.5.

3. Scour and Fill

3.1. Scour

Local scour depth, y_s around a bridge pier is obtained using the one-dimensional HEC 18 approach [19] as follows:

$$y_s = y_n K_4 \left[2K_1K_2K_3 \left(\frac{b}{y_n} \right)^{0.65} F^{0.43} \right] \quad (20)$$

where y_n is the height of the stream (derived from the Manning equation), K_4 is a bed material coefficient, K_1 is a correction factor for the pier nose shape, K_2 is a correction factor for the angle of attack of the flow, K_3 is a correction factor for bed condition, b is the pier width, and F is the Froude number. The local scour area, A_s, and wetted perimeter of the scour hole, P_{mh}, are calculated as

$$A_s \cong {y_s}^2 \quad (21)$$

$$P_{mh} \approx 2.33 y_s \quad (22)$$

When calculated, these values are added to the hydraulic area and wetted perimeter after each scouring event.

3.2. Fill

Haschenburguer [29] proposed a model to estimate the mean fill using both the Shields stress and the hydraulic radius, which are both considered discharge-dependent variables. The Haschenburguer fill model estimates the sediment deposit from a given intensity for small-scale events in stable channels. The model uses the Shields stress as the fundamental variable, and it does not limit the amount of accretion, a fact that could lead to unrealistic heights of the sediment layer. Such issue is also present in the HEC 18 equations for scour. Hence, it is considered to have a sufficient degree of approximation [30,31].

The scour–fill process is controlled by the transfer of material in the river bed. Thus, scour depths are time-dependent random variables. The underlying process which governs sediment transport has been a constant topic for research [32–34]. A Poisson distribution coupled with a homogeneous Poisson process (HPP) is used, so that the occurrence of the phenomenon can be expressed with a semiempirical PDF as [35]

$$f\left(y_f\right) = \phi e^{-\phi y_{yf}} \tag{23}$$

where $f\left(y_f\right)$ is the proportion of stream bed fill for a given flow depth, y_{yf}, which is the flow depth obtained from the minimum discharge at the start of the rainy season.

$$\phi = 3.3e^{-1.52\tau^*/\tau} \tag{24}$$

where the Shields mean stress, τ^*, is

$$\tau^* = \frac{\rho_w R_H S}{(\rho_s - \rho_w) D_{50}} \tag{25}$$

where ρ_w is the density of the sediment, ρ_s is the density of water at 20 °C, D_{50} is the mean size value of the particles, and R_H is the hydraulic ratio at the beginning and at the end of the rainy season. The hydraulic ratio is different from the one considered in the scour process, and S is the slope of the hydraulic grade line. A shear velocity criterion for incipient motion of sediment is used as follows [36]:

$$\tau \cong 0.215 + \frac{6.79}{D_{50}{}^{1.70}} - 0.75 \times exp\left(-2.61 \times 10^{-3} \times D_{50}\right) \tag{26}$$

The local fill area, A_{sf}, and the wetted perimeter, P_{mf}, are subtracted after each event and are calculated as

$$A_{sf} \cong f\left(y_f\right)^2 \tag{27}$$

$$P_{mf} \approx 2.33 f\left(y_f\right) \tag{28}$$

4. Scour Hazard Curves

4.1. Simulation of Random Events

Both the hydraulic radius (H_t) and the Froude number (F) change through time as scour and fill develop. Hence, an analytical solution to the stochastic process is difficult to obtain. A numerical approximation to the solution is used instead. Monte Carlo simulation (MCS) is a tool for the analysis of complex systems. The MCS approach used in this research simulates a large number of independent histories of occurrence–intensity, which are then used to estimate the time progression of scour. Each simulated history is one possible outcome in the probability space. Each history behaves differently because of the intrinsic stochastic nature of the state variables (scour depth, flow, Froude number, etc.). If a sufficient number of simulated histories is known, the probability function P of the phenomena can be approximated (law of large numbers).

Usually scour and fill in a bridge system (a collection of variables and components) are dealt with in a discrete time fashion, especially when a probabilistic approach is used. In this research, a discrete-event system is used, assuming that the state variables change instantly through discrete points in the simulation time. Since the form of the solution is unknown, a raw sampling approach was used.

4.2. Hazard Curves

Hazard curves show the intensity of scour depth and its corresponding probability of exceedance. The probabilistic scour assessment provides the scour intensity corresponding to a target hazard level (annual probability of exceedance). The proposed hazard curve for scour is site specific, which means that there is not a variable that deals with the source-to-site distance, which is a consequence of the fundamental approach to the phenomena. The location of the phenomenon does not have an influence on the model in the approach proposed. Therefore, the effects of the phenomenon govern the model, and its location is negligible. The annual rate of exceedance for a given scour depth can be estimated as follows [37]:

$$P[y_s(te) > y] = \frac{\lambda_e}{\xi_0}[\xi_{yt}] \tag{29}$$

where ξ_0 is the mean number of events that exceed a given scour threshold, (ψ_s), during an interval t. $P[y_s(te) > y]$ is the probability that one or more scour depth events $y_s(te)$ are greater than a threshold (y), and ξ_{yt} is the exceedance rate given a time interval t and is estimated as

$$\xi_{yt} = 1 - (1 - P[y_s(te) > y])^t \tag{30}$$

Thus, the annual rate of exceedance can be obtained by

$$\xi_0 = \frac{\left[1 - \int_{-\infty}^{x} exp\left[-\left(1 + \xi\frac{x-\mu}{\sigma}\right)^{-\frac{1}{\xi}}\right]dx\right] * \xi_{yt}}{\lambda_e} \tag{31}$$

5. Case Study

The bridge is located in the state of Oaxaca in one of the highest scour-prone zones in Mexico. The bridge is a four-span, simply supported bridge; each span has a length of 56 m, as shown in Figure 2a. Bridge piers and foundation piles are built with reinforced concrete in a multi-column bent type with circular columns of 1.2 m in diameter, as seen in Figure 2a. A simplified trapezoidal cross-section is used, as shown in Figure 2b.

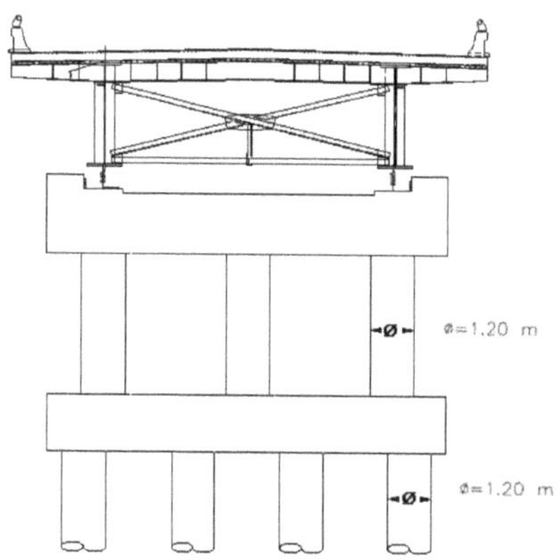

(a)

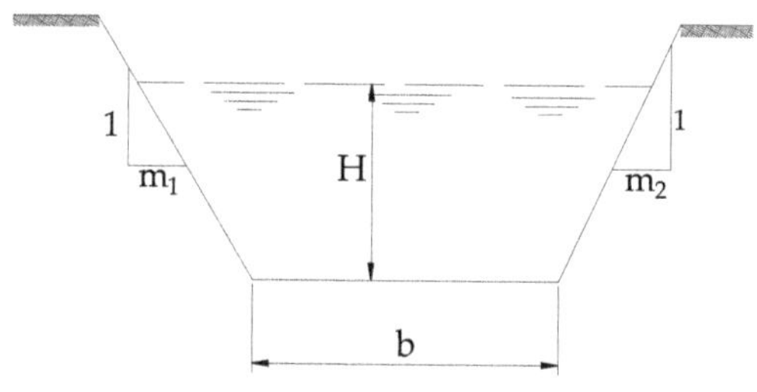

(b)

Figure 2. Study case: (**a**) typical pier of the bridge; (**b**) simplified cross-section of the river.

The properties of the simplified cross-section for a given value of H are as follows:

$$A_h = H\left\{b + \frac{H(m_1 + m_2)}{2}\right\} \tag{32}$$

$$P_m = b + H\left[\sqrt{1 + m_1^2} + \sqrt{1 + m_2^2}\right] \tag{33}$$

$$Q = \frac{1}{n}A_h R_h^{\frac{2}{3}} S_h^{\frac{1}{2}} \tag{34}$$

where A_h is the hydraulic area, P_m is the wetted perimeter, $R_h = A_{ht}/P_m$, and H is the height of the stream for a given discharge Q. Equation (34) can be written as

$$Q = \frac{1}{n}\left[\frac{\left(H\left\{b + \frac{H(m_1+m_2)}{2}\right\}\right)^{5/3}}{\left(b + H\left[\sqrt{1+m_1^2} + \sqrt{1+m_2^2}\right]\right)^{2/3}}\right] S_h^{\frac{1}{2}} \quad (35)$$

H_t is the new value of y_s, and in a similar way τ^* is deducted from the river discharge for recurring events; thus, two different hydraulic ratios are used to find the values of y_s and τ^*. The values used in Equation (35) are taken from the Monte Carlo simulation. For this case study $b = 80\ m$, $m_1 = 0.8$, $m_2 = 1$, $s = 0.0068$, and $n = 0.025$.

5.1. Data Augmentation

Figure 3 shows the river discharge, Q, obtained from the National Superficial Water Database for this river. Both the maximum annual record and the minimum annual discharge are presented. Scour phenomenon is controlled by events that are infrequent in nature and have a high magnitude; these are known as "extreme events". Extreme events are represented in the maximum annual discharge data base, and they are used to forecast scour depths. Unlike scour, fill is controlled by recurring events at the beginning and end of the rainy season; then, the minimum discharge is used to forecast sediment deposit depths. From the data set shown in Figure 3, a "complete" data set is estimated using the data augmentation algorithm described in Section 2. Results are shown in Figure 4.

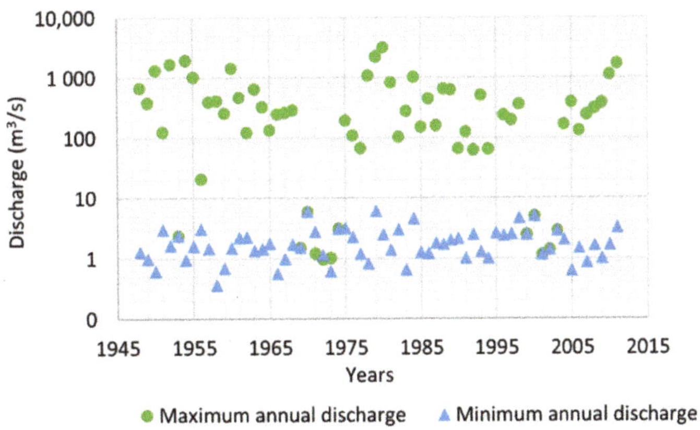

Figure 3. Discharge data.

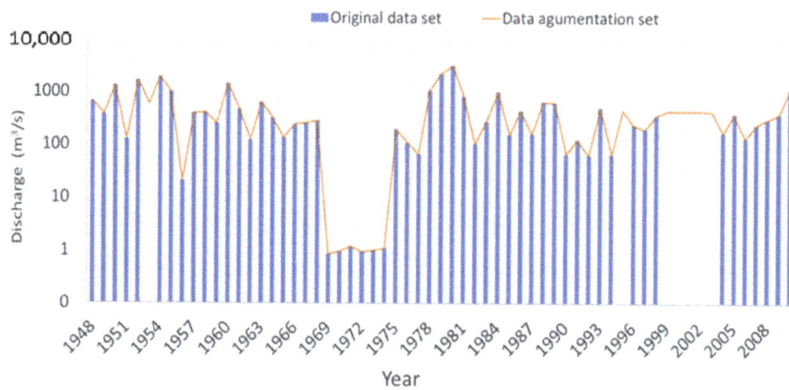

Figure 4. Comparison between original data set and data augmentation set.

5.2. Fit to a Probability Distribution

Different probability distribution functions (PDFs) are tested to find the appropriate PDF that best characterizes the maximum annual discharge and, in a similar fashion, the minimum annual discharge. Therefore, the PDF with the lowest AIC is selected. Figure 5a shows the AIC for different PDFs such as generalized extreme value (GEV), Weibull 2P, normal, exponential, and logistic. It is noticed that the general extreme value distribution is the best fit for the maximum annual discharge. Figure 5b shows different PDFs to characterize the minimum annual discharge data at the beginning and end of the rainy season. Figure 5b indicates that the lognormal distribution presents the minimum AIC; thus, it is the best fit for the minimum flow.

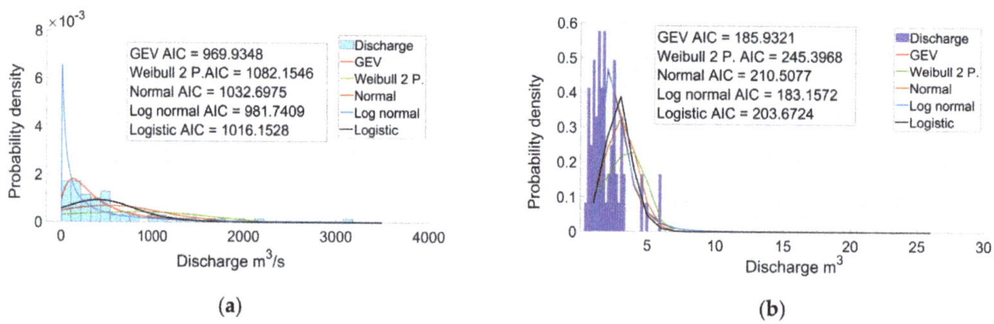

Figure 5. Different probability distribution functions: (**a**) data histogram of maximum annual flow; (**b**) histogram of minimum flow.

The selected PDFs that characterize the maximum and minimum annual discharge are represented in probability papers. The probability paper can be defined as a mathematical–graphic technique to verify if the data follow a PDF. Figure 6a shows the maximum annual discharge and the GEV PDF with a continuous line in probability paper. Figure 6b shows the probability paper of the minimum annual discharge with a lognormal PDF.

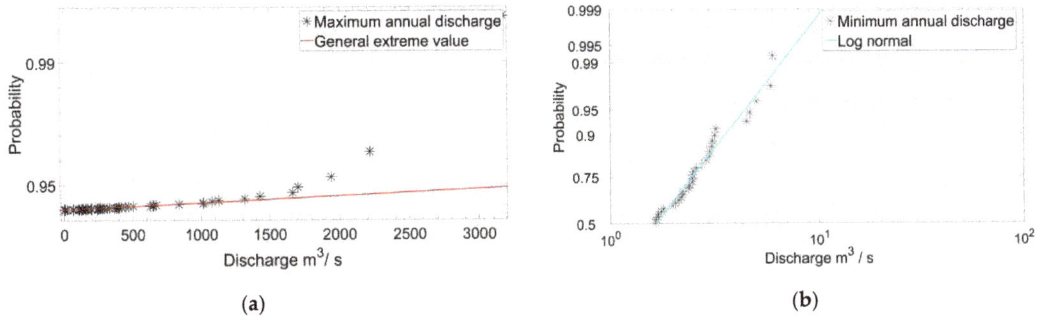

Figure 6. Probability papers for (**a**) maximum annual discharge; (**b**) minimum annual discharge.

Once the probability distribution for the intensity of events has been found, the mean number of events at a specific time interval is found, as described in Section 2.5, and then the joint probability distribution (JPD) is used with the simulations, as shown in Figure 7.

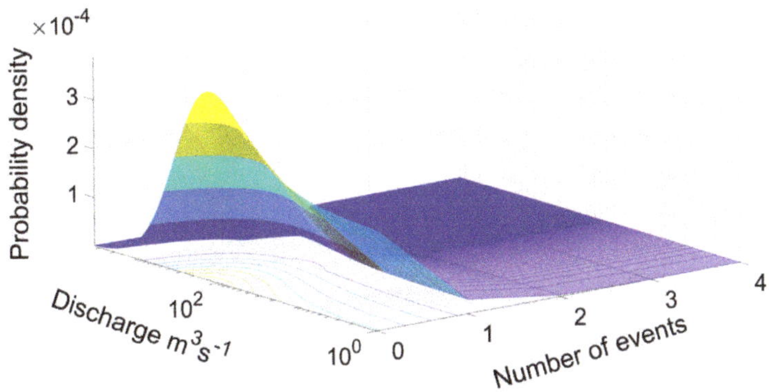

Figure 7. Joint PDF of maximum annual flow and number of events per year.

Table 1 shows all the constants and variables needed to carry out the simulations of scour and fill events; $k_1, k_2 = 1$ for the scour process.

Table 1. Variables and their distributions.

Variable	Description	Distribution	σ	μ	ξ	λ
ω_q	Maximum annual discharge	GEV/Poisson	230	209	0.512	$\cong$ 0.409
q_t	Minimum annual discharge	Log normal/Poisson	1.74	1.73	-	$\cong$ 0.818
d_{50}	Characteristic particle size (mm)		8	4		
k_3	Bed condition correction factor	Rand (Normal)	1.15			
d_a	Soil density kg/m^3	Normal	2000	250		

5.3. Simulation of Events

Monte Carlo simulations were carried out to generate random events that could trigger the scour and fill process. In this context, the necessary number of simulations to have a probability $\alpha \equiv 0.99$ providing a percentage error $\varepsilon \leq 5$ %, is 1060, which is the number of simulations carried out to solve the stochastic processes for each simulated time interval. Figure 8a,b shows the outcome of the simulations for a time interval equal to 170 years for the extreme and recurring events. The discharge extreme events are several orders of magnitude bigger than the recurring events. The simulations of occurrences and intensity are used for the estimation of scour and fill in a discrete time model.

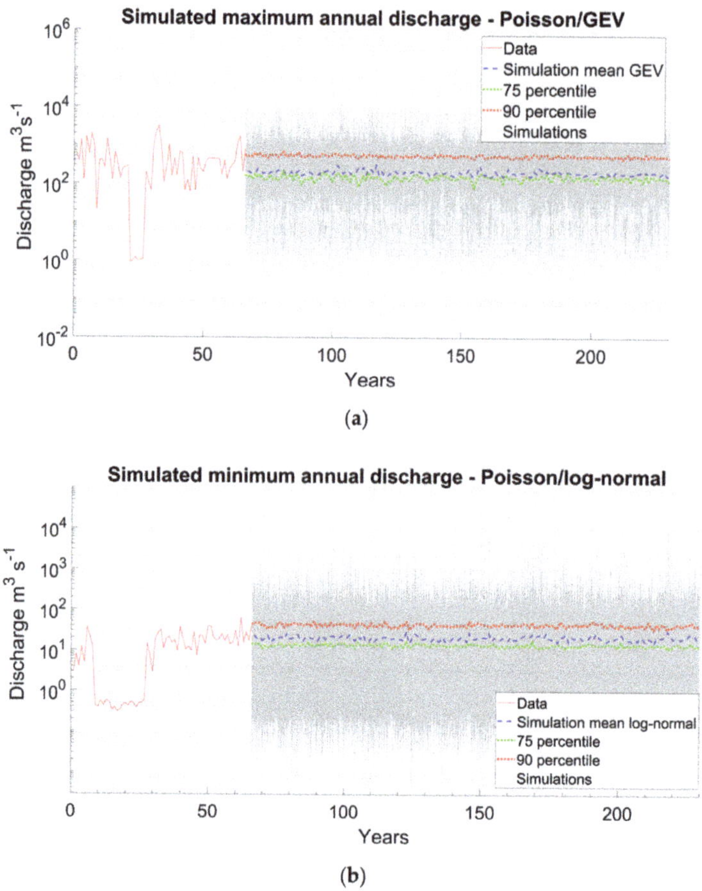

Figure 8. Simulation outcome of occurrences and intensity in accordance with both phenomena and its occurrence rates λ: (**a**) maximum annual streamflow; (**b**) minimum annual streamflow.

For comparison's sake, a set of 1024 simulations of 175 years of daily mean discharge is investigated using an ARIMA model as described in Section 2.6. The mean daily discharge data used are shown in Figure 9a. Figure 9b shows the outcome of the simulations. It can be noted that Figure 9a has outliers that have a negative impact on the forecasting precision. Since the data sets used for time series are generally much larger than that used to forecast extreme events (because of the BMA approach), outliers are more common and hard to deal with. This is a limitation on the use of time series such as ARIMA. In addition, when the data series is large and complex, advanced forecasting techniques such as wavelet multiresolution analysis should be used to improve accuracy.

5.4. Scour Survival Function

Considering that the bridges in Mexico are designed for a service life of 100 years, several simulated histories were developed for time intervals of 20, 30, 40, 60, 75, 90, 110, 130, 150, and 170 years, in order to track the degradation of thickness of the sediment layer. Figure 10a shows one simulation of the thickness of the sediment layer $y_{s(t)}$ in the river bed as a function of time; positive values denote scour, whereas negative values denote fill. Figure 10a shows the result of a single simulation for $t = 170$ years (simulation starts at the end of the recorded data year 64). Scouring events and fill events are added in the simulation. Since scour is controlled by extreme events, its magnitude is bigger than fill. Figure 10b shows the outcome of 1064 simulations.

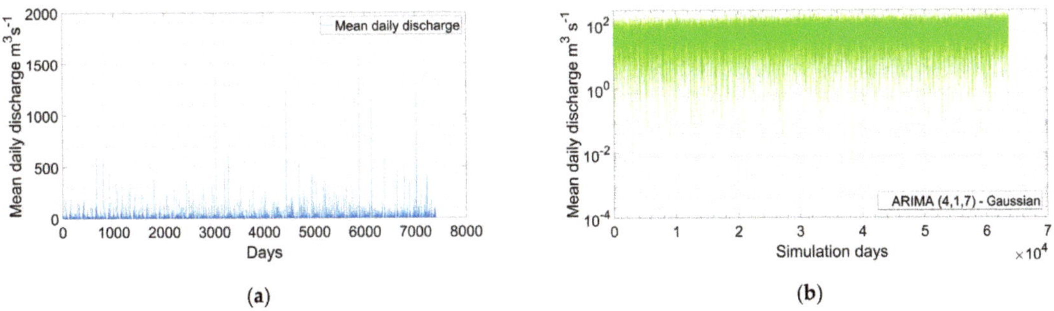

Figure 9. Mean daily discharge: (**a**) available readings of mean daily discharge at the bridge site; (**b**) 175 years of mean daily discharge simulation using ARIMA model with p:5, d:1, q:7 and a Gaussian innovation distribution.

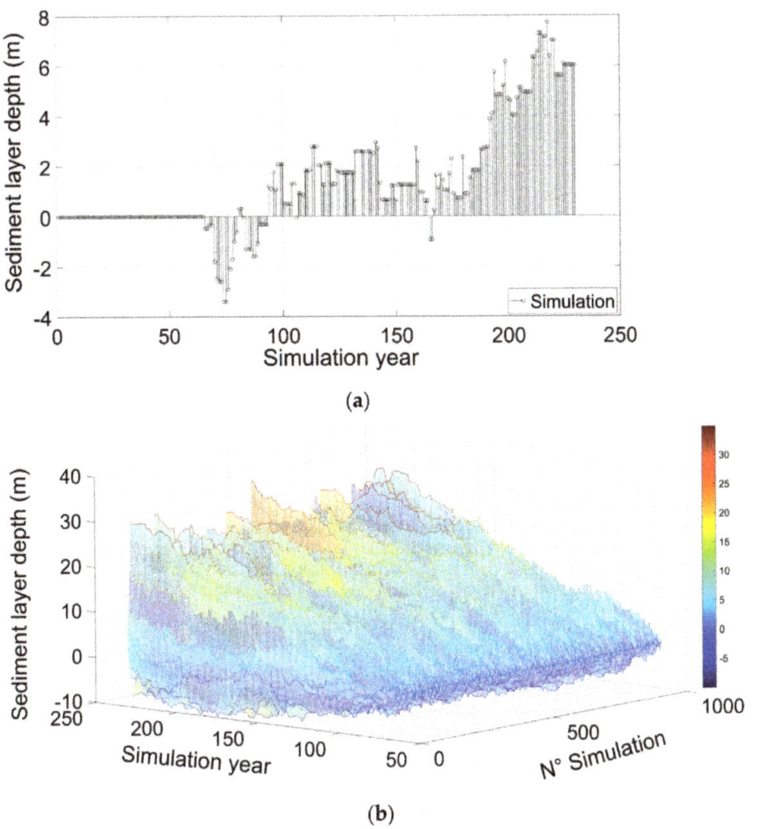

Figure 10. Simulation outcome for scour–fill depth over time (simulation starts at year 64): (**a**) one 170-year simulation; (**b**) 1064 simulations of 170 years.

The maximum scour depth value is obtained from each one of the 1064 simulations, and the best fit distribution is obtained as per 2.4 of this study. Once the probability function is found, the survival function is obtained as follows:

$$P[y_s(te) > y \mid t] = 1 - \int_{-\infty}^{t} exp\left[-\left(1+\xi\frac{t-\mu}{\sigma}\right)^{-\frac{1}{\xi}}\right]dt \tag{36}$$

Figure 11a shows the survivor function of the GEV function for different time intervals, and Figure 11b shows the PDF of the maximum scour depth.

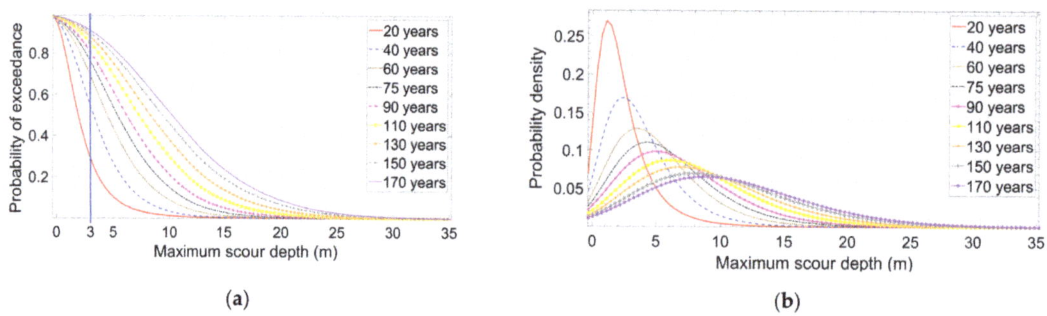

Figure 11. Outcome of the simulations for different time intervals of maximum scour depth: (**a**) GEV survivor function, 100-year return period scour depth is shown as a blue line; (**b**) GEV PDF.

Results show that, as expected, scour accumulates through time. Figure 11b shows the PDFs of several time intervals of the simulation; the distribution tends to zero skewness as time increases, and it has a negative skewness and positive kurtosis at lower time intervals. A scour depth of 2.67 m was computed using the HEC 18 methodology assuming a flood with a return period of 100 years, and it is used for comparison. Such a threshold has a notable probability of exceedance, as shown in Figure 11a. Multiple events of scour with low return periods can accumulate in time, and they can be larger than the design event when the bridge is not repaired. As time tends to infinity, the maximum scour depth increases because, with the approach proposed, scour depth is not limited by harder soils that are naturally found in deeper layers.

Figure 12a shows the scour survivor function of several time intervals using the SRICOS-EFA methodology [5]. Figure 12b shows a comparison between the proposed approach that takes in to account an estimation of sediment deposit, which lessens the severity of scour accumulation in time, and the SRICOS-EFA methodology.

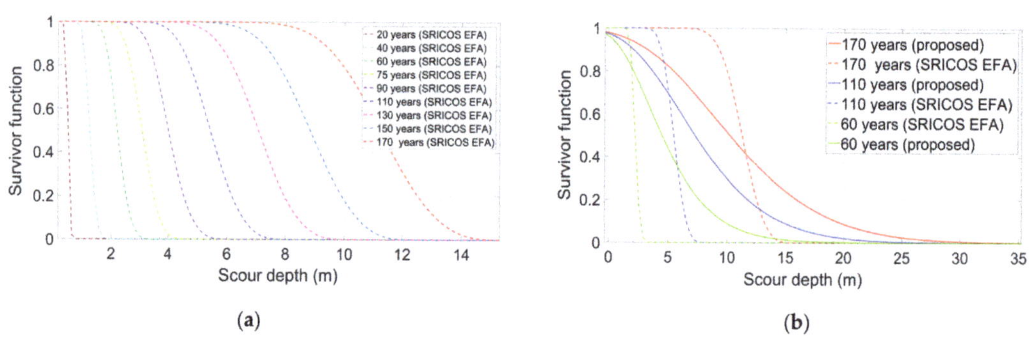

Figure 12. Simulations for different time intervals of maximum scour depth: (**a**) SRICOS EFA survivor function; (**b**) comparison between the proposed methodology and SRICOS EFA [5].

5.5. Discussion

The difference in results shown in Figure 12b are due to the differences in event simulation and in the methodology used to estimate the scour depths. Some notable differences are as follows: (**a**) SRICOS-EFA does not take in to account every event for the scour accumulation (flood has to be larger than flood to produce scour), and in a similar way a shear threshold is needed to produce scour. (**b**) It relies on average daily discharge produced by an ARIMA model, which has limitations in dealing with outliers. (**c**) It does not take in to account the accretion process (thus the sharp fall in the survivor function). (**d**) The SRICOS-EFA method uses equations obtained from flume test conducted at Texas A&M University [5], which are not equal to the HEC 18 equations for scour depth, and this leads to a natural spread on the comparison.

The proposed model has the following key aspects that differ from the SRICOS-EFA: (**a**) larger scour depths in the proposed methodology are likely due to the BMA approach on the maximum annual flow, and this approach leads to higher discharge and, thus, higher scour depths; (**b**) the addition of the accretion process leads to a heavy tail in the GEV distribution and a smoother fall in the survivor function; (**c**) in the proposed approach the hydraulic radius changes in every time step (unlike the SRICOS-EFA); (**d**)every discharge causes scour since the HEC 18 methodology does not establish a shear threshold for the scour development.

5.6. Scour Hazard

Scour hazard curves provide annual exceedance probabilities of scouring events having different intensities. Since scour is time-dependent, several time interval curves are needed because such curves are developed using all the available data of the events that could trigger the scour or fill. They reflect the behavior of the phenomena in a time-dependent fashion, and they can be used to evaluate both new and deteriorating bridges (Figure 13). The estimates of maximum scour are derived from simplified 1D equations. Such estimates are not exact and do not predict the future, but they provide a reasonable approximation to it.

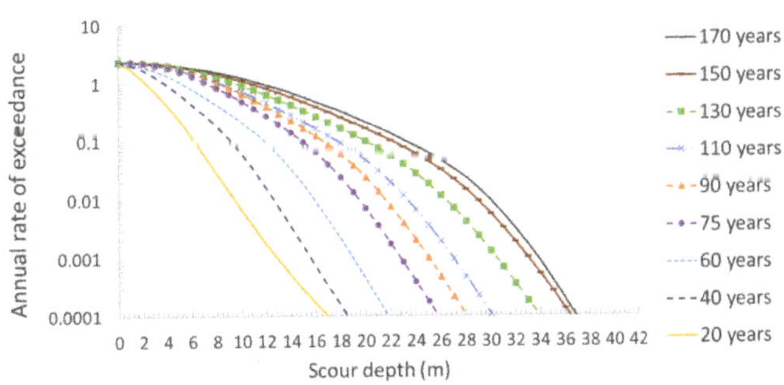

Figure 13. Scour hazard curves.

Figure 13 shows different exceedance rates due to scour or scour hazard curves for different time intervals. It is noticed that the scour depth increases as the time interval increases, which means that a certain scour depth takes different values for each time interval. Scour hazard curves may be improved if the next items are included in the model. First of all, a multi-layer approach is needed to improve the estimation of the scour process, and either a nonstationary stochastic process or an autoregressive integrated moving average with additive outliers in the ARIMA-AO model is needed to calculate the discharge forecast. In addition, it is important to consider climate change, which could impact the parameters related to discharge. It is also important to verify the computed scour depth with the actual bridge scour.

6. Concluding Remarks

A time-dependent scour hazard model that took into account missing data and the interaction between erosion and accretion was developed. The proposed model is straightforward, can be easily applied, and can be useful for the design and the re-design of new bridges prone to scour deterioration. In the case of existing bridge structures, the model can be used to estimate the time instant in which the structure could present an undesirable performance level caused by scour effects.

The proposed stationary stochastic model developed to study the scour and fill process shows that a single event with a high return period, such as $T_r = 100$ years, might not be adequate to represent the maximum scour that could be developed throughout the return period. Bridges subjected to continuous events of scour and fill without maintenance can acquire scour depths greater than the designed scour depth. This phenomenon makes bridges more vulnerable and partially explains why scour is the leading cause of bridge collapses in Mexico.

Scour hazard curves for different time intervals were estimated based on a reinforced concrete bridge located in one of the zones with high values of river discharge, such as Oaxaca. The scour hazard curves provide both an idea of an expected exceedance of scour associated with different time intervals and represent an important tool to estimate the reliability due to scour. Moreover, in zones with a high frequency of seismic occurrences, the scour hazard and seismic hazard can be treated from a multi-hazard point of view to estimate reliability indicators such as fragility curves, exceedance demand rates, mean annual rate of failure, or confidence factor. The above indicators lead engineers and decision-makers to re-design or repair the structural system in the case of new or existing bridges, respectively. The approach used to preprocess, characterize, and simulate events can be used for any phenomenon that has an "extreme" nature. In addition, the inclusion of missing data in the model improves the scour hazard estimation.

Author Contributions: Conceptualization, D.F.-V., D.T. and R.G.; methodology, D.F.-V., D.T. and R.G.; software, D.F.-V.; validation, D.F.-V., D.T. and R.G.; data curation, D.F.-V.; writing—original draft preparation, D.F.-V. and R.G.; writing—review and editing, D.F.-V., R.G. and D.T.; funding acquisition, R.G. and D.T. All authors have read and agreed to the published version of the manuscript.

Funding: This research was funded by Universidad Nacional Autónoma de México.

Institutional Review Board Statement: Not applicable.

Informed Consent Statement: Not applicable.

Data Availability Statement: Data are contained within the manuscript.

Acknowledgments: The first author gratefully acknowledges economic support from Instituto de Ingeniería of Universidad Nacional Autónoma de México during his PhD studies. The second author gratefully acknowledges the support provided by the Instituto de Ingeniería of Universidad Nacional Autónoma de México.The third author appreciates the support given by both Universidad Autónoma Metropolitana and Consejo Nacional de Ciencia y Tecnología through the Ciencia Básica Project CB 2017-2018 A1-S-8700.

Conflicts of Interest: The authors declare no conflict of interest.

References

1. Pizarro, A.; Manfreda, S.; Tubaldi, E. The science behind scour at bridge foundations: A review. *Water* **2020**, *12*, 374. [CrossRef]
2. Mahalder, B.; Schwartz, J.S.; Palomino, A.M.; Zirkle, J. Scour hole development in natural cohesive bed sediment around cylinder shaped piers subjected to varying sequential flow events. *Water* **2021**, *13*, 3289. [CrossRef]
3. Manfreda, S.; Link, O.; Pizarro, A. A theoretically derived probability distribution of scour. *Water* **2018**, *10*, 1520. [CrossRef]
4. Van Noortwijk, J.; Kok, M.; Cooke, R. Optimal maintenance decisions for the sea-bed protection of the Eastern-Scheldt barrier. In *Engineering Probabilistic Design and Maintenance for Flood Protection*; Cooke, R., Mendel, M., Vrijling, H., Eds.; Springer: Boston, MA, USA, 1997; pp. 25–56. [CrossRef]
5. Brandimarte, L.; D'Odorico, P.; Montanari, A. A probabilistic approach to the analysis of contraction scour. *J. Hydraul. Res.* **2006**, *44*, 654–662. [CrossRef]

6. Tubaldi, E.; Macorini, L.; Izzuddin, B.A.; Manes, C.; Laio, F. A framework for probabilistic assessment of clear-water scour around bridge piers. *Struct. Saf.* **2017**, *69*, 11–22. [CrossRef]
7. Pizarro, A.; Tubaldi, E. Quantification of modelling uncertainties in bridge scour risk assessment under multiple flood events. *Geosciences* **2019**, *9*, 445. [CrossRef]
8. Briaud, J.-L.; Brandimarte, L.; Wang, J.; D'Odorico, P. Probability of scour depth exceedance owing to hydrologic uncertainty. *Georisk Assess. Manag. Risk Eng. Syst. Geohazards* **2007**, *1*, 77–88. [CrossRef]
9. Liao, K.-W.; Hoang, N.-D.; Gitomarsono, J. A probabilistic safety evaluation framework for multi-hazard assessment in a bridge using SO-MARS learning model. *KSCE J. Civ. Eng.* **2018**, *22*, 903–915. [CrossRef]
10. Johnson, P.A.; Dock, D.A. Probabilistic bridge scour estimates. *J. Hydraul. Eng.* **1998**, *124*, 750–754. [CrossRef]
11. Kallias, A.N.; Imam, B. Probabilistic assessment of local scour in bridge piers under changing environmental conditions. *Struct. Infrastruct. Eng.* **2016**, *12*, 1228–1241. [CrossRef]
12. Bolduc, L.C.; Gardoni, P.; Briaud, J.-L. Probability of exceedance estimates for scour depth around bridge piers. *J. Geotech. Geoenviron. Eng.* **2008**, *134*, 175–184. [CrossRef]
13. Contreras-Jara, M.; Echaveguren, T.; Chamorro, A.; Vargas-Baecheler, J. Estimation of exceedance probability of scour on bridges using reliability principles. *J. Hydrol. Eng.* **2021**, *26*, 04021029. [CrossRef]
14. Gómez, R.; Flores, D.; Arenas, M.; Flores, R. *Vulnerabilidad de Estructuras de Puentes en Zonas de Gran Influencia de Ciclones Tropicales*; Technical report; CENAPRED: Mexico City, MC, Mexico, 2017. (in Spanish)
15. Wardhana, K.; Hadipriono, F.C. Analysis of recent bridge failures in the United States. *J. Perform. Constr. Facil.* **2003**, *17*, 144–150. [CrossRef]
16. Rubin, D. Inference and missing data. *Biometrika* **1976**, *63*, 581–592. [CrossRef]
17. Pawitan, Y. *All Likelihood: Statistical Modelling and Inference Using Likelihood*; Oxford University Press Inc.: New York, NY, USA, 2001; pp. 336–351.
18. Schafer, J.L.; Graham, J.W. Missing data: Our view of the state of the art. *Psychol. Methods* **2002**, *7*, 147–177. [CrossRef] [PubMed]
19. Arneson, L.A.; Zevenbergen, L.W.; Lagasse, P.F.; Clopper, P.E. *Evaluating Scour at Bridges*, 5th ed.; Publication No. FHWA-HIF-12-003; FHWA: Springfield, VA, USA, 2012.
20. Folch-Fortuny, A.; Arteaga, F.; Ferrer, A. Missing data imputation toolbox for MATLAB. *Chemometr. Intell. Lab. Syst.* **2016**, *154*, 93–100. [CrossRef]
21. Tan, M.T.; Tian, G.-L.; Ng, K.W. *Bayesian Missing Data Problems: EM, Data Augmentation and Noniterative Computation*; Chapman and Hall/CRC: Boca Raton, FL, USA, 2009; pp. 39–42.
22. Young, G.A.; Smith, R.L. *Essentials of Statistical Inference*; Cambridge University Press: Cambridge, UK, 2005; pp. 13–15.
23. Rubin, D.B. *Multiple Imputation for Nonresponse in Surveys*; Wiley & Sons. Inc.: New York, NY, USA, 1987; p. 258.
24. Takahashi, M. Statistical Inference in Missing Data by Mcmc and Non-Mcmc Multiple Imputation Algorithms: Assessing the Effects of between-Imputation Iterations. *Data Sci. J.* **2017**, *16*, 1–17. [CrossRef]
25. Schafer, J.L. *Analysis of Incomplete Multivariate Data*; Chapman and Hall/CRC: New York, NY, USA, 1997; pp. 55–59.
26. Akaike, H. Likelihood of a model and information criteria. *J. Econom.* **1981**, *16*, 3–14. [CrossRef]
27. Katz, R.W.; Parlange, M.B.; Naveau, P. Statistics of extremes in hydrology. *Adv. Water Resour.* **2002**, *25*, 1287–1304. [CrossRef]
28. Ferreira, A.; De Haan, L. On the block maxima method in extreme value theory: PWM estimators. *Ann. Stat.* **2015**, *43*, 276–298. [CrossRef]
29. Haschenburger, J.K. Scour and Fill in a Gravel-Bed Channel: Observations and Stochastic Models. PhD Thesis, University of British Columbia, Vancouver, BC, Canada, 1996.
30. Haschenburger, J.K. Channel scour and fill in coastal streams. In *Carnation Creek and Queen Charlotte Islands Fish/Forestry Workshop: Applying 20 Years of Coastal Research to Management Solutions*; BC Ministry of forest; Land management handbook No. 41; Hogan, D.L., Tschaplininski, P.J., Chatwin, S., Eds.; Crown Publications Inc.: Victoria, BC, CA, 1998; pp. 109–117.
31. Bigelow, P.E. Scour, fill, and salmon spawning in a California coastal stream. Master Thesis, Humboldt State University, Arcata, CA, USA, 2003.
32. Einstein, H.A. *The Bed-Load Function for Sediment Transportation in Open Channel Flows*; Technical report No. 1026; US Government Printing Office: Washington, DC, USA, 1950.
33. Todorovic, P.; Zelenhasic, E. A stochastic model for flood analysis. *Water Resour. Res.* **1970**, *6*, 1641–1648. [CrossRef]
34. Sirangelo, B.; Versace, P. Flood-induced bed changes in alluvial streams. *Hydrol. Sci. J.* **1984**, *29*, 389–398. [CrossRef]
35. Haschenburger, J.K. A probability model of scour and fill depths in gravel-bed channels. *Water Resour. Res.* **1999**, *35*, 2857–2869. [CrossRef]
36. Simões, F.J.M. Shear velocity criterion for incipient motion of sediment. *Water Sci. Eng.* **2014**, *7*, 183–193. [CrossRef]
37. Alamilla, J.L.; Campos, D.; Ortega, C.; Soriano, A.; Morales, J.L. Optimum selection of design parameters for transportation of offshore structures. *Ocean Eng.* **2009**, *36*, 330–338. [CrossRef]

water

MDPI

Article

Turbulent Flow through Random Vegetation on a Rough Bed

Francesco Coscarella [1,*], Nadia Penna [1], Aldo Pedro Ferrante [1], Paola Gualtieri [2] and Roberto Gaudio [1]

[1] Dipartimento di Ingegneria Civile, Università della Calabria, 87036 Rende, CS, Italy; nadia.penna@unical.it (N.P.); aldopedro.ferrante@unical.it (A.P.F.); gaudio@unical.it (R.G.)

[2] Dipartimento di Ingegneria Civile, Edile e Ambientale, Università degli Studi di Napoli "Federico II", 80125 Napoli, NA, Italy; paola.gualtieri@unina.it

* Correspondence: francesco.coscarella@unical.it; Tel.: +39-0984-496552

Abstract: River vegetation radically modifies the flow field and turbulence characteristics. To analyze the vegetation effects on the flow, most scientific studies are based on laboratory tests or numerical simulations with vegetation stems on smooth beds. Nevertheless, in this manner, the effects of bed sediments are neglected. The aim of this paper is to experimentally investigate the effects of bed sediments in a vegetated channel and, in consideration of that, comparative experiments of velocity measures, performed with an Acoustic Doppler Velocimeter (ADV) profiler, were carried out in a laboratory flume with different uniform bed sediment sizes and the same pattern of randomly arranged emergent rigid vegetation. To better comprehend the time-averaged flow conditions, the time-averaged velocity was explored. Subsequently, the analysis was focused on the energetic characteristics of the flow field with the determination of the Turbulent Kinetic Energy (TKE) and its components, as well as of the energy spectra of the velocity components immediately downstream of a vegetation element. The results show that both the vegetation and bed roughness surface deeply affect the turbulence characteristics. Furthermore, it was revealed that the roughness influence becomes predominant as the grain size becomes larger.

Keywords: rigid vegetation; bed roughness; turbulent flow; Turbulent Kinetic Energy (TKE); energy spectra

Citation: Coscarella, F.; Penna, N.; Ferrante, A.P.; Gualtieri, P.; Gaudio, R. Turbulent Flow through Random Vegetation on a Rough Bed. *Water* **2021**, *13*, 2564. https://doi.org/10.3390/w13182564

Academic Editor: Mouldi Ben Meftah

Received: 30 July 2021
Accepted: 15 September 2021
Published: 17 September 2021

1. Introduction

Vegetation exerts important effects on hydraulic resistance, turbulent structures, mixing processes and sediment transport in rivers [1–5]. For this reason, a large amount of both experimental and numerical researches has been devoted to the study of the impacts of vegetation on the flow characteristics, influencing mass and momentum exchange across the river section, together with geomorphology, water quality and aquatic biodiversity (e.g., [6,7]). The flow through emergent rigid vegetation has been widely investigated, neglecting impacts introduced by natural vegetation usually observed in many fluvial ecosystems [8]. In order to understand the flow evolution in the presence of emergent and submerged vegetation, which is founded on different specific aspects of the canopy flow (mean momentum balance, turbulence budget, exchange dynamics), several works have focused on flexible vegetation (e.g., [9–11]).

Maji et al. [12] compiled a state-of-the-art study that included works on flow dynamics and interactions between flow and vegetation. Most of them aimed only at the study of the flow–vegetation interactions on smooth beds (e.g., [1,2,13–21]). Nevertheless, special interest should be devoted to works on vegetated flows with rough beds, since the interactions between fluid, vegetation and bed sediment allow for reaching better knowledge of the turbulence characteristics in real rivers, which have a crucial role in sediment transport. In fact, with respect to a smooth bed, in the case of a rough bed a pronounced velocity spike occurs near the rough surface and immediately downstream of a stem [22]. In addition, a rough bed induces a decrease in the temporal-averaged streamwise velocity and an increase in the Turbulent Kinetic Energy (TKE) [23]. In rough conditions, the streamwise velocity also has a quasi-constant distribution in the layer where the flow is principally controlled by the vegetation [16]. Then, approaching the bed, it reduces logarithmically from the maximum constant magnitude toward zero. Moreover, the Reynolds shear stresses present very small values, practically vanishing in the region dominated by the vegetation [16,22].

Very recently, Penna et al. [24,25] studied the flow field around a rigid cylinder in three different rough bed conditions with a uniform pattern of stems that are regularly aligned. Penna et al. [24] analyzed the velocity, shear stress distributions, TKE and the energy spectra, showing that, in the region near the free surface, the flow is deeply affected by the stems. Moving toward the bed surface, the flow is influenced by both the vegetation and bed roughness effects. Penna et al. [25], using the so-called Anisotropy Invariant Maps (AIMs), investigated for the first time the turbulence anisotropy through uniform emergent rigid vegetation on rough beds. The study of the AIMs indicated that, approaching the bed surface, the combined impact of vegetation and bed roughness affects the turbulence evolution from the quasi-three-dimensional isotropy to axisymmetric anisotropy. This proved that, as the influence of the bed roughness decreases, the turbulence tends to isotropy.

Starting from the results of Penna et al. [24,25] in the case of a uniform vegetation pattern, the aim of this work is to study the impact of bed roughness and random vegetation pattern distribution on turbulence. In particular, in order to describe the flow domain, the time-averaged approaching flow velocity field, TKE, normal shear stresses and energy spectra of the velocity components were computed in an area centered on a single stem.

The paper is organized as follows: Section 2 describes the laboratory and the methodology applied for the data analysis; Section 3 illustrates and discusses the results; Section 4 reports the conclusions of the present work.

2. Laboratory Experiments and Methodology

The experimental study was conducted in a 9.6-m long, 0.485-m wide and 0.5-m deep tilting flume at the *Laboratorio "Grandi Modelli Idraulici" (GMI), Università della Calabria*, Italy. In order to reduce the influence of the pump on the turbulence characteristics of the flow, a stilling tank, an uphill slipway and honeycombs (10 mm in diameter) were placed at the inlet of the channel. At the outlet, a tank equipped with a calibrated Thomson weir to measure the flow discharge Q and with a tailgate to regulate the water depth h of the flow were placed (Figure 1).

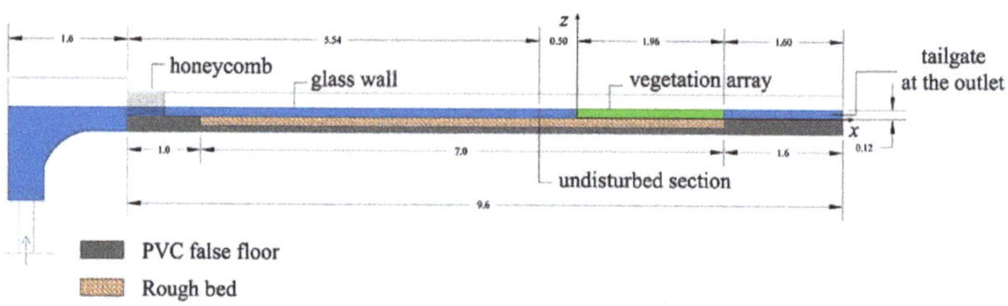

Figure 1. Schematic of the experimental facility (dimensions are expressed in meters).

The experiments were carried out with a flow depth $h \approx 0.12$ m, measured 50 cm upstream of the vegetation array (i.e., in undisturbed flow condition) by a point gauge with a decimal Vernier having an accuracy of ± 0.1 mm and a flow discharge Q equal to 19.73 L/s (measured with a Thomson weir). The approaching cross-section average flow velocity $U = Q/(Bh)$ was, hence, equal to 0.30 m s^{-1}, where B was the flume width. The longitudinal bottom slope of the flume, S, was fixed at 1.5‰ using a hydraulic jack.

The rigid vegetation was simulated with vertical, wooden and circular cylinders. The cylinder height and diameter were $h_c = 0.40$ m and $d = 0.02$ m, respectively. The stems were implanted into a 1.96-m long, 0.485-m wide and 0.015-m thick Plexiglas panel fixed to the channel bottom. A total of 68 cylinders were randomly arranged in the flume (Figure 2a). All the experiments were carried out in conditions of emergent vegetation. The frontal area per volume was $a = nd = 1.4\ \mathrm{m}^{-1}$, where $n = 71\ \mathrm{m}^{-2}$ was the number of cylinders per bed area, while the solid volume fraction occupied by the canopy els per bed area, while the solid volume fraction occupied by the canopy elements was $\phi = \pi ad/4 = n\pi d^2 4 = 0.02$, which is consistent with typical laboratory studies with vegetation [17,26,27]. Following Nepf [28], this vegetation distribution can be classified as dense. Focus was given to the evolution of the turbulence characteristics in a study area selected around a single stem (Figure 2b).

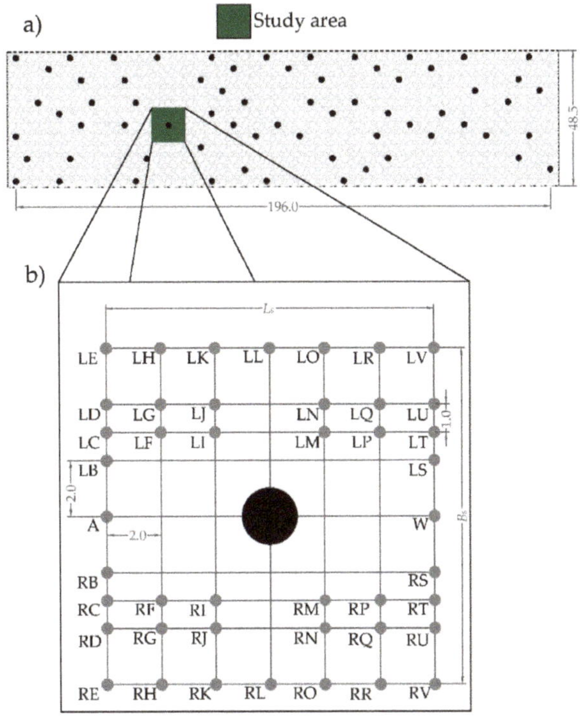

Figure 2. (**a**) Sketch of the cylinder array in the laboratory flume (dimensions are in cm); (**b**) the measurement verticals within the study area. Here, B_s and L_s are the width and length of the study area, respectively.

Three different types of bed roughness were simulated, employing very coarse sand ($d_{50} = 1.53$ mm), fine gravel ($d_{50} = 6.49$ mm) and coarse gravel ($d_{50} = 17.98$ mm), respectively (Figure 3). The grain size distributions were relatively uniform, i.e., as reported by Dey and Sarkar [29], with a geometric standard deviation $\sigma_g = (d_{84}/d_{16})^{0.5} < 1.5$, where d_{16} and d_{84} are the sediment sizes for which 16% and 84% by weight of sediment is finer, respectively. At the beginning of each run, the flume was filled in with the sediments, which were successively screeded to make the longitudinal bed slope equal to that of the flume bottom.

Figure 3. Sediments used in the experimental runs: (**a**) $d_{50} = 1.53$ mm; (**b**) $d_{50} = 6.49$ mm; (**c**) $d_{50} = 17.98$ mm.

The experimental conditions used in this research do not reproduce a specific situation of a real river. Nevertheless, they could represent a new dataset that, for instance, may be used for the calibration of advanced numerical models. In fact, as is known, vegetation, bed roughness or man-made structures acting as an obstruction for the flow generate turbulence and affect the entire flow velocity distribution, modifying the turbulence behavior [30,31].

In Table 1, the hydraulic conditions of the experimental study are reported. Along with the aforementioned characteristics, in Table 1 the following quantities are listed: the shear velocity (u_*), the critical velocity for the inception of sediment motion (U_c), the mean water temperature (T) measured with the Acoustic Doppler Velocimeter (ADV) integrated thermometer (having an accuracy of ±0.1 °C), the water kinematic viscosity (ν), computed as a function of the water temperature [32], the flow Froude number Fr [$=U/(gh)^{0.5}$], the flow Reynolds number Re ($=Uh/\nu$), the shear Reynolds number Re_* ($=u_*\varepsilon/\nu$, where ε is the Nikuradse equivalent sand roughness, equal to about $2d_{50}$) and the Reynolds number of the vegetation stems Re_d ($=Ud/\nu$). In accordance with Manes et al. [33] and Dey and Das [34], the shear velocity used to scale the flow statistics was determined as $u_* = (\tau_*/\rho)^{0.5}$, where τ_* is the total stress acting at the roughness tops. This can be obtained by extending linearly the distribution of the turbulent shear stress captured 50 cm upstream of the vegetation pattern (i.e., in correspondence with the undisturbed flow condition) from the region above the roughness elements to their tops. Thus, the shear velocity was evaluated at the sediment crest level as $\left(-\overline{u'w'}\right)^{0.5}$, where u' and w' are the fluctuations of the temporal velocity signal in the streamwise and vertical directions, respectively, and the symbol $\bar{\ }$ indicates the time averaging operation. The critical velocity for the inception of sediment motion U_c was established 50 cm upstream of the vegetation array through the well-known Neill formula [35], as follows:

$$U_c = \sqrt{2.5\left(\frac{h}{d_{50}}\right)^{0.2} g\Delta d_{50}} \quad (1)$$

where g is the gravitational acceleration, $\Delta = (\rho_s - \rho)/\rho$ is the relative submerged grain density, ρ_s is the grain density and ρ is the fluid density. All the experiments were performed in clear-water condition ($U < U_c$), which was also verified from the direct observation of the flow.

Table 1. Hydraulic conditions of the experimental study for the approaching flow.

Parameter (Units)	Run 1	Run 2	Run 3
d_{50} (mm)	1.53	6.49	17.98
h (m)	0.12	0.12	0.12
Q (l/s)	19.73	19.73	19.73
U (m/s)	0.34	0.34	0.34
u_* (m/s)	0.021	0.022	0.028
U_c (m/s)	0.39	0.69	1.04
S (‰)	1.50	1.50	1.50
T (°C)	16.67	18.06	18.70
ν (m^2/s)	1.09×10^{-6}	1.05×10^{-6}	1.03×10^{-6}
Fr	0.31	0.31	0.31
Re	37,431	38,857	39,612
Re_*	59	272	978
Re_d	6239	6192	6602

An ADV profiler with down-looking probe, four beams (Nortek Vectrino) and an automatic movement system (the Traverse System by HR Wallingford Ltd., Oxfordshire UK) was used to capture the instantaneous velocity components (streamwise u, spanwise v and vertical w) with an accuracy of ±5% (assessed in previous works). The instrument sampling frequency was 100 Hz, and the duration of a single sampling was 300 s for a total number of samples of 30,000 which, as reported by [34,36,37], is adequate for determining accurate turbulence statistics. The sampling volume was a 1 mm long cylinder with a diameter of 6 mm. The ADV receivers pointed at 50 mm below their own transmitter. Hence, the measurements were not performed near the free surface flow zone (i.e., 50 mm below the free surface). The spatial coordinates of the Traverse System had an accuracy of ±0.1 mm. Prior to the analysis of the ADV data, it was necessary to proceed with the spike detection. Firstly, the ADV raw data were prefiltered, discarding the values with correlation (COR) lower than 70% and a signal-to-noise ratio (SNR) lower than 15 dB [34]; secondly, the contaminated velocity records were cleaned using the phase-space thresholding method and each spike was replaced with a cubic polynomial through 12 points on either side of itself [38]. The de-spiking method resulted in a rejection of less than 5% of the original velocity time series.

For all the runs, 44 vertical profiles were captured, as shown in Figure 2b. The vertical spatial resolutions were 3 mm for $z \leq 15$ mm and 5 mm above, where z is the vertical axis starting from the maximum crest level in the study area.

To describe the undisturbed flow 50 cm upstream of the vegetation array, the velocity vertical distribution was captured at the centerline of the laboratory flume during each run. In Figure 4a, the undisturbed profiles of the dimensionless time-averaged velocity in the streamwise direction $\hat{u}_{UP}$ ($= \overline{u}_{UP}/u_*$, where $\overline{u}$ is the time-averaged velocity in the same direction) for the three experimental runs are reported. The vertical axis $\hat{z}$ was made dimensionless by dividing the elevation z by the local water level that, for the undisturbed distributions, was equal to the flow depth h reported in Table 1. As the elevation z increases, the streamwise velocities u increase; instead, near the sediment grains, in the so-called roughness sublayer, they tend to zero owing to the bed roughness (this is typical in the open-channel flow condition) [39]. In particular, for Run 3, owing to higher roughness dimension, the streamwise velocity profile tended rapidly to zero starting from elevation $z = 0.1h$ [40]. In addition, $\hat{u}$ in the three runs shows different values at a fixed $\hat{z}$. This happens owing to the different bed roughness conditions that lead to an increase of the shear velocity as d_{50} increases and, consequently, to a decrease in $\hat{u}$. The distributions of the dimensionless turbulent shear stresses $\hat{\tau}_{uw}$ ($= -\overline{u'w'}/u_*^2$) and of the dimensionless viscous shear stresses $\hat{\tau}_\nu$ [$= \nu(d\overline{u}/dz)/u_*^2$] along $\hat{z}$ are represented in Figure 4b,c, respectively. In particular, above the roughness surface, the prevalence of the Reynolds shear stresses can be noted, while the viscous shear stresses are practically negligible as $\hat{z}$ increases. The viscous shear stresses achieve their maximum values near the grain crests for each experimental run. Conversely, the turbulent shear stresses reach the peak above the crest level and then they reduce as the vertical distance increases.

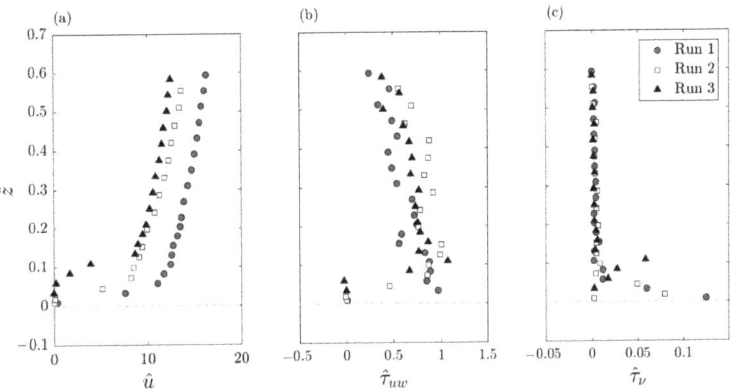

Figure 4. Vertical profiles of (**a**) dimensionless time-averaged velocity, (**b**) dimensionless Reynolds shear stress and (**c**) dimensionless viscous shear stress for the undisturbed flow condition (50 cm upstream to the vegetation array) in Run 1, Run 2 and Run 3.

3. Results and Discussion

3.1. Time-Averaged Flow

To investigate the time-averaged flow velocity field with respect to the approaching flow velocity in the spatial flow domain (i.e., in the upstream plane section identified with letters from LE to RE in the study area), Figure 5 shows the colormaps of the dimensionless time-averaged accelerated and decelerated flow fields upstream of the investigated stem in all the runs in the plane $\hat{y}$-$\hat{z}$. The abscissa, represented by $\hat{y}$, was made dimensionless by dividing y by the study area width (B_s = 12 cm). The time-averaged flow is accelerated if $(\hat{u}_{UP} - \hat{u}) < 0$ (blue values in the colormaps) and decelerated if $(\hat{u}_{UP} - \hat{u}) > 0$ (red values in the colormaps), where $\hat{u}$ is the dimensionless time-averaged streamwise velocity.

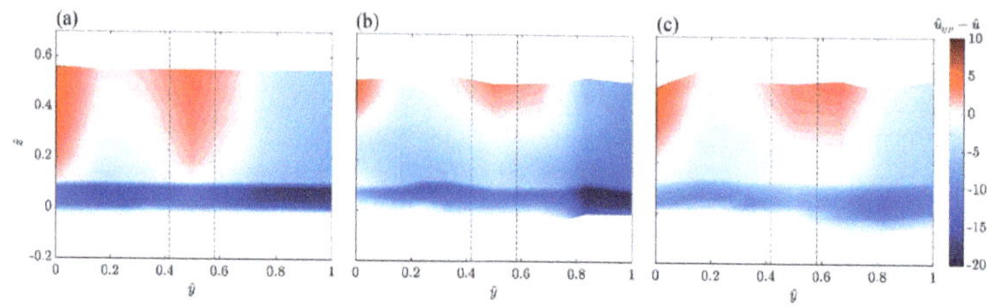

Figure 5. Contours of the dimensionless time-averaged accelerated $(\hat{u}_{UP} - \hat{u}) < 0$ and decelerated $(\hat{u}_{UP} - \hat{u}) > 0$ flow field in the upstream plane section (from LE to RE) for (**a**) Run 1, (**b**) Run 2 and (**c**) Run 3. The black dashed lines show the cylinder position.

A strong spanwise variation of $\hat{u}$ can be noted with respect to the approaching flow profile. Specifically, in correspondence with the cylinder ($0.42B_s < y < 0.58B_s$), in all the runs the flow is decelerated. This probably occurs owing to the incidence of both the analyzed stem and another stem immediately upstream of the former (Figure 2a). An examination of the contours also reveals a decelerated flow zone on the left side of the cylinder ($0< y < 0.2B_s$) in all the experiments. This happens owing to the presence of another stem immediately upstream of the study area (Figure 2a). Instead, the flow fields are accelerated both on the right side of the analyzed cylinder and, to a lesser extent, on the left one. Moving toward the bed, it is evident that the flow velocity is also influenced by the bed roughness. As d_{50} decreases, this zone becomes strongly accelerated. Conversely, as d_{50} increases, in the near-bed layer, the flow field results to be influenced by the bed roughness and much more by the vegetation, with a minor acceleration intensity.

A similar behavior can be appreciated in the downstream flow domain in Figure 6 (i.e., in the downstream plane section identified with letters from LV to RV in the study area). In particular, immediately downstream of the cylinder ($0.42B_s < y < 0.58B_s$), the flow is decelerated in all the runs. Conversely, Run 2 shows an accelerated flow both on the right and on the left sides of the cylinder, while Run 1 and Run 3 show an accelerated flow mostly on the right side of the cylinder. A sensible difference is clear in the values of $\hat{u}_{UP} - \hat{u}$ immediately upstream (vertical A; $y = 0.5B_s$) and downstream (vertical W; $y = 0.5B_s$) of the cylinder. Specifically, the time-averaged streamwise velocity $\overline{u}$ is reduced by about 30% and 50% with respect to the time-averaged streamwise velocity of the undisturbed profile, $\overline{u}_{UP}$, at the verticals A and W, respectively, in all the runs. It is possible to assure that a major deceleration is obtained beyond a cylinder. In fact, as observed in the downstream section at y = 0.5Bs (vertical W) with respect to the upstream section at y = 0.5 Bs (vertical A), the velocity reduction is practically due to the vicinity of the studied stem (6 cm upstream). On the contrary, a minor deceleration is visible at vertical A, although it is between two cylinders (an upstream stem at 9 cm and a downstream stem at 6 cm).

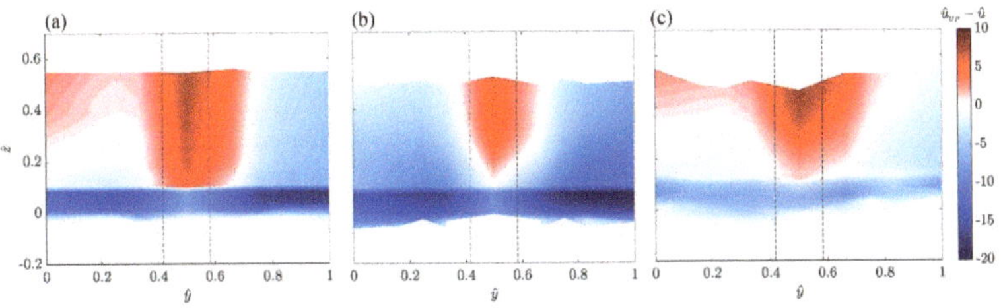

Figure 6. Contours of the dimensionless time-averaged accelerated $(\hat{u}_{UP} - \hat{u}) < 0$ and decelerated $(\hat{u}_{UP} - \hat{u}) > 0$ flow field in the downstream plane section (from LV to RV) for (**a**) Run 1, (**b**) Run 2 and (**c**) Run 3. The black dashed lines show the cylinder position.

In order to investigate the longitudinal evolution of the flow field, the contours of the dimensionless time-averaged accelerated and decelerated flow field are analyzed. For the sake of simplicity, they are reported in Figure 7 only for the extreme right vertical plane identified with letters from RE to RV in the plane $\hat{x}$-$\hat{z}$. The abscissa $\hat{x}$ was made dimensionless by dividing x by the study area width (L_s = 12 cm). In Figure 7, it is evident that the presence of vegetation has a very visible effect on the flow field: the flow is accelerated in all the runs with higher streamwise velocities than in the undisturbed flow profiles along the whole water depth. At each measurement location, the vegetation causes the velocity profile to maintain a constant value (for $z > h_l$). Moving toward the bed ($z < h_l$), the influence of vegetation decreases, and the flow field becomes more accelerated, owing to the presence of the bed.

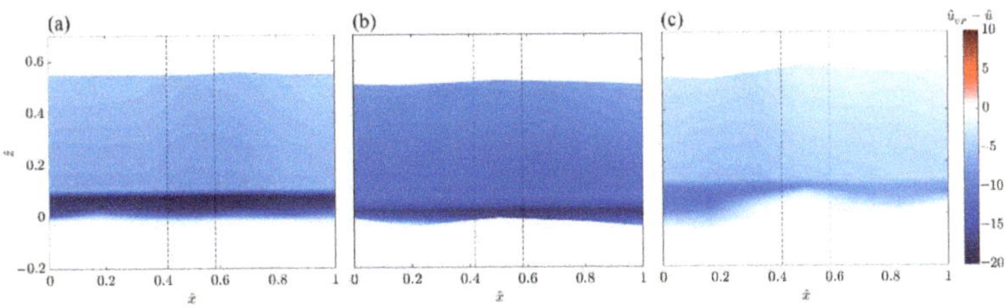

Figure 7. Contours of the dimensionless time-averaged accelerated $(\hat{u}_{UP} - \hat{u}) < 0$ and decelerated $(\hat{u}_{UP} - \hat{u}) > 0$ flow field in the longitudinal plane section (from RE to RV) for (**a**) Run 1, (**b**) Run 2 and (**c**) Run 3. The black dashed lines show the cylinder position.

3.2. TKE and Normal Stresses

The computation of the spatial distributions of the TKE is very significant in the assessment of the energetic process in open-channel flows. Since the time-averaged spatial fluctuations influence mechanical dispersion [41] and, in turn, this latter may be influenced by the vegetation stems and the bed roughness, the topic is of considerable interest. In fact, the presence of vegetation adds a further turbulence production in the wakes of the plant elements [42]. The TKE is defined as half the sum of the variances of the velocity components:

$$\text{TKE} = \frac{1}{2}\left[\overline{(u')^2} + \overline{(v')^2} + \overline{(w')^2}\right]. \tag{2}$$

where v' is the velocity fluctuation of the spanwise velocity v.

The colormaps of the dimensionless TKE (i.e., TKE divided by u_*^2) on the transversal upstream plane section (identified with the letters from LE to RE in the spanwise direction) are shown in Figure 8 for each experimental run. The highest values of the TKE are located in front of the cylinder and at the vertical LE ($y = 0$).

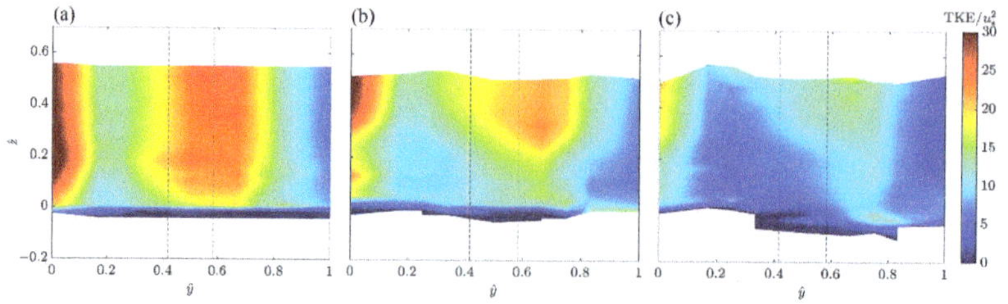

Figure 8. Contours of the dimensionless TKE in the upstream plane section (from LE to RE) for (**a**) Run 1, (**b**) Run 2 and (**c**) Run 3. The black dashed lines show the cylinder position.

To analyze this behavior, in Figure 9 we consider, for all the runs, the dimensionless normal stresses in the streamwise, spanwise and vertical directions, $\overline{u'u'}/u_*^2$, $\overline{v'v'}/u_*^2$ and $\overline{w'w'}/u_*^2$, respectively, which are essentially the three addends in Equation (2). It is possible to observe that the high value on the left part of TKE contours ($y = 0$ in Figure 8a) is practically ascribable to the streamwise and spanwise effects (Figure 9) owing to the presence upstream of the studied stem, both of an empty zone without vegetation, which influences $\overline{u'u'}$, and of a wake vortex, that affects $\overline{v'v'}$. In contrast, the high magnitude of the TKE immediately behind the vegetation element ($0.42B_s < y < 0.58B_s$) is mostly due to the spanwise fluctuations, as a consequence of the circumvention of the obstacle.

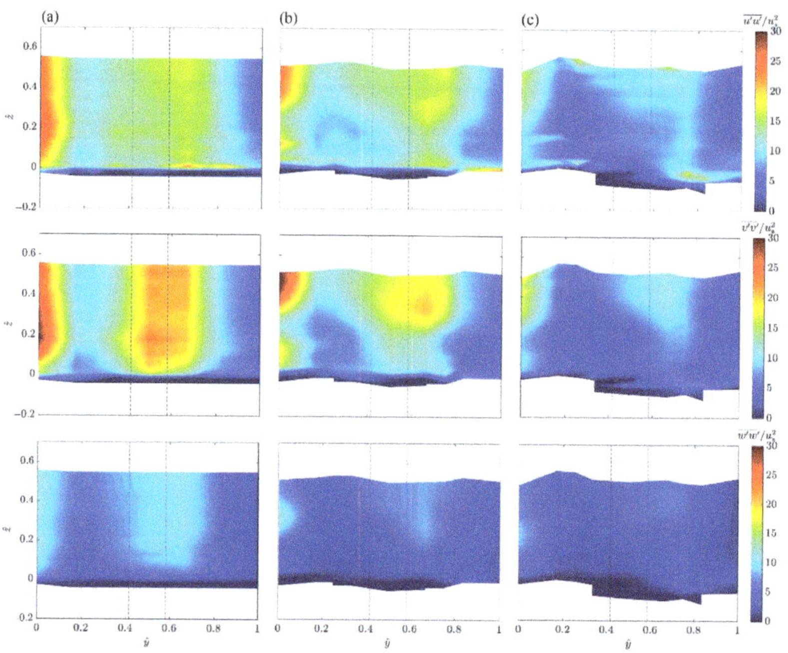

Figure 9. Contours of the dimensionless normal stresses (**a**) $\overline{u'u'}/u_*^2$, (**b**) $\overline{v'v'}/u_*^2$ and (**c**) $\overline{w'w'}/u_*^2$ in the upstream plane section (from LE to RE) for (**a**) Run 1, (**b**) Run 2 and (**c**) Run 3. The black dashed lines show the cylinder position.

Conversely, downstream of the studied vegetation element, the dimensionless TKE shows higher magnitudes immediately beyond the stem, whereas no high TKE value is detectable at $y = 0$ (Figure 10).

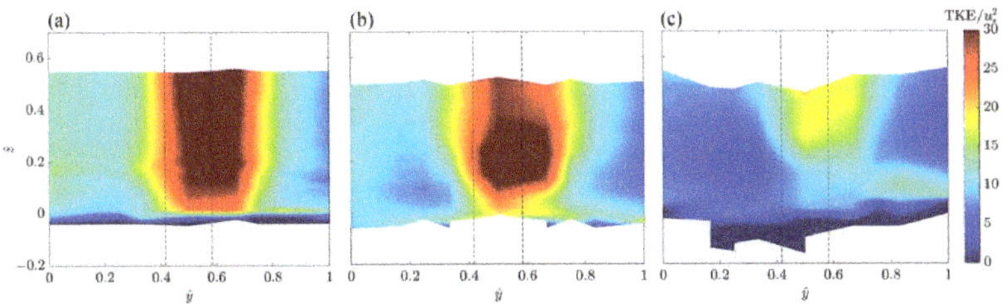

Figure 10. Contours of the dimensionless TKE in the downstream plane section (from LV to RV) for (**a**) Run 1, (**b**) Run 2 and (**c**) Run 3. The black dashed lines show the cylinder position.

Analogously to the upstream flow field, from the normal shear stress distributions shown in Figure 11, it is possible to evaluate the contributions of the TKE patterns. Specifically, the higher kinetic energy values are influenced by both the streamwise, $\overline{u'u'}$, and spanwise, $\overline{v'v'}$, normal stresses, which increase owing to the presence of the von Kármán wake vortex. This behavior, although with different magnitudes, is displayed in all the runs and, consequently, is clearly a vegetation effect. Furthermore, from a comparison between the TKE and the normal stress lateral distributions (Figures 8 and 10, and Figures 9 and 11, respectively), it is evident a predominant effect of the vegetation element in the downstream plane section and a contribution of the vertical normal stresses in the vertical W (downstream of the cylinder) greater than in the vertical A (upstream of the cylinder).

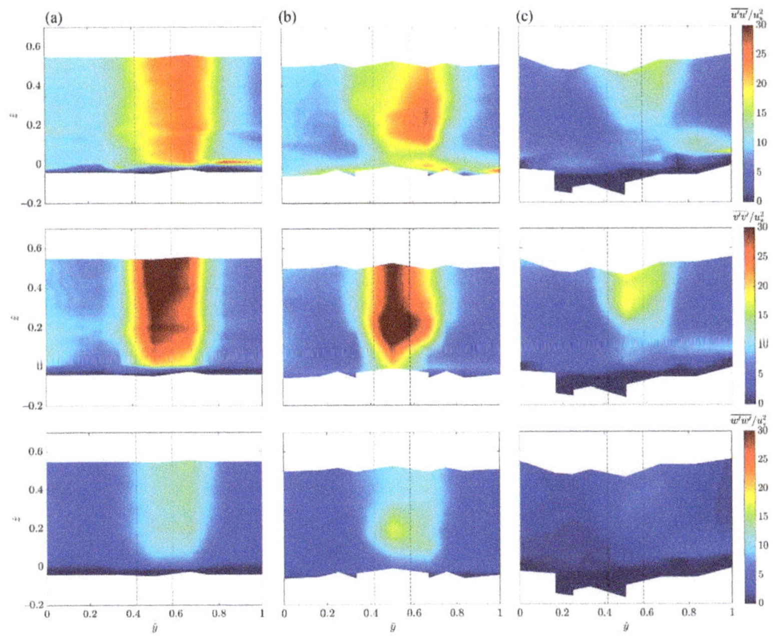

Figure 11. Contours of the dimensionless normal stresses (**a**) $\overline{u'u'}/u_*^2$, (**b**) $\overline{v'v'}/u_*^2$ and (**c**) $\overline{w'w'}/u_*^2$ in the downstream plane section (from LV to RV) for (**a**) Run 1, (**b**) Run 2 and (**c**) Run 3. The black dashed lines show the cylinder position.

The distributions of the dimensionless TKE in the longitudinal extreme right vertical plane (identified with letters from RE to RV) are illustrated in Figure 12. High TKE magnitudes are observed in the near-bed flow zone, where the bed roughness surface causes higher fluctuations of the velocity components. However, the TKE reduces progressively moving upwards from $z > 0$, owing to the inhibition of u', v', and w'. The TKE contours are slightly spatially nonuniform for all the experiments. This is probably due to the random vegetation array and the spatial irregular bed sediment.

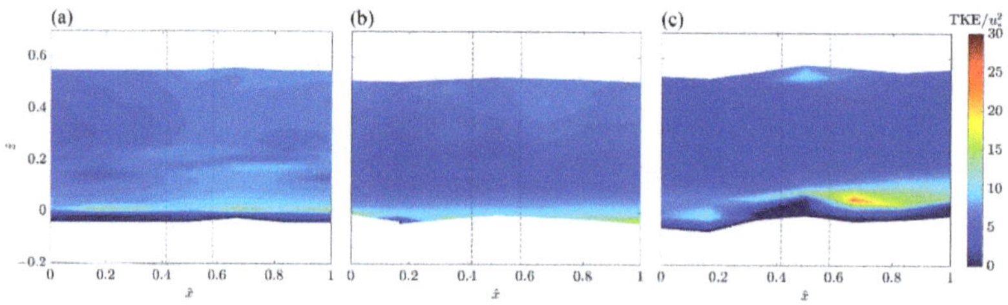

Figure 12. Contours of the dimensionless TKE in the longitudinal plane section (from RE to RV) for (**a**) Run 1, (**b**) Run 2 and (**c**) Run 3. The black dashed lines show the cylinder position.

3.3. Energy Spectra

In order to further analyze the turbulent characteristics, the measured velocity data were explored through the energy spectra of the velocity fluctuations. In particular, the energy spectra are shown in Figure 13 for the vertical W, i.e., immediately downstream of the studied cylinder for all the runs and three different elevations ($z = 0$, $z = 0.2\ h$ and $z = 0.4\ h$), as a function of the Strouhal number of the cylinder ($St = fd/\overline{u}$, where f is the frequency with a resolution equal to F_S/N, and N is the number of samples equal to 30,000 for an acquisition time of 300 s). The energy spectra were determined by employing the discrete fast Fourier transform of the autocorrelation function.

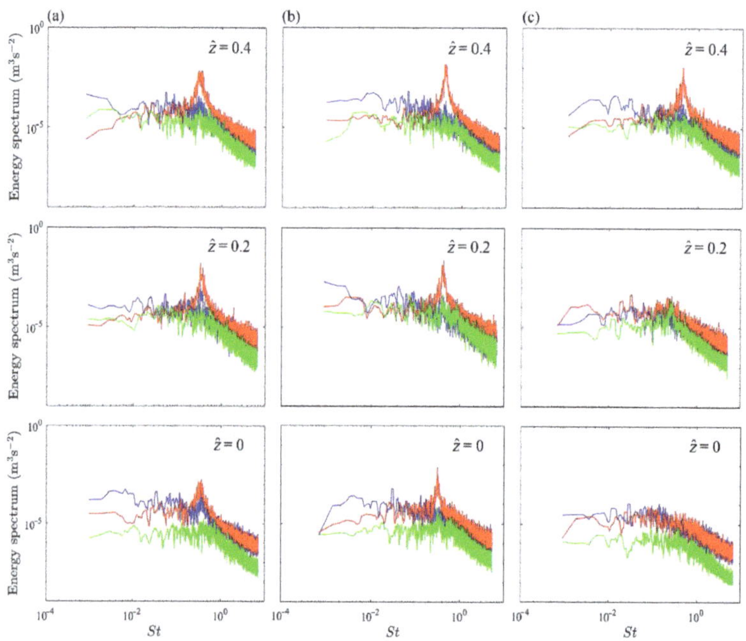

Figure 13. Energy spectra for (**a**) Run 1, (**b**) Run 2 and (**c**) Run 3 of streamwise (blue lines), spanwise (red lines), and vertical (green lines) velocity fluctuations at three different levels $\hat{z}$ of the vertical W.

From a comparison of the spectra, the spanwise velocity revealed the presence of large-scale coherent structures, evident as a peak located in the energy-containing range of the energy spectra (Figure 13). Specifically, these peaks are present in all the runs, with the exception of two spectra of Run 3, where it is evident that the peak recedes as z decreases. This may be due to a roughness effect, which in Run 3 is explicated with a medium sediment diameter equal to 17.98 mm, i.e., comparable with the stem diameter, equal to 20 mm. Moreover, the peaks were observed at the same Strouhal number in the $u'v'$ cross-spectra (not shown here for the sake of brevity), demonstrating that the identified coherent structures were responsible across the vegetation for the lateral momentum transport [5].

4. Conclusions

In the present work, an experimental study was carried out to investigate the impact of different uniform bed roughness on the flow characteristics through randomly arranged emergent rigid vegetation. In particular, focus was given to the evolution of the turbulence characteristics in a study area selected around a single stem. The principal results are summarized below.

In the sections, respectively, upstream and downstream of the generic stem, the dimensionless time-averaged accelerated and decelerated flow field is deeply affected by the vegetation and, to a lesser extent, by the bed roughness. In particular, in the flow field downstream of the studied cylinder it is possible to point out these zones: a decelerated one, in correspondence to the stem ($0.42B_s < y < 0.58B_s$), another decelerated zone, on the left side of the stem ($0 < y < 0.2B_s$, only for Runs 1 and 3) and an accelerated one elsewhere.

From a comparison perspective, it is possible to evaluate a velocity reduction of about 50% in the downstream plane immediately behind the analyzed stem. Conversely, downstream of the cylinder, on the left side ($0< y < 0.2B_s$), the decelerated zones fade about 30%. This is clearly attributable to the random distribution of the vegetation pattern. Instead, in the longitudinal plane the flow is decelerated in all the runs with higher streamwise velocities than in the undisturbed flow profiles along the whole water depth.

The analysis of the TKE distribution clearly shows the effects of the vegetation, with high magnitudes immediately upstream and downstream of the investigated stem. This indicates that the velocity oscillations get excited by the cylinders, producing an increased turbulence intensity in the proximity of the stem. Moving toward the free zone (i.e., without vegetation), this influence vanishes, causing a decrease in the TKE value. In addition, the behavior observed in TKE colormaps was examined by comparing the streamwise, spanwise and vertical normal stress contours. The results revealed a strong influence of the u' and v' fluctuations on the energy distribution and highlighted the influence of the so-called von Kármán vortices. In addition, from the analysis of the TKE distributions, an effect of the cylinder greater in the downstream plane section than in the upstream one is manifested, with an increased mean value of 25% at the abscissa $y = 0.5\ B_s$. The longitudinal TKE distribution revealed high values in the near-bed flow zone. This suggests that the velocity oscillations get excited by the rough bed, producing an increase of the turbulence level in the vicinity of the sediments. Mowing toward the free surface this effect disappears, inducing a decrease in the TKE.

The evaluation of the energy spectra of the velocity fluctuations showed a clear influence of both vegetation and roughness. In particular, the spanwise velocity component revealed energetic peaks that indicate the presence of large-scale coherent structures due to the von Kármán wake vortex along the flow depth in the case of lower roughness (Run 1 and 2) and only in the upper part of the water depth in the case of higher roughness (Run 3). In fact, it is interesting to point out that, when the median sediment diameter is comparable with the stem diameter (i.e., Run 3) going towards the bed, the energetic peak is lowered. This involves a consideration of the nature of coherent structures of turbulence, which are significantly influenced by the characteristic scales of the flow conditions, such as vegetation diameter, water depth and roughness size.

Further study is necessary to describe more deeply the turbulence structures in the presence of both vegetation and sediments. In fact, the results referring to a single generic cylinder are representative only for such a case and cannot be generalized, owing to the lack of further data relating to different vegetation elements. For this purpose, in future works we intend to perform other laboratory experiments, intensifying the measurements and adopting advanced techniques, such as the Particle Image Velocimetry (PIV), in order to better characterize the two-dimensional (2D) turbulence structures.

Author Contributions: Conceptualization, F.C., N.P., A.P.F., R.G.; methodology, F.C., N.P., P.G., R.G.; formal analysis, F.C., N.P.; data curation, F.C.; writing—original draft preparation, F.C.; writing—review and editing, F.C., N.P., A.P.F., P.G., R.G.; supervision, R.G.; funding acquisition, R.G. All authors have read and agreed to the published version of the manuscript.

Funding: This research was funded by the "*SILA-Sistema Integrato di Laboratori per l'Ambiente*-An Integrated System of Laboratories for the Environment" PONa3_00341.

Data Availability Statement: Publicly available datasets were analyzed in this study. This data can be found here: http://www.ingegneriacivile.unical.it/lgmi/turbulent-flow-through-random-vegetation-on-a-rough-bed/, accessed on 14 September 2021.

Acknowledgments: The authors would like to thank Erik Calabrese Violetta for his valuable work during the performance of the experimental runs.

Conflicts of Interest: The authors declare no conflict of interest.

References

1. White, B.L.; Nepf, H.M. A vortex-based model of velocity and shear stress in a partially vegetated shallow channel. *Water Resour. Res.* **2008**, *44*, 1–15. [CrossRef]
2. Meftah, M.B.; De Serio, F.; Mossa, M. Hydrodynamic behavior in the outer shear layer of partly obstructed open channels. *Phys. Fluids* **2014**, *26*, 065102. [CrossRef]
3. Caroppi, G.; Gualtieri, P.; Fontana, N.; Giugni, M. Vegetated channel flows: Turbulence anisotropy at flow–rigid canopy interface. *Geosciences* **2018**, *8*, 259. [CrossRef]
4. Rowiński, P.M.; Västilä, K.; Aberle, J.; Järvelä, J.; Kalinowska, M.B. How vegetation can aid in coping with river management challenges: A brief review. *Ecohydrol. Hydrobiol.* **2018**, *18*, 345–354. [CrossRef]
5. Caroppi, G.; Västilä, K.; Järvelä, J.; Rowiński, P.M.; Giugni, M. Turbulence at water-vegetation interface in open channel flow: Experiments with natural-like plants. *Adv. Water Resour.* **2019**, *127*, 180–191. [CrossRef]
6. Kemp, J.; Harper, D.; Crosa, G. The habitat-scale ecohydraulics of rivers. *Ecol. Eng.* **2000**, *16*, 17–29. [CrossRef]
7. Nepf, H.M. Hydrodynamics of vegetated channels. *J. Hydraul. Res.* **2012**, *50*, 262–279. [CrossRef]
8. Västilä, K.; Järvelä, J. Modeling the flow resistance of woody vegetation using physically based properties of the foliage and stem. *Water Resour. Res.* **2014**, *50*, 229–245. [CrossRef]
9. Nepf, H.M.; Vivoni, E.R. Flow structure in depth-limited, vegetated flow. *J. Geophys. Res. Oceans* **2000**, *105*, 28547–28557. [CrossRef]
10. Järvelä, J. Effect of submerged flexible vegetation on flow structure and resistance. *J. Hydrol.* **2005**, *307*, 233–241. [CrossRef]
11. Li, C.W.; Xie, J. FNumerical modeling of free surface flow over submerged and highly flexible vegetation. *Adv. Water Resour.* **2011**, *34*, 468–477. [CrossRef]
12. Maji, S.; Hanmaiahgari, P.R.; Balachandar, R.; Pu, J.H.; Ricardo, A.M.; Ferreira, R.M. A Review on Hydrodynamics of Free Surface Flows in Emergent Vegetated Channels. *Water* **2020**, *12*, 1218. [CrossRef]
13. Ricardo, A.M.; Franca, M.J.; Ferreira, R.M. Turbulent flows within random arrays of rigid and emergent cylinders with varying distribution. *J. Hydraul. Eng.* **2016**, *142*, 04016022. [CrossRef]
14. Nezu, I.; Sanjou, M. Turburence structure and coherent motion in vegetated canopy open-channel flows. *J. Hydro-Environ. Res.* **2008**, *2*, 62–90. [CrossRef]
15. Yang, W.; Choi, S.U. A two-layer approach for depth-limited open-channel flows with submerged vegetation. *J. Hydraul. Res.* **2010**, *48*, 466–475. [CrossRef]
16. Shimizu, Y.; Tsujimoto, T. Numerical analysis of turbulent open-channel flow over a vegetation layer using a k-ε turbulence model. *J. Hydrosci. Hydraul. Eng.* **1994**, *11*, 57–67.
17. Nepf, H. Drag, turbulence, and diffusion in flow through emergent vegetation. *Water Resour. Res.* **1999**, *35*, 479–489. [CrossRef]
18. Righetti, M.; Armanini, A. Flow resistance in open channel flows with sparsely distributed bushes. *J. Hydrol.* **2002**, *269*, 55–64. [CrossRef]
19. Choi, S.U.; Kang, H. Numerical investigations of mean flow and turbulence structures of partly-vegetated open-channel flows using the Reynolds stress model. *J. Hydraul. Res.* **2006**, *44*, 203–217. [CrossRef]
20. Poggi, D.; Krug, C.; Katul, G.G. Hydraulic resistance of submerged rigid vegetation derived from first-order closure models. *Water Resour. Res.* **2009**, *45*, W10442. [CrossRef]

21. Gualtieri, P.; De Felice, S.; Pasquino, V.; Doria, G. Use of conventional flow resistance equations and a model for the Nikuradse roughness in vegetated flows at high submergence. *J. Hydrol. Hydromech.* **2018**, *66*, 107–120. [CrossRef]
22. Liu, D.; Diplas, P.; Fairbanks, J.D.; Hodges, C.C. An experimental study of flow through rigid vegetation. *J. Geophys. Res. Earth Surf.* **2008**, *113*, F04015. [CrossRef]
23. De Serio, F.; Ben Meftah, M.; Mossa, M.; Termini, D. Experimental investigation on dispersion mechanisms in rigid and flexible vegetated beds. *Adv. Water Resour.* **2018**, *120*, 98–113. [CrossRef]
24. Penna, N.; Coscarella, F.; D'Ippolito, A.; Gaudio, R. Bed roughness effects on the turbulence characteristics of flows through emergent rigid vegetation. *Water* **2020**, *12*, 2401. [CrossRef]
25. Penna, N.; Coscarella, F.; D'Ippolito, A.; Gaudio, R. Anisotropy in the free stream region of turbulent flows through emergent rigid vegetation on rough bed. *Water* **2020**, *12*, 2464. [CrossRef]
26. Petryk, S. Drag on Cylinders in Open Channel Flow. Ph.D. Thesis, Colorado State University, Fort Collins, CO, USA, 1969.
27. Stone, B.M.; Shen, H.T. Hydraulic resistance of flow in channels with cylindrical roughness. *J. Hydraul. Eng.* **2002**, *128*, 500–506. [CrossRef]
28. Nepf, H.M. Flow and transport in regions with aquatic vegetation. *Annu. Rev. Fluid Mech.* **2012**, *44*, 123–142. [CrossRef]
29. Dey, S.; Sarkar, A. Scour downstream of an apron due to submerged horizontal jets. *J. Hydraul. Eng.* **2006**, *132*, 246–257. [CrossRef]
30. Penna, N.; Coscarella, F.; Gaudio, R. Turbulent Flow Field around Horizontal Cylinders with Scour Hole. *Water* **2020**, *12*, 143. [CrossRef]
31. Pu, J.H.; Hussain, A.; Guo, Y.K.; Vardakastanis, N.; Hanmaiahgari, P.R.; Lam, D. Submerged flexible vegetation impact on open channel flow velocity distribution: An analytical modelling study on drag and friction. *Water Sci. Eng.* **2019**, *12*, 121–128. [CrossRef]
32. Julien, P.Y. *Erosion and Sedimentation*; Cambridge University Press: Cambridge, UK, 1998.
33. Manes, C.; Pokrajac, D.; McEwan, I. Double-averaged open-channel flows with small relative submergence. *J. Hydraul. Eng.* **2007**, *133*, 896–904. [CrossRef]
34. Dey, S.; Das, R. Gravel-bed hydrodynamics: Double-averaging approach. *J. Hydraul. Eng.* **2012**, *138*, 707–725. [CrossRef]
35. Neill, C.R. Mean-velocity criterion for scour of coarse uniform bed material. In Proceedings of the International Association of Hydraulic Research 12th Congress, Fort Collins, CO, USA, 11–14 September 1967; pp. 46–54.
36. Strom, K.B.; Papanicolaou, A.N. ADV measurements around a cluster microform in a shallow mountain stream. *J. Hydraul. Eng.* **2007**, *133*, 1379–1389. [CrossRef]
37. Afzalimehr, H.; Moghbel, R.; Gallichand, J.; Jueyi, S.U.I. Investigation of turbulence characteristics in channel with dense vegetation. *Int. J. Sediment Res.* **2011**, *26*, 269–282. [CrossRef]
38. Goring, D.G.; Nikora, V.I. Despiking acoustic Doppler velocimeter data. *J. Hydraul. Eng.* **2002**, *128*, 117–126. [CrossRef]
39. Dey, S. *Fluvial Hydrodynamics*; Springer: Heidelberg, Germany, 2014.
40. Dey, S.; Sarkar, S.; Ballio, F. Double-averaging turbulence characteristics in seeping rough-bed streams. *J. Geophys. Res. Earth Surf.* **2011**, *116*. [CrossRef]
41. Tanino, Y.; Nepf, H. Lateral dispersion in random cylinder arrays at high Reynolds number. *J. Fluid Mech.* **2008**, *600*, 339–371. [CrossRef]
42. Defina, A.; Bixio, A.C. Mean flow and turbulence in vegetated open channel flow. *Water Resour. Res.* **2005**, *41*, W07006. [CrossRef]

water

MDPI

Article

A Study on the Friction Factor and Reynolds Number Relationship for Flow in Smooth and Rough Channels

Yeon-Moon Choo, Jong-Gu Kim and Sang-Ho Park *

Department of Civil and Environmental Engineering, Pusan National University, Busandaehak-ro, 63beon-gil, Geumjeong-gu, Busan 46241, Korea; chooyean@naver.com (Y.-M.C.); koreaws@empas.com (J.-G.K.)
* Correspondence: sangogo@nate.com; Tel.: +82-051-510-7654

Abstract: The shear velocity and friction coefficient for representing the resistance of flow are key factors to determine the flow characteristics of the open-channel flow. Various studies have been conducted in the open-channel flow, but many controversies remain over the form of equation and estimation methods. This is because the equations developed based on theory have not fully interpreted the friction characteristics in an open-channel flow. In this paper, a friction coefficient equation is proposed by using the entropy concept. The proposed equation is determined under the rectangular, the trapezoid, the parabolic round-bottomed triangle, and the parabolic-bottomed triangle open-channel flow conditions. To evaluate the proposed equation, the estimated results are compared with measured data in both the smooth and rough flow conditions. The evaluation results showed that R (correlation coefficient) is found to be above 0.96 in most cases, and the discrepancy ratio analysis results are very close to zero. The advantage of the developed equation is that the energy slope terms are not included, because the determination of the exact value is the most difficult in the open-channel flow. The developed equation uses only the mean velocity and entropy M to estimate the friction loss coefficient, which can be used for maximizing the design efficiency.

Keywords: friction coefficient; open-channel flow; entropy; Reynolds number

Citation: Choo, Y.-M.; Kim, J.-G.; Park, S.-H. A Study on the Friction Factor and Reynolds Number Relationship for Flow in Smooth and Rough Channels. *Water* **2021**, *13*, 1714. https://doi.org/10.3390/w13121714

Academic Editor: Mouldi Ben Meftah

Received: 1 June 2021
Accepted: 16 June 2021
Published: 21 June 2021

1. Introduction

Head loss, h_f, is a very important physical parameter for both the experimental and the theoretical analyses of fluid phenomena. The mechanism of the head loss in the open-channel flow is very complex and is not clearly explained yet. Usually, friction losses can be calculated whenever the friction coefficient, f, is clearly defined by the Darcy-Weisbach equation [1].

The friction coefficient is a key factor to determine the flow velocity in channel flows, which is also important to ensure the optimum hydraulic design. Furthermore, most studies on the friction coefficient f are not performed in an open channel but in a circular pipe flow. However, theoretical research for the open-channel flow was performed in the case of a relatively slow flow with no secondary current and small distribution. The theoretical analysis of the pipe flow was relatively easy compared to that of the open-channel flow [2].

The experimental data from Bazin [3,4] in the late 1800s and Varwick [5] in the early 1900s showed relationships between the friction coefficient and Moody curves in a pipe flow. For the open-channel flow, similar results by Bazin and Varwick [5] were presented. These results showed that the flow characteristics in the laminar, the transition, and the turbulence flows were similar to those in the pipe flow.

The Bazin, Manning [6], and Ganguillet-Kutter [7] equations were developed by using experimental data performed by Bazin and Varwick [5]. Many researchers, including Chezy [8] and Manning [6], developed empirical equations by using observed data in the open channel. However, these equations do not adequately represent the flow characteristics in the open channel. In order to improve the accuracy for representing the flow characteristics in an open channel, a sufficient study on the adequate and reliable analyses is required.

The Manning's equation [6] was less accurate, even in well-flowing controlled man-made waterways. Chow [9] also recommended the adjustment of Manning's roughness coefficient as a function of depth. Studies on the calculation of the friction coefficient and the head losses in the past are still in use or are not actively underway because of difficulties in the physical solution. This is based on studies done about 100 years ago. Additionally, Choo [10] used the entropy concept to derive the mean velocity using Chiu's velocity formula [11,12], which was also used in this paper to derive the friction coefficient.

For the safe design, operation, and management of an increasingly developed and complex water resource facility, research on the calculations of a more accurate friction coefficient should be based on this study.

Therefore, this paper proposes a new theoretical equation to reflect the probabilistic entropy concepts and hydraulic properties. This article proposes a theory using the two-dimensional velocity formula and the probabilistic entropy to get the equation of the friction coefficient calculations. Equations based on these theories can be expected to be much more reliable than the empirical equations. The relationship between smooth and rough waterways is shown by means of comparison of the measured and estimated values to verify the accuracy of the determined values.

The results from the developed equation are based on a theoretical background. The friction coefficient was calculated directly based on the guidance equation combined with the physical factors. We found that it can be used to calculate the friction coefficients with very high accuracy without using energy slopes.

2. Theoretical Background

2.1. Existing Friction Loss Coefficient

The Hagen-Poiseule equation is shown as Equation (1), which calculates the friction head loss and can be written as:

$$h_L = \frac{8\mu\overline{u}}{\omega_0 R_H^4} l \tag{1}$$

where h_L is the friction head loss, μ is the fluid viscosity coefficient, ω_0 is the water unit weight, R_H is the hydraulic radius of a pipe, and l is the pipe length.

In this case, the friction head loss coefficient takes form by rearranging Equation (1) with $R_H = d/4$, generating Equation (2) (Darcy-Weisbach [1]):

$$h_L = f\frac{l}{d}\frac{\overline{u}^2}{2g} \tag{2}$$

The representative equation for the friction loss coefficient in uniform flow can be written as Equation (3):

$$f = \frac{8gR_hS}{V^2} \tag{3}$$

where f is the friction coefficient, V is the mean velocity, g is the acceleration of gravity, R_h is the hydraulic radius, $S = h_f/L$ is an energy slope, h_f is the friction head losses, and L is the length of a given section.

This equation can be applied in streams close to the uniform flow, because it is difficult to calculate the value accurately in a nonuniform or unsteady flow, because S is an energy slope.

In a smooth pipe flow, the relationship can be expressed by using the Blasius [13] equation, written as Equation (4):

$$f = \frac{0.223}{Re^{0.25}} \tag{4}$$

where Re is the Reynold's number.

The equation is limited to valid values of Re between 750 and 25,000. For higher values, von Karman [14] developed a general expression modified by Prandtl [15], which matches the data measured by Nikuradse more accurately [16]. The resulting Prandtl-von Karman formula [5,14,15] can be written as Equation (5):

$$\frac{1}{\sqrt{f}} = 2\log\left(Re\sqrt{f}\right) + 0.4 \tag{5}$$

Therefore, it is possible to establish a relationship between f and Re by an experiment in an open-channel flow using the above equations. However, the relation factors to f are different between the open-channel flow and the pipe flow, because it is affected by multiple factors, such as free water surfaces in the waterways, hydraulic radius, and water surface slope. Figure 1 shows the relationship between smooth and rough channel flows by analyzing various overseas experimental data [17].

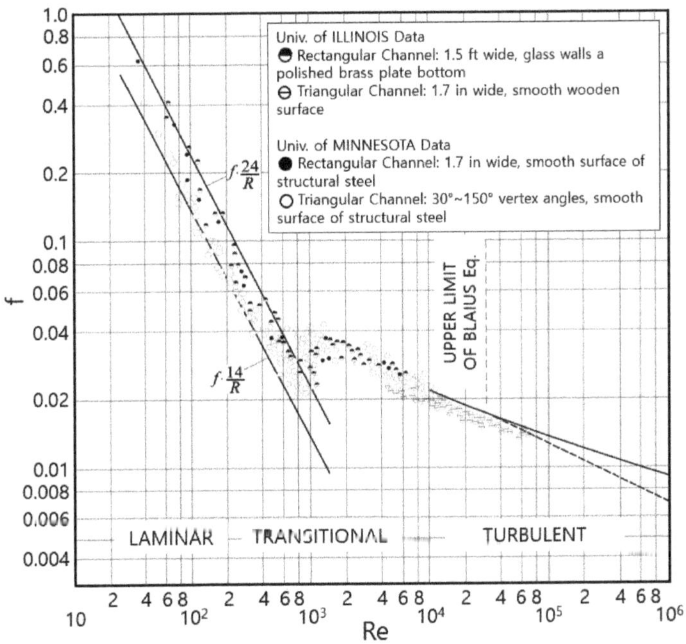

Figure 1. Relationship between f and Re for smooth channels (Flow conditions: Laminar, Transitional, and Turbulent [14]).

The shortcomings of the above studies are that it is difficult to calculate the energy slopes correctly in an open-channel flow. In addition, equations should be applied differently according to the scope of the Re. For the example, the Prandtl-von Karman's equation has to consider an uncertainty when using Equation (3). This is because it is hard to obtain an accurate flow velocity at the bottom of the channel.

2.2. New Friction Coefficient Using Entropy

Shannon [18] first defined entropy by function $H(x)$ and can be written as Equation (6):

$$H(x) = -\int_{-\infty}^{+\infty} p(x)\ln p(x)dx \tag{6}$$

where $p(x)$ is the probability density function, and $\ln p(x)$ is dimensionless but dx has dimension.

Equation (6) means maximizing the entropy that represents the uncertainty for x, given $p(x)$ for the continuous state variable x. Applying this concept to the water velocity can be written as Equation (7):

$$H(u) = -\int_0^{u_{max}} p(u) \ln p(u) du \tag{7}$$

The following constraints are used, such as the average value and probability, which are available information about u, which is instantaneous (point) velocity and can be written as Equations (8) and (9):

$$\int_0^{u_{max}} p(u) du = 1 \tag{8}$$

$$\int_0^{u_{max}} u p(u) du = \overline{u} \tag{9}$$

Arranging the independent constraint conditions can be given as Equation (10):

$$\int_a^b \Phi_i(u, p) du \quad i = 1, 2 \tag{10}$$

where a is the minimum value of u, b is the maximum value of u, i is the constraint number ($i = 1$ is Equation (8) and $i = 2$ is Equation (9)).

Therefore, $p(u)$, which maximizes the entropy, can be obtained using the method of Lagrange as Equations (11)–(13):

$$\frac{\partial I(u,p)}{\partial p} + \sum_{i=1}^{2} \lambda_i \frac{\partial \phi_i(u,p)}{\partial p} = 0 \tag{11}$$

$$I(u,p) = -p(u) lnp(u) \tag{12}$$

where $\phi_1(u,p) = p(u)$, $\phi_2(u,p) = u \times p(u)$.

$$\frac{\partial \phi_1(u,p)}{\partial p} = 1, \quad \frac{\partial \phi_2(u,p)}{\partial p} = u \tag{13}$$

where λ_1 and λ_2 are Lagrange multipliers.

Substituting Equations (12) and (13) into Equation (11) can be constructed as the following Equation (14):

$$-1 - lnp(u) + \lambda_1 + \lambda_2 u = 0 \tag{14}$$

where $\lambda_1 - 1 = a_1$ and $\lambda_2 = a_2$ are the Lagrange multipliers.

Differentiating Equation (14) with respect to $p(u)$ results in the velocity as Equation (15):

$$p(u) = e^{a_1 + a_2 u} \; (0 \le u \le u_{max}) \tag{15}$$

Equation (15) and $M = a_2 u_{max}$ (entropy coefficient) are substituted into Equation (8) to obtain Equation (16):

$$u_{max} e^{a_1} = \frac{M}{(e^M - 1)} \tag{16}$$

Then Equation (15) and $M = a_2 u_{max}$ are substituted into Equation (9) to obtain Equation (17) (This is the two-dimensional average velocity equation, which is Chiu's velocity equation [11,12]):

$$\frac{\overline{u}}{u_{max}} = \phi(M) = \left(\frac{e^M}{(e^M - 1)} - \frac{1}{M} \right) \tag{17}$$

Equation (17) can be used to restructure Equation (16) to obtain Equation (18):

$$e^{a_1} = \frac{\phi(M)M}{(e^M - 1)\overline{u}} \quad (18)$$

The shear stress is the product of the dynamic viscosity (kinematic viscosity is the dynamic viscosity divided by density) and the velocity gradient, which can be expressed as Equation (19):

$$\tau = \rho v \left(\frac{du}{\overline{h}_\xi d\xi} \right) \quad (19)$$

where τ is the shear stress, ρ is the density of the fluid, v is the kinematic viscosity of the fluid, $\overline{h}_\xi$ is the mean value of h_ξ, and h_ξ is the scale factor, which has the length dimensions.

The shear stress at the channel boundary (bottom) is the shear stress when ξ is ξ_0, as in Equation (20):

$$\tau_0 = \rho v \left(\frac{du}{\overline{h}_\xi d\xi} \right)_{\xi=\xi_0} = \rho g R_h S_f \quad (20)$$

where τ_0 is the waterway boundary shear stress, g is the gravitational acceleration, and S_f is the energy gradient.

The velocity cumulative probability of u is suggested by Chiu [11,12] as Equation (21):

$$P(u) = \int_0^u p(u)du = \frac{\xi - \xi_0}{\xi_{max} - \xi_0} \quad (21)$$

where ξ is the spatial coordinates ($0 \le \xi \le 1$), u is the velocity at ξ, ξ_0 is the minimum value of ξ (occurring at the channel boundary where $u = 0$), and ξ_{max} is the maximum value of ξ (where u is at its maximum (i.e., u_{max})) (see Figure 2).

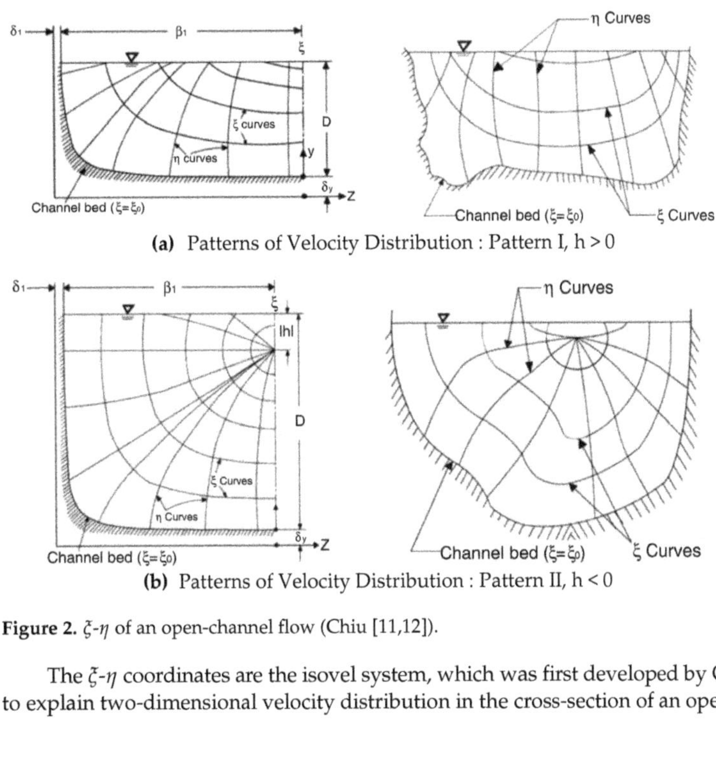

Figure 2. ξ-η of an open-channel flow (Chiu [11,12]).

The ξ-η coordinates are the isovel system, which was first developed by Chiu [11,12] to explain two-dimensional velocity distribution in the cross-section of an open channel.

In Equation (21), if ξ is ξ_0 at the bottom of the channel, u is 0 and $\xi_{max} - \xi_0$ is 1, and by differentiating the velocity gradient in the channel bed, where $p(u) = e^{a_1}$, Equation (22) can be obtained:

$$\left(\frac{du}{\overline{h}_\xi d\xi}\right)_{\xi=\xi_0} = \frac{1}{\overline{h}_\xi e^{a_1}} \tag{22}$$

For a wide channel, $\xi/\xi_{\max}$ can be simplified to y/D and, hence, $h_\xi = D$. For the hydraulic radius, $R_h \cong D$. Therefore, substituting Equation (22) into Equation (20), Equation (23) is obtained:

$$e^{a_1} = \frac{\nu}{gR_h^2 S_f} \tag{23}$$

Equating Equations (18) and Equation (22) expresses the average water velocity in the open-channel flow as Equation (24):

$$\overline{u} = \frac{gR_h^2 S_f}{\nu F(M)} \tag{24}$$

where $F(M) = \left(e^M - 1\right)/(\phi(M)M)$.

For the friction velocity (u_*), the relationship between the average water velocity and the friction velocity is shown as Equation (25):

$$u_* = \sqrt{\frac{\tau_0}{\rho}} \tag{25}$$

To calculate the friction term in Equation (25), Choo's mean velocity distribution [10] is used for Equation (26):

$$u = \frac{\overline{u}}{K(M)} \ln\left[1 + \left(e^M - 1\right)\left(\frac{\xi - \xi_0}{\xi_{max} - \xi_0}\right)\right] \tag{26}$$

where $K(M) = \phi(M)M$.

Choo's mean velocity was used earlier for calculating the discharge. However, in this paper, it will be used for converting friction velocity, since it has already been modified for the average water velocity in the open-channel flow [10].

The water velocity slope $du/\overline{h}_\xi d\xi$ is differentiated from Equation (26), and $\xi_0 \simeq 0$ and $\xi_{max} - \xi_0 = 1$ are applied as Equation (27):

$$\left(\frac{du}{\overline{h}_\xi d\xi}\right) = \frac{\overline{u}(e^M - 1)}{R_h K(M)(1 + \xi(e^M - 1))} \tag{27}$$

where, because the bottom boundary layer $\xi_0 = 0$, Equation (27) is equal to Equation (28):

$$\left(\frac{du}{\overline{h}_\xi d\xi}\right)_{\xi=\xi_0} = \frac{\overline{u}\left(e^M - 1\right)}{R_h K(M)} \tag{28}$$

Equation (28) is inserted into Equation (20), which is the shear stress at the channel boundary, to obtain Equation (29):

$$\tau_0 = \rho\nu\left(\frac{du}{\overline{h}_\xi d\xi}\right)_{\xi=\xi_0} = \rho\nu\frac{\overline{u}\left(e^M - 1\right)}{R_h K(M)} \tag{29}$$

The relationship between the average water velocity and the friction velocity of the friction loss coefficient of the pipe flow is shown in Equation (30):

$$\frac{\overline{u}}{u_*} = \sqrt{\frac{8}{f}} \tag{30}$$

Equations (25) and (30) are substituted into Equation (29) to obtain Equation (31):

$$\tau_0 = \frac{8\rho v^2}{f}\left(\frac{(e^M - 1)}{R_h K(M)}\right)^2 \quad (31)$$

Therefore, if Equation (31) is substituted for Equations (24) ($F(M) = (e^M - 1)/(\phi(M)M)$) and (26) ($K(M) = \phi(M)M$) and use $Re = (\overline{u}d)/v$, Equation (32) can be obtained:

$$f = \frac{8dF(M)}{ReR_h} \quad (32)$$

Equation (32) can be used to estimate the frictional loss coefficient (f) of the open-channel flow, which reflects its entropy. Equation (32) does not require the hydraulic factors used in the existing equations, such as shear velocity (u_*) or energy gradient (S_f). In addition, the friction loss coefficient (f) can be expressed with only the average water velocity and the entropy M, which are easy to obtain. In addition, there is also an advantage in that the energy gradient (S_f) can be estimated by using Equation (32).

Therefore, in this study, we proposed the friction loss coefficient of Equation (32) in the open-channel flow by using the concept of entropy, which has been used in many fields recently. The data used to demonstrate the utility of the equation were obtained by Yuen [19] and Babaeyan-Koopaei [20] for each stream of water. It is shown in the Figure 1 that the estimated friction loss coefficient was compared with the measured friction loss coefficient.

3. Experimental Data

To evaluate the accuracy of the proposed equation, we calculated the friction coefficient based on the data measured at the rectangular channel. The estimated results were compared with the measured data, as shown in Figure 1. First, the data measured by Yuen [19] at the trapezoidal section were used. Then, the data were measured by Babaeyan-Koopaei [20] at the trapezoid, the parabolic round-bottomed triangle, and the parabolic-bottomed triangle trapezoidal channel.

Yuen obtained data in a fully developed turbulent flow of the smooth trapezoidal open-channel flow. The ranges of the data were $0.5 < Fr < 3.5$, $1.9{\cdot}10^4 < Re < 6.2{\cdot}10^5$ and $0.3 < 2b/H < 15$ (where $2b/H$ was the aspect ratio). The subcritical flow was also studied for the compound trapezoidal channel, which ranged in depths of $0.05 < Dr < 0.5$. Here, Dr is the relative depth ratio ($\frac{(H-h)}{H}$, where H is the flow depth of the channel, and h is the depth of the lower main channel). Several series of experiments were undertaken by using the Preston tube technique. These experiments were performed in a 21.26-m-long tilting channel with a working cross-section of 0.615 m wide × 0.365 m deep. A total of three sets were measured under the equivalent conditions, varying the bed slope at 0.001, 0.004, 0.009, 0.015, and 0.023. In addition, the point velocities were measured across the whole cross-section for the selected flow depths. Particular attention was focused on understanding the Reynolds and Froude number effects on these distributions.

Babaeyan-Koopaei [20] measured the data in the trapezoidal, parabolic, round-bottomed triangle, and parabolic-bottomed triangle channel. For each section, the measured data were used with the changes in the flow velocity and water levels under three flow conditions: 1 m^3/s, 10 m^3/s, and 100 m^3/s (see Babaeyan-Koopaei for more information).

The values of the measured Re data are shown in Table 1. It can be seen that the measured data in the rectangular section reflected the transition zone and the turbulence zone. In the trapezoidal, parabolic, round-bottomed triangle, and parabolic-bottomed triangle sections, the measured data reflected the full turbulence zone.

Table 1. The range of Reynolds numbers with the cross-section shape and the channel slope.

Data	Cross-Section Shape	Channel Slope	Reynolds Number Range
Yuen [19]	Rectangular	0.001	16,920~156,400
		0.004	45,770~160,900
		0.009	71,450~358,000
		0.015	108,600~335,000
		0.023	124,400~618,300
Babaeyan-Koopaei [20]	Trapezoidal	0.001	167,000~4,474,000
	Parabolic	0.001	135,000~4,630,000
	Round-bottomed triangle	0.001	167,000~4,684,000
	Parabolic-bottomed triangle	0.001	167,000~4,630,000

4. Estimation of the Entropy Parameter, M

An estimate of the entropy parameter M is needed to use Equation (29). For the estimation of entropy parameter M, most researches used Equation (15) to calculate the entropy parameters in which the equation required, essentially, the maximum flow velocity in an open channel.

However, the maximum velocity occurred at the center of the pipe flow, but the location of the maximum velocity was unclear at the open-channel flow. Additionally, a lot of manpower, time, and effort were required to measure the maximum velocity in the open-channel flow. Moramarco [21] calculated the M values by using Equation (15). For that, he used data obtained from the average and maximum velocities at the upper river basin. Moramarco [22] proposed an equation for calculating $\phi(M)$ by substituting Chiu's theory and the Manning and Prandtl-von Karman equations. However, the disadvantage of these equations were that it was difficult to clearly identify the point where the maximum velocity occurred, y_{max}, and imaginary distance, y_0, where the velocity was zero in the riverbed.

This study determined the entropy parameter M by using the expression developed by Choo [23]. The advantage of this method was that the entropy parameters in the stream could be obtained at any time without using the uncertain maximum velocity.

The entropy parameters M and the Re were calculated by using the same characteristics as those shown in Figures 3 and 4. As the entropy parameter, M was increased, and the Re was also increased. On the other hand, as the friction coefficient f increased, Re decreased. Based on the value of Re, the two flows were identified as turbulent flows.

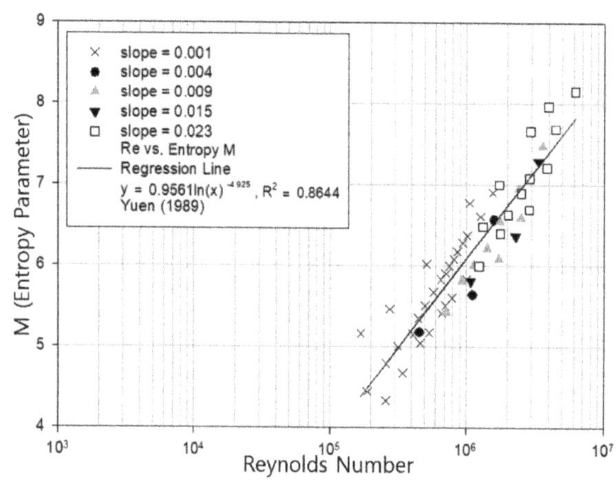

Figure 3. Relationship between M and Re calculated using Yuen's data.

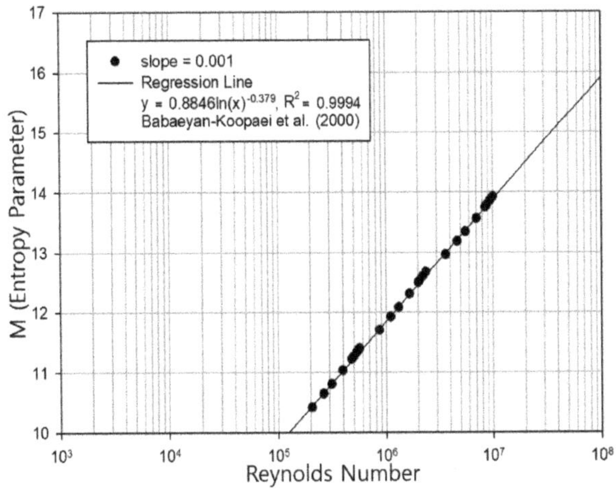

Figure 4. Relationship between M and Re calculated using Babaeyan-Koopaei's data.

5. Results Analysis

The entropy parameter M, defined in Section 4, was used in Equation (32) to calculate the coefficient of friction f in an open-channel flow. The relationship between f and Re is shown in Figures 5 and 6.

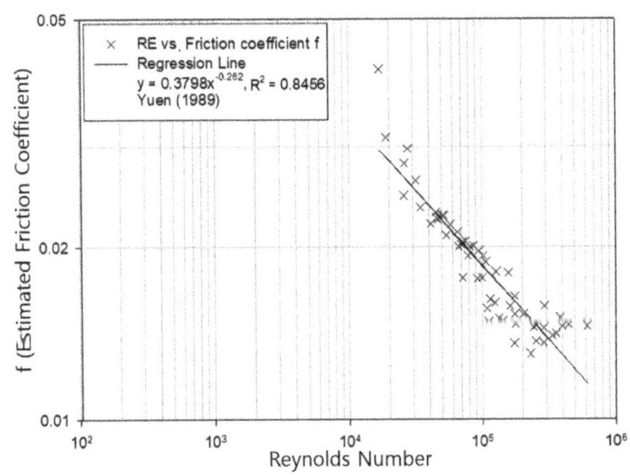

Figure 5. The relationship between f and Re calculated using Yuen's data.

The coefficient of friction f in Figures 5 and 6 shows the same trend as in Figure 1, where f tends to decrease as Re increases. In addition, in Figures 7 and 8, the friction coefficient, f, shows a tendency to decrease as the discharge increases. The discrepancy ratio is the error ratio between the measurement and calculated values, separated by a range. The proportions on the y-axis are the ratio of the total comparison quantity to the range of the x-axis. Figures 9–12 show that the discrepancy ratio results of the proposed equation were all distributed near 0.

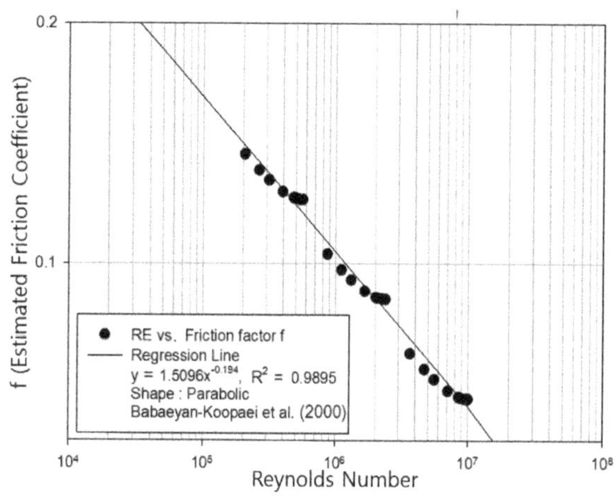

Figure 6. The relationship of f and Re calculated using Babaeyan-Koopaei's data.

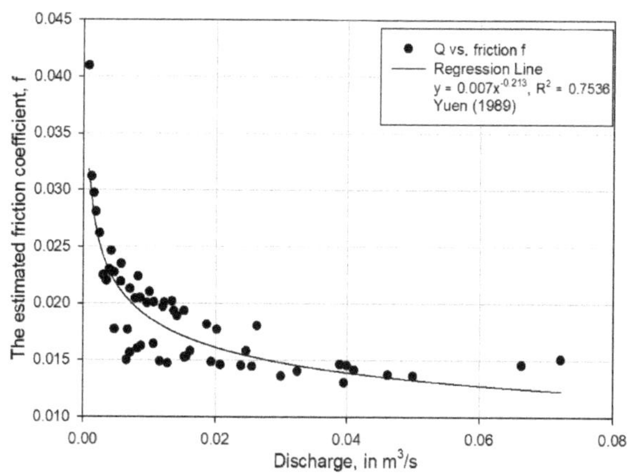

Figure 7. The relationship of f and Q calculated using Yuen's data.

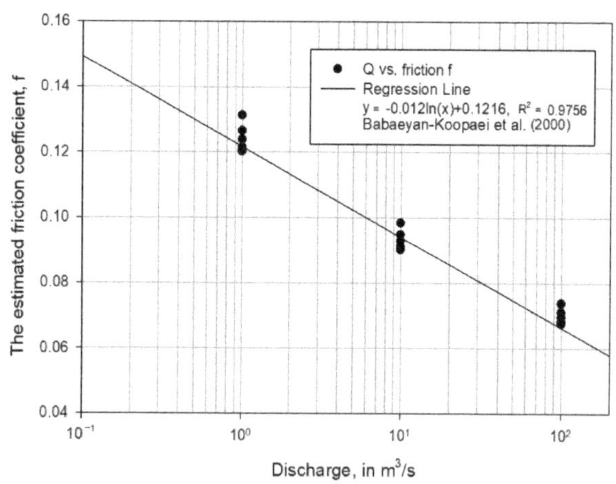

Figure 8. The relationship of f and Q calculated using Babaeyan-Koopaei's data.

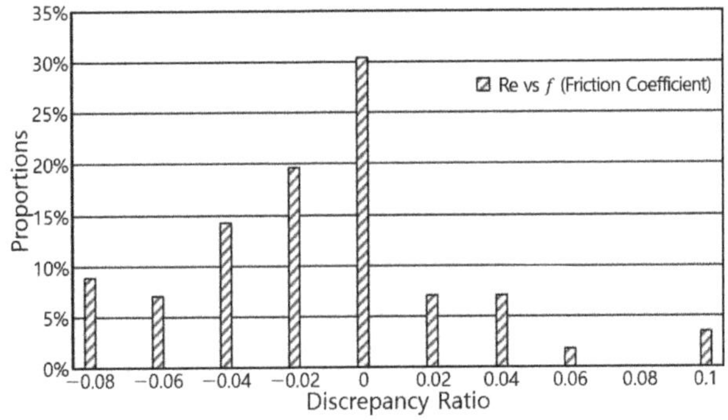

Figure 9. The discrepancy ratio for f and Re calculated using Yuen's data.

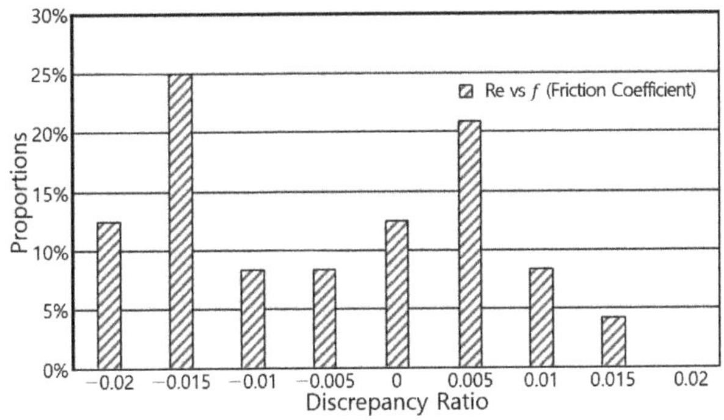

Figure 10. The discrepancy ratio for f and Re calculated using Babaeyan-Koopaei's data.

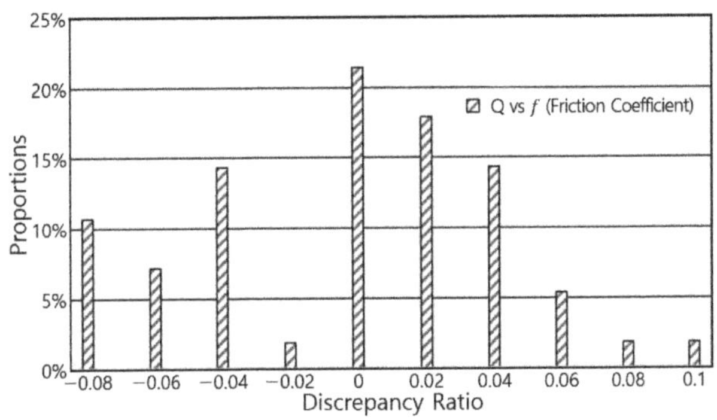

Figure 11. The discrepancy ratio for f and Q calculated using Yuen's data.

The above results are summarized as follows: The entropy parameter M is a linear function of $Log(Re)$, which is an increasing function of Re. The friction coefficient f is a linear decreasing function of $Log(Re)$, which is a decreasing function of Re. Figure 13 shows the above results calculated using the proposed Equation (32), along with the relationship between the friction coefficients f and Re. Comparing the scale with a previous empirical study resulted in rough flows; the coefficient of determination was observed to be 0.8 within the range of the flow of the rectangular channel and 0.75 within the range of the flow of the compound trapezoidal channel.

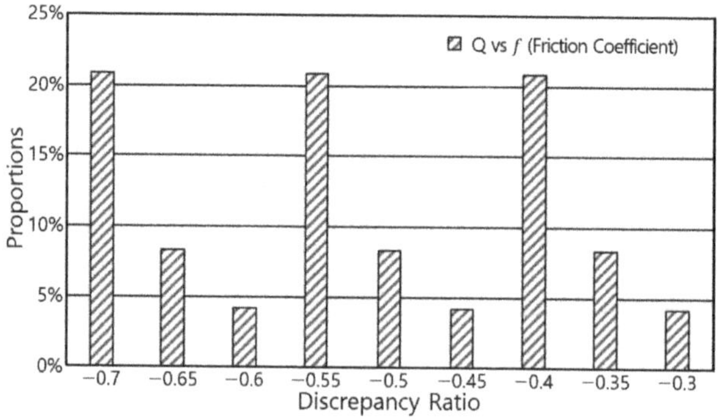

Figure 12. The discrepancy ratio for f and Q calculated using Babaeyan-Koopaei's data.

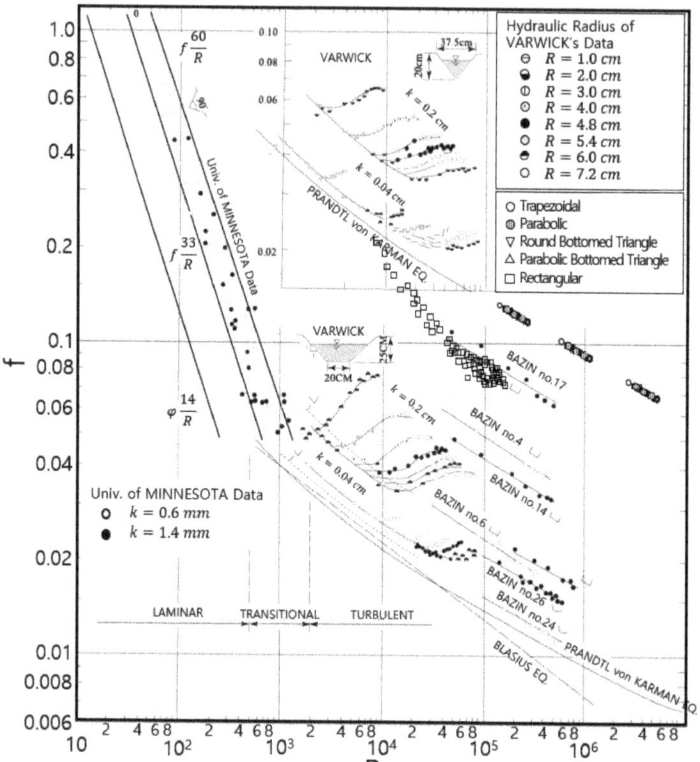

Figure 13. The comparison between the empirical and calculated friction coefficients using Yuen and Babaeyan-Koopaei (rough channel).

Figure 10 shows the results determined with Yuen's data measured in the rectangular and trapezoidal channels. Particularly, the calculated f value was expressed as $1.7 \times 10^5 < Re < 6.18 \times 10^6$, which exceeded the existing Re value. This range means that the proposed equation can represent the actual phenomenon in the natural stream.

The picture on the right shows the results determined with Babaeyan-Koopaei's data measured in the trapezoidal, parabolic, round-bottomed triangle, and parabolic-bottomed triangle channels. In this case, the calculated f value was expressed as $13.5 \times 10^4 < Re < 46.84 \times 10^5$, which was close to the maximum range of the previous *Re* value. The important thing is that even the compound trapezoidal sections that were similar to the natural section, which were not available in the existing graph, as shown in Figure 1, can easily be calculated for f and *Re*. In Tables 2 and 3, the regression equations from the Yuen [19] and Babaeyan-Koopaei [20] data show the relationship between M, f, *Re*, and Q.

Table 2. The results for the relationship between M, f, *Re*, and Q using Yuen's data.

Division		Regression Equation	R^2
M and *Re*	Yuen [19]	$y = 0.9561 \ln(x)^{-4.925}$	$R^2 = 0.8644$
f and *Re*	Yuen [19]	$y = 0.3798x^{-0.262}$	$R^2 = 0.8456$
f and Q	Yuen [19]	$y = 0.007x^{-0.213}$	$R^2 = 0.7536$

Table 3. The results for the relationship between M, f, *Re*, and Q using Babaeyan-Koopaei's data.

Division		Regression Equation	R^2
M and *Re*	Babaeyan-Koopaei [20]	$y = 0.8846 \ln(x)^{-0.379}$	$R^2 = 0.9994$
f and *Re*	Babaeyan-Koopaei [20]	$y = 1.5096x^{-0.194}$	$R^2 = 0.9895$
f and Q	Babaeyan-Koopaei [20]	$y = -0.012 \ln(x)^{0.1236}$	$R^2 = 0.9533$

Comparing the scale with a previous empirical study results in rough flows; the coefficient of determination was observed to be 0.8 within the range of the flow of the rectangular channel and 0.75 within the range of the flow of the compound trapezoidal channel.

As can be seen in Figure 13, the values of the coefficient of friction determined from past experiences are properly correlated with the measured data. There is no bouncing value on the graph. For the rectangular channel, this seems expressed fairly well as Bazin no. 17. Other types of channels matched well with the extended lines of the Prandtl-von Karman equation. In other words, it is meaningful that the values of the friction coefficient determined from proposed equation can be accurately estimate based on a theoretical formula rather than on an empirical parameter under a number of conditions.

6. Conclusions

The results of a study conducted approximately 100 years ago are still used to estimate the friction coefficient in an open-channel flow. However, as with the pipe flow (perfusion), the friction coefficient must be correctly determined in order to interpret the correct flow. This paper proposes a new form of friction coefficient calculation by using the two-dimensional velocity formula of Chiu [11,12] and probabilistic entropy.

The advantage of this equation is that it eliminates the terms of energy slopes, which are difficult to measure or calculate in an open-channel flow, making their application simple and very accurate on a theoretical basis.

In uniform flow conditions, a channel bed gradient may be the same or almost the same as an energy slope or water surface gradient. The normal depth is maintained as long as the slope, cross-section, and the surface roughness of the channel remains unchanged; thus, the average flow velocity remains constant. However, in natural flow and human-made open channels, such as irrigation systems and sewer lines, there are mostly nonuniform or unsteady flows. Unlike uniform flow conditions, these varied flows do not share the same energy slope, bed gradient, and water surface gradient.

Based on the data measured in the rectangular section, the proposed equation was used to determine the entropy parameter M and the friction coefficient f. The induced entropy parameters were shown to be a linear function of $Log(Re)$, and the friction coefficient was the decreasing function of $Log(Re)$.

If this study is to be carried out continuously by hydraulic data measured in various channel shapes, laboratory channels, and natural streams, the friction coefficient value estimated from the proposed equation will be actively used in the flow analysis and the design of hydraulic structures.

Author Contributions: Y.-M.C. and S.-H.P. carried out the survey of the previous studies. Y.-M.C. wrote the manuscript. S.-H.P. conducted all the simulations. Y.-M.C., J.-G.K., and S.-H.P. conceived the original idea of the proposed method. All authors have read and agreed to the published manuscript.

Funding: This research was funded by the Institute of Industrial Technology, Pusan National University.

Institutional Review Board Statement: Not applicable.

Informed Consent Statement: Not applicable.

Acknowledgments: This research was supported by the Institute of Industrial Technology, Pusan National University.

Conflicts of Interest: The authors declare no conflict of interest.

References

1. Darcy, H.; Bazin, H. *Recherches Hydrauliques*; Enterprises par M.H. D'Arcy; Imprimerie Nationale: Paris, France, 1865.
2. Yu, K.K. Particle Tracking of Suspended-Sediment Velocities in Open-Channel Flow. Ph.D. Thesis, Civil and Environmental Engineering, University of Iowa, Iowa City, IA, USA, 2004.
3. Bazin, H.E. *Recherches Experimentales sur Lecoulement de Leau dans les Canaux Decouverts*; Memoire Presentes par Divers Savants al Academie des Sciences: Paris, France, 1865.
4. Bazin, H.E. Etude d'une nouvelle formule pour calculer le debit des canaux decouverts. In *Annales des Ponts et Chaussées*; P. Vicq-Dunod: Paris, France, 1897; Volume 14, pp. 20–70.
5. Varwick, F. *Zur Fließformel für offene Künstliche Gerinne*; Dresden University: Dresden, Germany, 1945. (In German)
6. Manning, R.; Griffith, J.P.; Pigot, T.F.; Vernon-Harcourt, L.F. *On the Flow of Water in Open Channels and Pipes*; Transaction of the Institution of Civil Engineers of Ireland: Dublin, Ireland, 1890; Volume 20.
7. Ganguillet, E.; Kutter, W.R. An investigation to establish a new general formula for uniform flow of water in canals and rivers. *Z. Oesterreichischen Ing. Archit. Ver.* **1869**, *21*, 6–25.
8. Chezy, A. *Thesis on the Velocity of the Flow in a Given Ditch*; des Ponts et Chaussees, Library in France: Paris, France, 1775; Volume 847.
9. Chow, V.T. *Open-Channel Hydraulics*; Mcgrow-Hill Civil Engineering Series; McGraw-Hill: Tokyo, Japan, 1959.
10. Choo, T.H.; Yoon, H.C.; Lee, S.J. An estimation of discharge using mean velocity derived through Chiu's velocity equation. *Environ. Earth Sci.* **2013**, *69*, 247–256. [CrossRef]
11. Chiu, C.L. Velocity distribution in open channel flow. *J. Hydraul. Eng.* **1988**, *115*, 576–594. [CrossRef]
12. Chiu, C.L. Application of entropy concept in open-channel flow study. *J. Hydraul. Eng.* **1991**, *117*, 615–628. [CrossRef]
13. Blasius, H. *Das Ähnlichkeitsgesetz bei Reibungsvorgängen in Flüssigkeiten*; Forschungsheft des Vereins Deutscher Ingenieure: Berlin, Germany, 1913; Volume 131.
14. Kårmån, T. Mechanische Ähnlichkeit und turbulenz. In Proceedings of the 3rd International Congress for Applied Mechanics, Stockholm, Sweden, 24–29 August 1930; Volume 1, pp. 85–93.
15. Prandtl, L. >*The Mechanics of Viscous Fluids*; Durand, B., Ed.; Aerodynamic Theory Springer: Berlin, Germany, 1935; Volume III, division G. 142.
16. Nikuradse, J. *Gesetzmassigkeiten der turbulenten Stromung in glatten Rohren*; Ver Deutsch. Ing. Forschungsheft: Berlin, Germany, 1932; Volume 356.
17. Ryu, H.J.; Kim, J.H.; Lee, B.H.; Lee, W.H.; Jang, I.S. *Choisin Hydrography*; Donghwa, Korea, 2007.
18. Shannon, C.E. A mathematical theory of communication. *Bell Syst. Tech. J.* **1948**, *27*, 379–423. [CrossRef]
19. Yuen, W.H. A Study of Boundary Shear Stress, Flow Resistance and Momentum Transfer in Open Channels with Simple and Compound Trapezoidal cross Section. Ph.D. Thesis, Department of Civil Engineering, University of Birmingham, Birmingham, UK, 1989.
20. Babaeyan-Koopaei, K.; Valentine, E.M.; Swailes, D.C. Optimal design of parabolic-bottomed triangle canals. *J. Irrig. Drain. Eng.* **2000**, *126*, 408–411. [CrossRef]
21. Moramarco, T.; Saltalippi, C.; Singh, V.P. Estimation of mean velocity in natural channels based on Chiu's velocity distribution equation. *J. Hydrol. Eng.* **2004**, *9*, 442–450. [CrossRef]
22. Moramarco, T.; Singh, V.P. Formulation of the entropy parameter based on hydraulic and geometric characteristics of river cross sections. *J. Hydraul. Eng.* **2010**, *15*, 852–858. [CrossRef]
23. Choo, Y.M.; Yun, G.S.; Choo, T.H.; Kwon, Y.B.; Sim, S.Y. Study of shear stress in laminar pipe flow using entropy concept. *Environ. Earth Sci.* **2017**, *76*, 1–27. [CrossRef]

water

MDPI

Article

Hydrodynamic Characteristics of Flow in a Strongly Curved Channel with Gravel Beds

Ying-Tien Lin [1,2,3], Yu Yang [2], Yu-Jia Chiu [4,*] and Xiaoyan Ji [5]

1 State Key Laboratory of Hydraulic Engineering Simulation and Safety, Tianjin University, Tianjin 300350, China; kevinlin@zju.edu.cn
2 Ocean College, Zhejiang University, Hangzhou 310058, China; yangyu168@zju.edu.cn
3 The Engineering Research Center of Oceanic Sensing Technology and Equipment, Ministry of Education, Zhoushan 316021, China
4 Hydrotech Research Institute, National Taiwan University, Taipei 106, Taiwan
5 China National Environmental Monitoring Center, Beijing 100012, China; jixy@cnemc.cn
* Correspondence: yujiachiu@ntu.edu.tw; Tel.: +886-2-3366-2609

Citation: Lin, Y.-T.; Yang, Y.; Chiu, Y.-J.; Ji, X. Hydrodynamic Characteristics of Flow in a Strongly Curved Channel with Gravel Beds. *Water* **2021**, *13*, 1519. https://doi.org/10.3390/w13111519

Academic Editor: Mouldi BEN MEFTAH

Received: 6 April 2021
Accepted: 26 May 2021
Published: 28 May 2021

Publisher's Note: MDPI stays neutral with regard to jurisdictional claims in published maps and institutional affiliations.

Abstract: This study experimentally and numerically investigated the hydrodynamic characteristics of a 180° curved open channel over rough bed under the condition of constant downstream water depth. Three different sizes of bed particles (the small, middle and big cases based upon the grain size diameter D_{50}) were selected for flume tests. Three-dimensional instantaneous velocities obtained by the acoustic Doppler velocimeter (ADV) were used to analyze hydrodynamic characteristics. Additionally, the Renormalization-Group (RNG) turbulence model was employed for numerical simulations. Experimental results show that rough bed strengthens turbulence and increases turbulent kinetic energy along curved channels. The power spectra of the longitudinal velocity fluctuation satisfy the classic Kolmogorov $-5/3$ law in the inertial subrange, and the existence of rough bed shortens the inertial subrange and causes the flow reach the viscous dissipation range in advance. The contributions of sweeps and ejections are more important than those of the outward and inward interactions over a rough bed for the middle case. Flow-3D was adopted to simulate flow patterns on two rough bed settings with same surface roughness (skin drag) but different bed shapes (form drag): one is bed covered with thick bottom sediment layers along the curved part of the flume (the big case) as the experimental condition, and the other one is uniform bed along the entire flume (called the big case_flat only for simulations). Numerical simulations reveal that the secondary flow is confined to the near-bed area and the intensity of secondary flow is improved for both rough bed cases, possibly causing more serious bed erosion along a curved channel. In addition, the thick bottom sediments (the big case), i.e., larger form drag, can enhance turbulence strength near bed regions, enlarge the transverse range of secondary flow, and delay the shifting of the core region of maximum longitudinal velocity towards the concave bank.

Keywords: acoustic Doppler velocimeter; rough bed; secondary flow; turbulent bursting; turbulence kinetic energy

1. Introduction

Curved channels are commonly found in natural rivers around the world. Flow in curved open-channel follows a helicoidal path and induces a secondary flow, which can redistribute mass, momentum, boundary shear stress and thereby plays an important role in hydraulic engineering [1,2]. Considerable studies focused on the curved channel characteristics in different ways, including theoretical derivation, field observations and measurements, physical model experiments, and numerical simulations. For example, Zeng et al. [3] showed that the turbulent kinetic energy in the channel bend is significantly larger than that at the entrance and the exit according to laboratory experiments. Blanckaert [4] and Blanckaert and Vriend [5] found that water surface gradient and streamwise curvature variations are the major factors for velocity redistribution in sharp bend channels. Vaghefi et al. [6] indicated that there are two persistent clockwise vortexes from the streamlines drawn in cross sections and the maximum shear stress occurs from the entrance of the bend to the bend apex area near the inner wall. However, these studies only focus on flow over smooth curved channels, and the hydrodynamic characteristics over a curved and rough bed, more consistent with practical field conditions, has been paid less attention and is not well-understood [7].

Over the past few decades, most of studies focus on the flow patterns over a straight and rough bed open channel. Grass et al. [8] found that the typical scale of near-wall vortical structure is directly proportional to the bed roughness size under fully developed flow conditions. Ferro [9,10] revealed that the bed roughness can change the mean velocity profile near the wall and the friction coefficient. Nikora et al. [11,12] applied the double-averaging method to large and organized roughness element bed flows and showed that the intrinsic averaged streamwise velocity profile is linear inside the bed roughness with high submergence. Mignot et al. [13] concluded that the strongest turbulence activity occurs at gravel crest levels $z_c/h = 0.1$, where z_c is the distance from the bed and h is the flow depth, and the turbulent diffusion also reaches a maximum at this elevation. Dey et al. [14] showed that the friction factor decreases with bed-load transport substantiating the concept of reduction of flow resistance and the downward vertical fluxes of the *TKE* values increase in presence of bed-load transport. Qi et al. [15] used the Particle Image Velocimetry (PIV) techniques to measure the near-wall rough regions and performed that the *TKE* value firstly increases and then decreases as water depth increases. Most of these literatures aim to understand the effect of rough bed on straight open channels. In order to better know the flow patterns in natural rivers, the study of curved open channel flow over rough bed is highly required.

The flow patterns of curved open channel flow over rough bed are more complicated than those over smooth bed. Up to now, the related studies are relatively fewer in comparison with the straight and rough bed channels. The studies can be divided into two types. The first one focuses on the roughness of the side wall in curved channels. In this type, the longitudinal velocity decreases and the shear stress increases in the bank vicinity [16]. In addition, a reversed secondary flow occurs for all rough outer bank cases is found and the reversed secondary flow becomes is stronger as the roughness of the outer bank increases [17]. Hersberger et al. [18] suggested that the macroroughness placed on the outer side wall causes the maximum velocity cell move towards the center of the channel and also reduces the sediment transport capacity in the bend. The other one is that the roughness is due to bed sediment particles over curved channels. Jamieson et al. [19] showed that the magnitude and distribution of three-dimensional Reynolds stresses increase through a 135° channel bend with a mobile sand bed. Pradhan et al. [20] revealed that the resistance caused by the curvature is larger on smoother channels than that on fine gravel channels at the meander bends. These studies have provided some useful information on the flow patterns over a rough and curved channel. However, these results may not be able to fully reflect some practical situations. For example, on the purpose of navigation the water level at the downstream usually need to maintain constant, which is the main focus in this study.

In this paper, laboratory experiments and numerical simulations were conducted to investigate the flow field in a 180 degree U-shaped curved and rough bed flume under the condition of constant downstream water level. Experimental investigations were carried out to study the profile of longitudinal velocity, turbulent kinetic energy, and turbulent bursting. Gravel sediment particles with different grain sizes were used to present the level of bed roughness. ADV was used to obtain three dimensional instantaneous velocities at different cross sections along the curved channel. In addition, the Renormalization-Group (RNG) turbulence model in the FLOW-3D software was then utilized to discuss flow patterns on two rough bed settings with same surface roughness (skin drag) but different bed shapes (form drag): one is bed covered with a thick bottom sediment layer along the curved part of the flume as the experimental condition, and the other one is uniform bed along the entire flume. The paper was organized as follows: in Section 2 we detailed the experimental set-up and the numerical modelling. In Section 3, the experimental results including longitudinal velocity distribution, turbulent kinetic energy and power spectral density, and turbulent bursting under different bed roughness were performed. In addition, numerical results including water depth, longitudinal velocity, *TKE* (k) and secondary flow were presented and used to discuss the effects of thick sediment layers. Finally, conclusion and future prospects were given in Section 4.

2. Methodology

2.1. Experimental Setup

The experiments were carried out in a recirculating Plexiglas flume of 0.40 m wide, 0.40 m deep and 33 m long in the Ocean Engineering Laboratory in Zhoushan Campus, Zhejiang University, China. The flume consists of a 12 m straight inflow reach, a 180-degree U-shaped curved reach with the radius of curvature R in the centre position of the channel equal to 1.4 m, and a 16 m straight outflow reach with a constant slope of 0.5 % of entire flume. A series of honeycomb grids were installed at the entrance of the channel to stabilize the flow and prevent the formation of large-scale flow disturbances. Three different discharges Q were set at 0.015 $m^3\ s^{-1}$, 0.025 $m^3\ s^{-1}$ and 0.030 $m^3\ s^{-1}$, respectively. The flume has an adjustable tailgate at the downstream end of the flume, where the water level was set to 35 cm, to regulate the flow depth. The test sections were located in the curved region of the flume, where the flow was in fully developed turbulent regimes.

Nortek Vectrino ADV (Nortek AS, Vangkroken 2, N-1351 Rud, Norway) was used to measure three-dimensional instantaneous velocities (x—longitudinal direction, y—transverse direction, and z—vertical direction), i.e., longitudinal velocity (u), transverse velocity (v) and vertical velocity (w). In the experiment, the x and y direction was relative to the curved channel. The ADV sampling volume is 0.09 cm^3 and the actual measurement point is 5 cm below the tip of the probe. For each cross-section, the ADV beam with red marking (x-axis indicator) was pointed to the streamwise direction of the curved channel. In order to ensure the quality of the measured data, SNR (signal to noise ratio) should be guaranteed to be above 20 and Correlation coefficient should be above 70 for the measured data [21]. Additionally, the measured velocity signals were despiked using the method of Goring and Nikora [22]. In the experiments, the sampling time was set to 60 s to ensure the accuracy and reliability of the data [6,23,24] with the sampling frequency of 25 Hz. The 60 s records were compared to a 10 min record both for the bare case and the rough bed case, and the differences for the time-averaged velocities and the *TKE* values were <0.9% and <1.1%, respectively. Thus, the 60 s ADV records were sufficient to obtain stable first (time-averaged) and second (variance) moments of turbulent statistics. Velocity measurements were conducted at five selected cross-sections (0°, 45°, 90°, 135° and 180°) with 5 or 7 vertical locations for each cross-section (5 cm, 10 cm, 15 cm, 20 cm, 25 cm, 30 cm and 35 cm from the inner wall) and 15 measuring points at each vertical line (measurements carried out at every 1 cm at 0 cm to 10 cm from the bed, and every 2 cm at 10 cm to 20 cm from the bed). The surface of the rough bed was uneven, so the first measuring point (1 cm from the bed) at the vertical line was confirmed by the average distance of three sections (y/B = 0.25, y/B = 0.5 and y/B = 0.75). The flow depth was measured using water level gauges at three vertical locations (y/B = 0.25, y/B = 0.5 and y/B = 0.75) for each cross-section along the flume. For the bare case (smooth bed), five vertical locations were enough to obtain the hydrodynamics characteristics of flow. For rough bed cases, velocities at seven vertical locations were measured to provide the detailed flow fields. The scheme of experiment setup is shown in Figure 1.

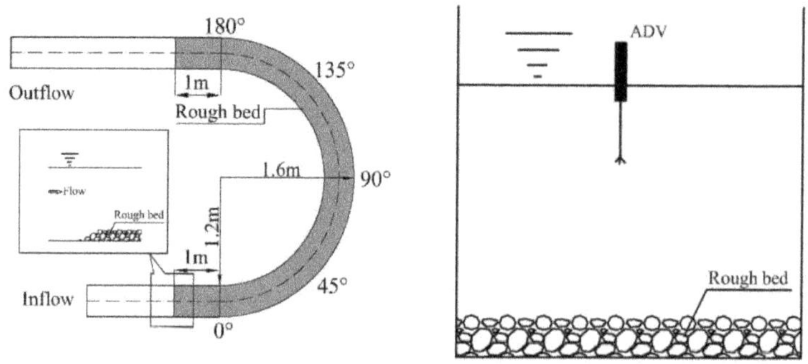

(a) (b)

Figure 1. Scheme of experiment setup: (**a**) The shadow shows rough bed regions and the inset figure represents rough bed configurations; (**b**) ADV measurement setup.

The rough bed in the experiments was made by placing a thick layer of sediments throughout the curved region (Figure 1a). The thickness of sediment layers was around 2~3D_{50}, where D_{50} is the grain size diameter D at 50% in the cumulative distribution of particle size [25]. Three different sizes of bed particles were chosen: D_{50} = 1.5 cm (refer to the small case hereafter), 3.0 cm (refer to the middle case hereafter) and 5.0 cm (refer to the big case hereafter) in the study. Sediment particles were placed in the regions of 1 m near the inlet and outlet of the curved channel to prevent flow instability as shown in Figure 1.

2.2. Fundamental Flow Conditions

For determining the flow conditions in the experiments, Reynolds numbers *Re* and Froude numbers *Fr* are respectively calculated by

$$Re = 4U_{m0}R_h/v \tag{1}$$

and

$$Fr = U_{m0}/\sqrt{gh_0} \tag{2}$$

where U_{m0} is mean flow velocity of the 0° section for each case; $R_h(=Bh/(B+2h_0))$ is the hydraulic radius; $B(= 0.4 \text{ m})$ is the channel width; h_0 is the water depth of the 0° section; $v = 10^{-6}$ m^2 s^{-1} is the kinematic viscosity of water; and $g = 9.81$ m^2 s^{-1} is the gravitational acceleration. As listed in Table 1, the Reynolds numbers ranged from 64,516 to 148,148, indicating the flow conditions were turbulent. Additionally, the Froude numbers for all experimental conditions were less than 1, meaning that the flows belong to the subcritical flow regime.

For rough bed cases, viscous sublayer is absent [20], and thus the hydrodynamic characteristics of flow is solely affected by the bed roughness. Yalin [26], and Schlichting and Gersten [27] classified the rough bed by the means of friction velocity u^* (m s^{-1}) as depicted below:

$$0 < \frac{u^* k_s}{v} < 5 \tag{3}$$

$$5 < \frac{u^* k_s}{v} < 70 \tag{4}$$

$$70 < \frac{u^* k_s}{v} \tag{5}$$

where $u^* = \sqrt{gR_hS}$ is the friction velocity of 0° section for each case, $S = 0.005$ denotes the bed slope and k_s refers to the roughness height in meters. Equation (3) refers to the condition of flow over a hydraulically smooth bed, while Equations (4) and (5) represent flow over transition region and hydraulically rough surfaces, respectively [20]. In these cases, D_{50} is used to define the grain size. The roughness height for the Perspex sheet is determined as 0.0001 m [20]. For non-uniform bottom sediment, k_s is expressed as a characteristic grain diameter, i.e., $k_s = D_{50}$ [28]. The detailed experimental parameters are listed in Table 1. As shown in Table 1, $0 < (u^* k_s)/v < 5$ for the bare case, representing a hydraulically smooth bed. For the rough bed cases, $(u^* k_s)/v \gg 70$ represents hydraulically rough surfaces.

Table 1. Parameters of the experiment.

Cases	Q (m^3 s^{-1})	D_{50} (cm)	k_s (m)	u^* (m s^{-1})	$\frac{u^* k_s}{v}$	*Re*	*Fr*
Bare case	0.015 0.025 0.030	0	0.0001	0.0744	1.26	64,516 102,248 129,588	0.088 0.129 0.178
Small case	0.015 0.025 0.030	1.5	0.015	0.0747	1120.50	71,260 119,616 146,340	0.115 0.196 0.249
Middle case	0.015 0.025 0.030	3.0	0.03	0.0749	2247.00	73,172 121,952 148,148	0.124 0.207 0.258
Big case	0.015 0.025 0.030	5.0	0.05	0.0750	3750.00	74,076 123,456 146,340	0.129 0.215 0.249

2.3. Numerical Analysis Using FLOW-3D

FLOW-3D is a commercial CFD software that specializes in solving transient, free-surface problems. The finite volume method (FVM) in a Cartesian, staggered grid is employed in FLOW-3D to solve the Reynolds' average Navier–Stokes (RANS) equations. FLOW-3D contains a powerful meshing capability through the Fractional Area/Volume Obstacle Representation (FAVOR), which is used to illustrate the complex boundaries of the computational domain. In FLOW-3D, the Volume of Fluid (VOF) is used to simulate the fluid with the free surface [29–32]. In this study, the water depth, longitudinal velocity and secondary flow with and without the thick layer of sediments were numerically studied using FLOW-3D. The numerical domains were based on the experimental channel mentioned above. In order to make the numerical results more accurate, the model geometry was lengthened 20 m in the straight inflow and outflow reach, respectively. The total grid number reached up to 1.17 million, and the grid size was 2.5 cm in the curved reach and 5 cm in the straight inflow and outflow reach. The irrelevance of the number of grids was verified by means of reducing the grid size, i.e., increasing the grid number. The grid size was increased to 2 cm, and the difference of velocity profiles in comparisons with the 2.5 cm grid size was less than 3%.

The boundary conditions for the inflow and outflow reach were set as mean flow velocity and water depth measured in flume experiments. For free water surface, the atmospheric pressure was assigned. Two inter-block junctions of straight and curved reach were defined as the symmetrical condition [33]. In addition, a no-slip condition was applied to wall boundaries. The simulation ran for 1000 s to ensure the reach of steady state conditions. In FLOW-3D, the Renormalization-Group (RNG) turbulence model known to describe low intensity turbulence flows and flows having strong shear regions more accurately [34–36], was selected. The RNG model systematically removes all small scales of motion from the governing equations by considering their effects in terms of larger scale motion and a modified viscosity [35].

Three cases numerically simulated in the study were: (i) the bare case (the same as experiment); (ii) the big case (the same as experiment): the rough bed region was set as a 10 cm high and solid step (=$2D_{50}$), simplified conditions, to represent the thick sediment layer, which is the same as the experiment setup. Additionally, a 20 cm long slope before and after the step was used to connect the step and flat regions of the flume. The solid step should be set to porous media in accordance with the experimental settings. However, for the FLOW-3D software it is unable to set the solid step as porous media as well as assign a surface roughness on the surface of porous media. Instead, we calibrated the surface roughness k_s and ensured the simulations close to flume measurements; (iii) the big case_flat, in which no thick sediment layer was covered, i.e., no step was set up in numerical tests. The entire flume bottom is uniform, but bed roughness was the same as the big case. Cases (ii) and (iii) were used to compare the effect of thick sediment layers on secondary flows along a curved channel. The roughness height k_s in the FLOW-3D model was a tuning factor for the three cases.

3. Results and Discussion

3.1. Water Depth

The transverse gradient at the channel curve caused by the centrifugal force is small and negligible due to the small quantity of flow and small radius of curvature. Therefore, the average depth of three sections ($y/B = 0.25$, $y/B = 0.5$ and $y/B = 0.75$) is calculated and shown in Figure 2 for four cases.

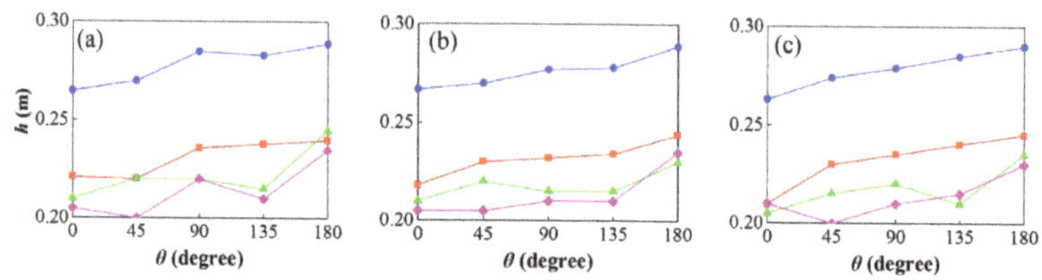

Figure 2. Water depth variations along the curved reach for four cases: (**a**) $Q = 0.015$ m^3 s^{-1}, (**b**) $Q = 0.025$ m^3 s^{-1}, and (**c**) $Q = 0.030$ m^3 s^{-1} (• Bare case; ■ Small case; ▲ Middle case; ♦ Big case).

It can be seen from Figure 2 that the water depth changes less obviously along the curved reach for the bare case (smooth bed). As bed particles become larger, the sediment layer over the curved channel is thicker. At the conditions of the constant downstream water level at the tailgate, the thicker sediment layers lead to shallow water depth over the curved reach, i.e., $h_{big\ case} > h_{middle\ case} > h_{small\ case}$ in general. Therefore, the bed with the largest particle sizes leads to fastest mean flow velocity when the water depth at the tailgate is constant. In addition, the smaller discharge as shown in Figure 2a leads to less water depth differences for rough bed cases. As discharge becomes larger, water depth differences for rough bed cases are more apparent as provided in Figure 2c.

3.2. Longitudinal Velocity Distribution

Figure 3 shows the longitudinal velocities along the depth of each section for smooth beds (bare case) and rough beds (small case, middle case and big case) for $Q = 0.030$ m^3 s^{-1}. For smooth bed (bare case, Figure 3a), the longitudinal velocity along the water depth at 0° and 45° sections follows logarithmic distribution, similar to the case in straight open channels. At the cross sections of 90° and 135°, the flow is affected by the channel bend. The longitudinal velocity gradually increases in the bottom region ($z/h < 0.4$) and reduces in the top region ($z/h > 0.4$), thus breaking the logarithmic distribution law. Consequently, the longitudinal velocity as shown in Figure 3a exhibits an approximate "constant" distribution along the water depth as Blanckaert [37] and Barbhuiya and Talukdar [38] found.

For rough bed cases (Figure 3b–d), the longitudinal velocity profile is significantly different as the bare case. For the bare case, longitudinal velocity ranges from 0.2 m s^{-1} to 0.35 m s^{-1}, while the values vary from 0.18 m s^{-1} to 0.45 m s^{-1} for the rough bed cases.

Since the water depth at the tailgate is constant for all cases, the water depth with larger bed sediment particles becomes shallower, leading to greater longitudinal velocity. At 0° sections, the longitudinal velocity along the water depth also follows distinct logarithmic distribution, but the longitudinal velocity near the bed greatly decreases due to the bottom friction from rough bed, causing the steeper slope of the longitudinal velocity profile and larger induced bottom shear stress as other studies [15,23,39,40] mentioned. For the middle and big cases, the slope of the longitudinal velocity profile has little difference, which possibly means that the longitudinal velocity profile changes a little when the grain size reaches a certain level. At the 45° section, the longitudinal velocity profile near convex bank is affected by the channel bend firstly, implying that the longitudinal velocity gradually increases in the bottom and reduces near the water surface. However, the velocity profile near concave bank keeps following the logarithmic distribution. At the cross sections of 90° and 135°, the flow in the whole section is affected by the channel bend. Under the both effects of channel bend and rough bed, the longitudinal velocity distribution exhibits a trend of increasing near the bed and then decreasing near the surface along the water depth. It also can be found that the maximum longitudinal velocity at the central line occurs between 0.2 to 0.4 dimensionless depth from the bed. The grain size of bed particle only changes the magnitude of velocity, making the phenomena more remarkable. The trends of longitudinal velocity profiles for different rough bed cases are similar.

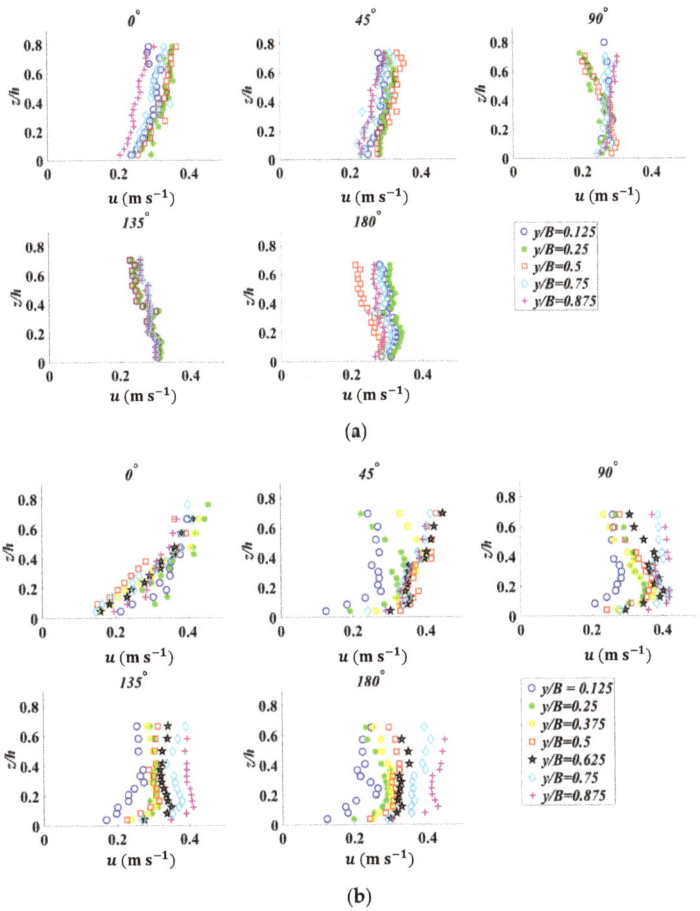

Figure 3. *Cont.*

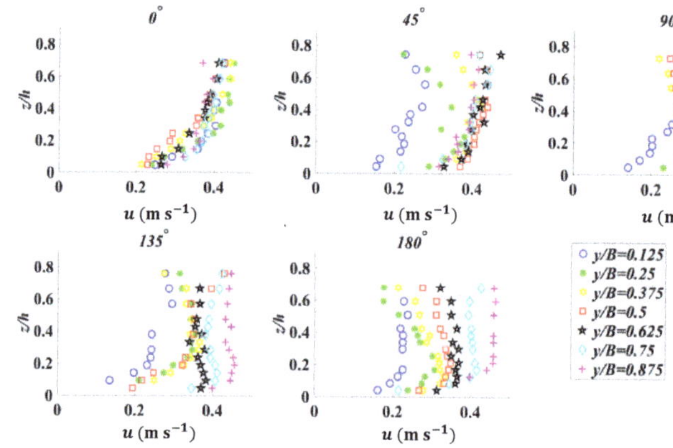

(c)

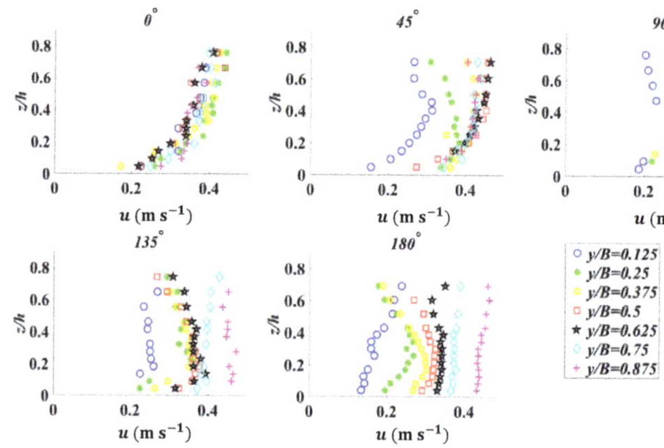

(d)

Figure 3. Longitudinal velocity profiles for four cases at five cross sections: (**a**) Bare case, (**b**) Small case, (**c**) Middle case, and (**d**) Big case (Q = 0.030 $m^3\ s^{-1}$).

The velocity gradient between convex bank and concave bank for the bare case is unobvious. On the other hand, for the rough bed cases the transverse velocity gradient cannot be neglected, especially for the big case. The longitudinal velocity of concave bank is greater than that of convex bank for all rough bed cases, which indicates that the rough bed can enlarge the centrifugal effect of channel bend. Longitudinal velocity profiles for different flow discharge are similar and not discussed in detail here.

3.3. Turbulent Kinetic Energy and Power Spectral Density Analysis

In the study, *TKE* (k), represented as the turbulence characteristics, is defined as

$$k = \frac{1}{2}\left(\overline{u'^2} + \overline{v'^2} + \overline{w'^2}\right) \tag{6}$$

where $u' = u - \overline{u}$, $v' = v - \overline{v}$ and $w' = w - \overline{w}$ are the fluctuation velocities of u, v and w, respectively.

Taking Q = 0.030 $m^3\ s^{-1}$ as an example, the *TKE* profiles, normalized by the overall mean velocity $U = Q/Bh$, under different tough bed conditions are shown in Figure 4. For the bare case (Figure 4a), the normalized k values show a constant distribution along the water depth for each cross section, but the normalized k values gradually increase from inlet (0° section) to the apex of bend (90° section) and decrease from the apex to outlet (180° section). The intense turbulence mainly concentrates in the cross section of 90°–135°. The reason is that the friction of the wall and the momentum exchange between the water flow and the wall leads to the increase of k values [4].

For the rough bed cases (Figure 4b–d), the normalized k values are smaller near the free surface, similar to the bare case. However, the normalized k values are larger near the bed region at the 0° section, implying that rough bed will enhance fluid mixing and strengthen turbulence as the results found in straight open channels [15,41]. The effect of bed roughness is not only confined for near bed regions but can be gradually extended to the whole water depth. As the particle size increases, the normalized k values near the bed increase a little because of the greater friction from the rough bed. Yet, the normalized k values near the surface remain basically unchanged. The slope of the variation of the normalized k values, i.e., $\frac{\Delta(z/h)}{\Delta(\text{normalized } k \text{ values})}$, decreases, which means that at the same difference $\Delta\left(\frac{z}{h}\right)$, $\Delta(\text{normalized } k \text{ values})$ becomes larger. As a result, the rate in loss of k for larger particles becomes faster. After the 45° section, the normalized k values are also affected by the channel bend. Similar to the bare case, the normalized k values of the rough bed on the convex bank are larger than those for concave bank. Additionally, the vertical gradient of the variation of normalized k value increases, indicating that the rate in loss of k becomes slow. The k distributions for three rough bed cases are similar in curved channels. Therefore, the bed roughness and curved channel both improve turbulence along the water depth, especially near the bed regions. The effect of roughness is essential for the transport of the turbulent kinetic energy along the vertical direction.

Figure 5 presents the variations in the power spectral density (PSD) of longitudinal velocity fluctuation $u\prime$ with frequency f for different experimental conditions measured at the mid-width of the flume (20 cm from inner bank). Since the flow discharge has little effect on PSD, the discharge Q equal to 0.030 $m^3\ s^{-1}$ is used as an example to reveal the variations of PSD. The spectral analysis used here is based on the Welch method with Hamming type windowing [42]. In Figure 5a,c,e,g, the feature point was measured at mid-depth (12 cm from the bed). In order to explore the effect of distance from the channel bed on PSD, three characteristic positions, i.e., 2 cm (bottom), 12 cm (middle) and 16 cm (top) from bed along the vertical direction, are provided at the 90° section in Figure 5b,d,f,h. Due to the limitation of ADV frequency, the energy input, inertial subrange and part of dissipation range can be observed. However, the higher frequency (part of dissipation range) is unable to present. Therefore, some of the small-scale turbulent structure can still be investigated.

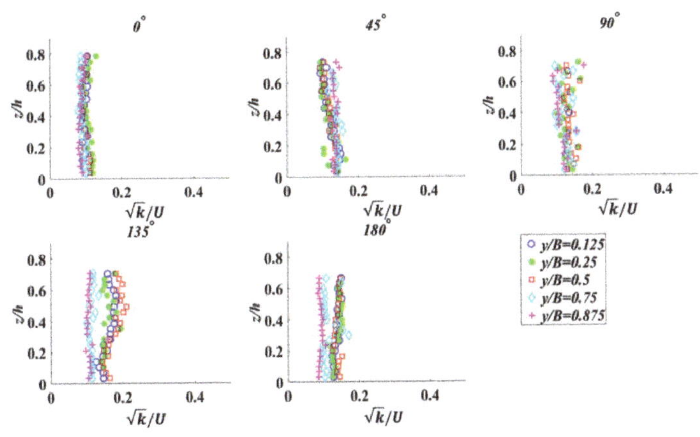

(a)

Figure 4. *Cont.*

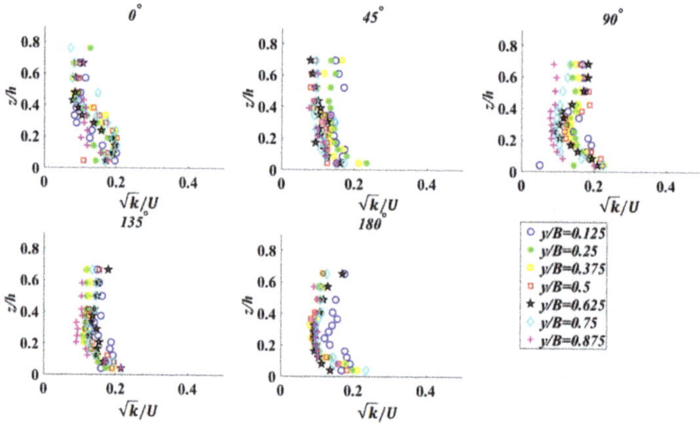

(b)

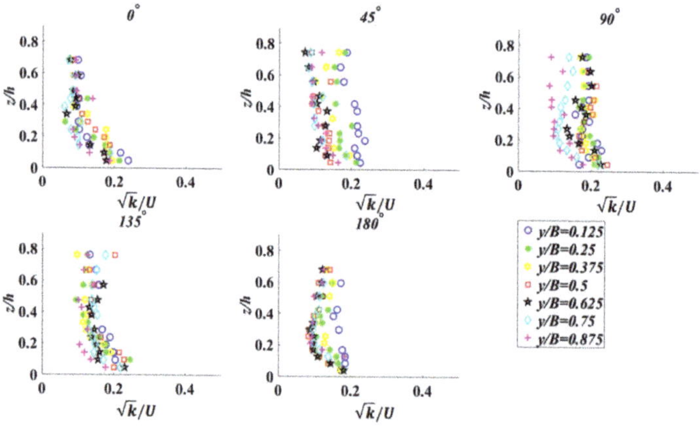

(c)

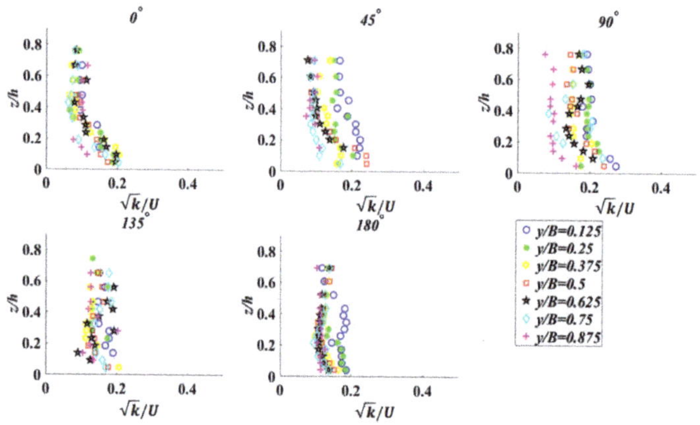

(d)

Figure 4. TKE (k) profiles for four cases at five cross sections: (**a**) Bare case, (**b**) Small case, (**c**) Middle case, and (**d**) Big case ($Q = 0.030$ m^3 s^{-1}).

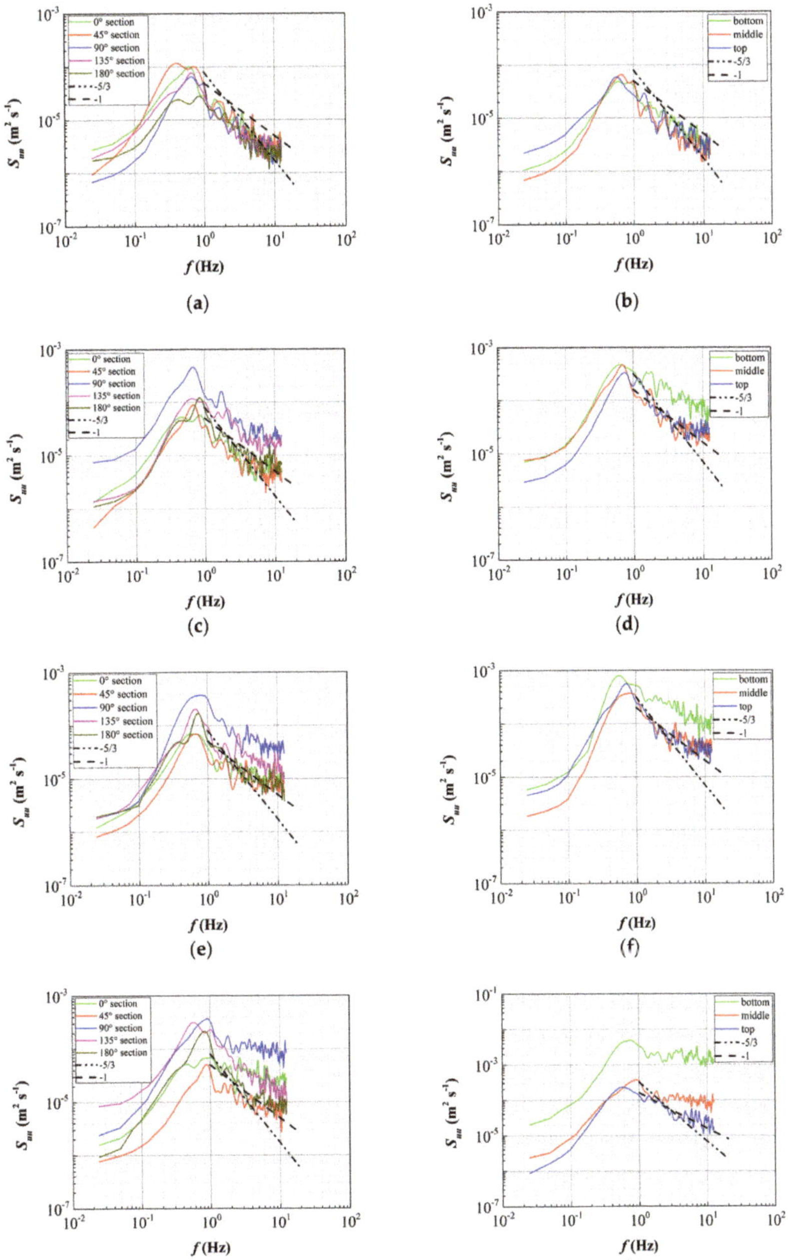

Figure 5. Power spectral density for different cases: (**a**) bare case, (**b**) different depth (2 cm, 12 cm, 16 cm) from bed for the bare case of 90° section, (**c**) small case, (**d**) different depth (2 cm, 12 cm, 16 cm) from bed for the small case of 90° sectiol (**e**) middle case, (**f**) different depth (2 cm, 12 cm, 16 cm) from bed for the middle case of 90° section, (**g**) big case, (**h**) different depth (2 cm, 12 cm, 16 cm) from bed for the big case of 90° section.

For the bare case (Figure 5a), the PSD value for the high-frequency part at the 45° section is the greatest, while the PSD value at the 180° section is the smallest, indicating that the turbulence becomes strong, and the high-frequency (small-scale) vortex increases due to the influence of channel bend. For Figure 5b, there are similar trends for different positions for the bare case, which means the structures of turbulent vortex along vertical directions are approximately identical over smooth bed.

For the rough bed cases (Figure 5c,e,g), the PSD value for the high-frequency part at the 90° section is obviously larger than other sections. For Figure 5d,f,h, the PSD values at near-bed positions are larger than those at middle and top positions, and the trend is more prominent with the increasing bed particle size, implying that the turbulent mixing near the bed regions are more intense. The PSD of the longitudinal velocity fluctuations satisfy the classic Kolmogorov $-5/3$ law in the inertial subrange. Comparing to the bare case, the existence of bed roughness shortens the inertial subrange, causing the energy access to the viscous dissipation in advance. The peak frequency for each rough bed case is similar and about 0.8 Hz, irrespective of bed particle sizes.

3.4. Turbulent Bursting

It is found that fluid motion near a bed is not completely chaotic in nature, but it is a clear "sequence of ordered motion" [43]. Such coherent motions are called the bursting process [43]. To quantify the intermittent instantaneous Reynolds stresses as well as identify turbulence structures within a turbulent bursting sequence, one of the widely used conditional sampling techniques is the quadrant analysis of the Reynolds shear stress [44–46]. Here, quadrant analysis was conducted to study the effect of rough bed on turbulent bursting. In the quadrant analysis, the Reynolds stress has four types of contributions according to the signs of the instantaneous velocity fluctuations [42]. Accordingly, turbulent events are defined by the four quadrants as outward interactions ($i = I,\ u\prime > 0,\ w\prime > 0$), ejections ($i = II,\ u\prime < 0,\ w\prime > 0$), inward interactions ($i = III,\ u\prime < 0,\ w\prime < 0$), and sweeps ($i = IV,\ u\prime > 0,\ w\prime < 0$). Results in quadrants *II* and *IV* mean the positive downward momentum flux and are involved in turbulence near-bed bursting [43].

In order to describe the turbulent event accurately, the hole (instead of zero) concept is used to eliminate smaller Reynolds stresses. The hole is formed by four hyperbolas $|u'(t)w'(t)| = G_0|u'w'|$, where G_0 is a threshold value. By using the threshold, small values can be ignored in the *i*-th quadrant [47]. The contribution of each quadrant can be represented as S_k (k = *I, II, III, IV*, indicate the four quadrants), where

$$S_k = \begin{cases} 1, & |u'(t)w'(t)| > G_0|u'w'|,\ [u'(t),\ w'(t)]\ in\ the\ same\ quadrant \\ & 0,\ \ otherwise \end{cases} \tag{7}$$

In the study, the threshold G_0 was set to 1.0 [48]. Then, the revised occurrence frequency, f_k, for the four turbulent events is given as:

$$f_k = \frac{\sum_{t=0}^{T} S_k}{\sum_{t=0}^{T} S_I + \sum_{t=0}^{T} S_{II} + \sum_{t=0}^{T} S_{III} + \sum_{t=0}^{T} S_{IV}} \tag{8}$$

where T is the length of measurement time.

Figure 6 shows characteristic distributions of the occurrence frequency f_k in different cases for Q = 0.030 m^3 s^{-1} as an example. The feature point is measured at the distance of 0.01 m from the bed in the vertical direction and 0.20 m from the inner bank (center of the flume) in the transverse direction in each section to investigate the turbulent event near bed. For the bare case (Figure 6a), it can be observed that sweep and ejection events (event *II* and *IV*) occur with comparable frequency and are much higher than outward and inward interactions at the 0° section, indicating that upwards motion of low-speed fluid and downwards motion of high-speed fluid are dominant. After the 45° section, the frequency of occurrence of events of *I* and *III* quadrants become higher. At the 180°

section, the frequency of occurrence for four events is similar. This reveals that high-speed fluid reflects by the bottom and low-speed fluid is pushed back as the influence of channel bend, showing a strong interaction of the turbulent structures with the main flow. For rough bed cases, it is noted that higher magnitude of sweep and ejections (event *II* and *IV*) contributions as shown in Figure 6b (small case) similar to the bare case at the 0° section. When flow enters the curved channel, the frequency of occurrence for event *II* and *IV* increases, indicating that sweep and ejection are the most dominant processes, which can potentially influence the sediment transport in the stream, causing the exchange of energy and momentum in the flow and the bed formation. The middle case (Figure 6c) performs the similar trend as the small case. However, for the big case (Figure 6d) the frequency of occurrence for event *II* and *IV* decreases first (0° section to 90° section) and then increases (90° section to 180° section). Therefore, the contributions of sweeps and ejections are more important than those of the outward and inward interactions over a rough bed, which are the same results obtained by other researches in straight channels [43,49]. Bed roughness is the leading factor to the turbulent bursting in curved channels when the grain size of bed particles is moderate (the small case and middle case), but for the big case the turbulent bursting is weaken before the apex of the bend. The possible explanation is the larger grain size of bed particles possible disturbs the bottom boundary layer, changing turbulent structure near the bed.

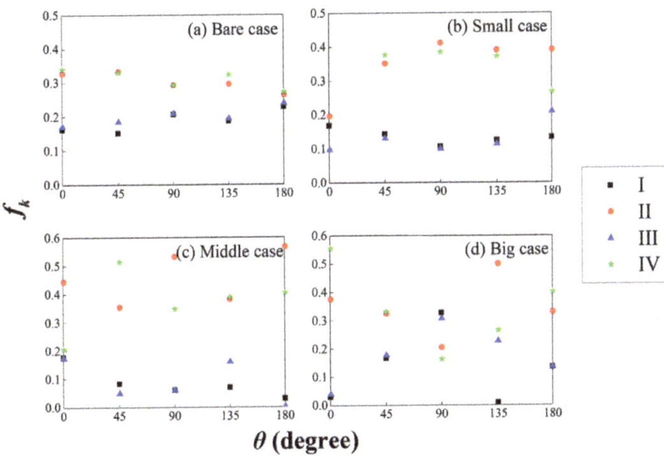

Figure 6. Frequency of occurrence of coherence turbulent events for different cases (Q = 0.030 m^3 s^{-1}).

3.5. Comparisons with Previous Experimental Results

All of the experiment cases in this paper were done in a 180 degree U-shaped curved and rough bed flume under the condition of constant downstream water level, close to the practical situations on the purpose of navigation. Thus, the experimental results in comparison with previous studies may have some differences. Firstly, since the downstream water level is fixed, as the bed particles become larger, the water depth along the curved channel decreases, leading to greater velocity compared with that over smooth curved channels. Without the constant downstream water depth, the water depth along the curved channel may keep similar [18]. Secondly, the longitudinal velocity distribution is relatively different. Pradhan et al. [20] found that the velocity remains higher towards the inner wall for rough bed. In our experimental results, longitudinal velocity near the concave bank is greater than that near the convex bank for all rough bed cases, which indicates that the rough bed can enlarge the centrifugal effect of channel bend. The reason may be the different shape of cross-section of channel (trapezoidal channel for Pradhan et al. [20]). Thirdly, sweeps and ejections are more important than those of the outward and inward interactions over a rough bed when the grain size of bed particles is intermediate [43,49]. For the big case, the proportion of the outward and inward interactions is higher than that of the sweeps and ejections in 90° section, which was not shown in the results of Najafabadi et al. [43] because of the limits of the grain size in the study.

3.6. Numerical Results

Numerical tests aim to understand the effect of thick sediment layers on secondary flow over a curved channel. The roughness height k_s of the FLOW-3D simulation was adjusted until the numerical results close to experimental data. The roughness height k_s = 0.005 m for the bare case and the roughness height k_s = 0.05 m for the big case can match well with the experimental water depths and longitudinal velocity as shown in Figures 7 and 8. The differences between the simulation and measurements can be possibly attributed to three aspects: (1) the isotropic assumptions in the RNG turbulence model are not suitable for curved channel flows, (2) the uneven bottom surface from sediment particles and flows through pores between sediment particles are unable to reflect in the numerical tests, where the solid step is used to represent the sediment layers, and (3) the slope of the entire channel may not be constant. For big case_flat, the water depth increases a little due to the bottom friction from rough bed, and the water depth of each section is approximately the same, i.e., uniform flow conditions, meaning that the friction exerted by the rough bed mainly balances gravity due to sloping bed.

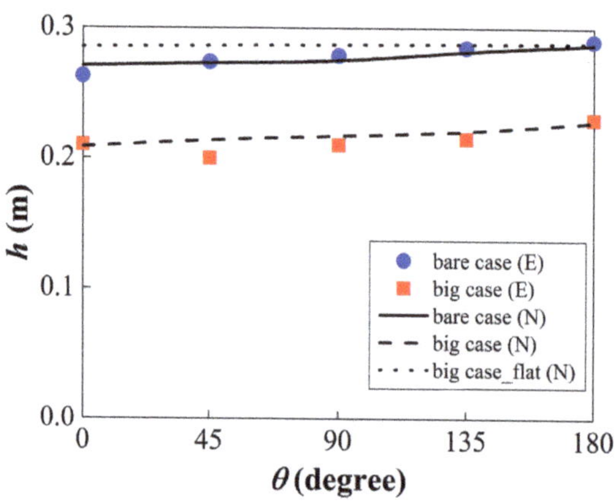

Figure 7. Comparisons of simulated and measured water depth for Q =0.030 m^3 s^{-1}.

Figure 8 performs the longitudinal velocity profiles at the mid-width of the flume (20 cm from inner bank) under different cases. The simulated velocity distribution at the 90° and 135° section does not fit very well with the experimental results, especially for the big case. For big case_flat, the longitudinal velocity near bed decreases due to the bed friction when compared with the bare case. On the other hand, the longitudinal velocity for the big case_flat is smaller than that for the big case owing to larger water depths.

The distributions of longitudinal velocity in three typical cross-sections (0°, 90° and 180°) are shown in Figure 9. For the bare case (Figure 9a), the core region of maximum longitudinal velocity has obviously shifted toward the concave bank at the 90° and 180° section in comparison with that at the 0° section. This is because of the advective transport of streamwise momentum by the cross-stream circulation in open-channel bends. At the bend entrance, the shorter distance in the inner bend than that in the outer bend would lead to longitudinal velocity u decreasing from the convex toward the concave bank. The flow acceleration/deceleration is induced by streamwise pressure gradients related to the sudden transverse tilting of the water surface in the bend. In addition, the sudden disappearance of the transverse tilting of the water surface leads to pronounced flow accelerations/deceleration in the concave/convex bend at the bend exit [3,50]. For the big case (Figure 9b), the core region of maximum longitudinal velocity stays around the center region of the cross-section at the 0° and 90° sections. At the 180° section, the core region shifts to the concave bank. For the big case_flat (Figure 9c), numerical results are similar to that for the bare case at the 0° and 90° sections. However, at the 180° section, the maximum longitudinal velocity for the big case_flat concentrates on the right middle region rather than the lower right region for the bare case. Based upon the results, it can be concluded that thick sediment layers delay the shifting of the core of maximum longitudinal velocity towards the concave bank. The bed roughness only reduces the longitudinal velocity magnitude.

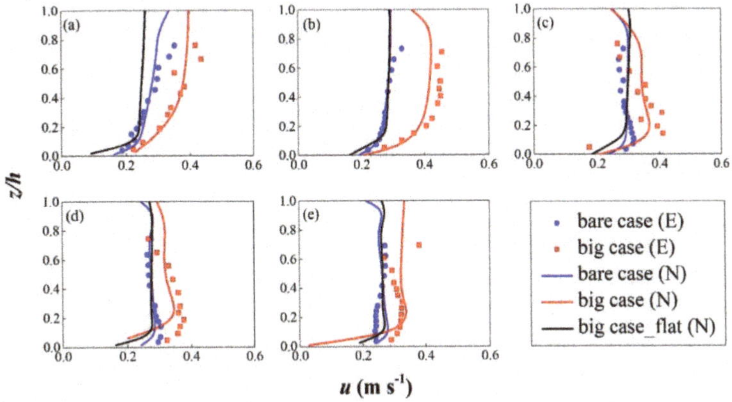

Figure 8. Comparisons of simulated and measured longitudinal velocity profiles on 5 cross sections: (**a**) 0° section, (**b**) 45° section, (**c**) 90° section, (**d**) 135° sectionInd (**e**) 180° section.

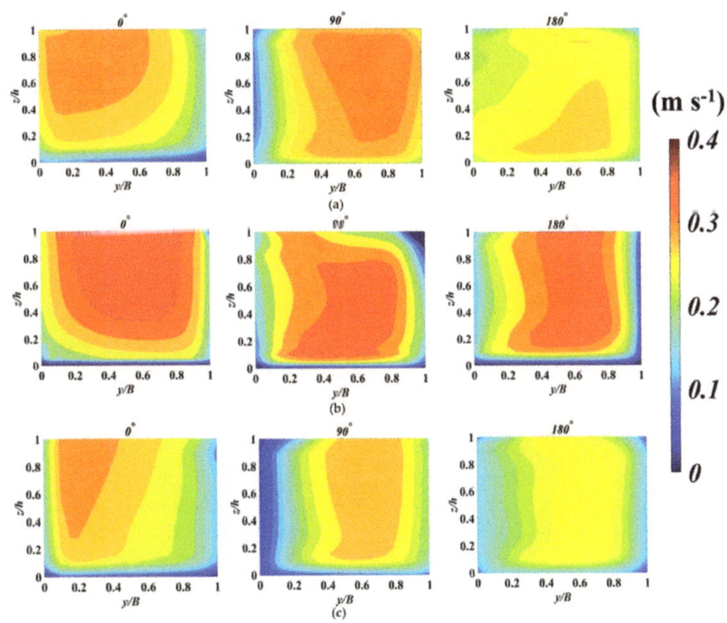

Figure 9. The distributions of the longitudinal velocity at the 0°, 90° and 180° section: (**a**) Bare case, (**b**) Big case, and (**c**) Big case_flat.

The distributions of k at the mid-width of the flume (20 cm from inner bank) at three typical cross-sections (0°, 90° and 180°) are shown in Figure 10. For the 0° section (Figure 10a), the k values of three cases are smaller near the free surface but become larger near the rough bed regions, implying that rough bed will enhance fluid mixing and strengthen turbulence. Additionally, the k values of the big case_flat are slightly larger than those of the big case. For the 90° and 180° sections (Figure 10b,c), the k values firstly increase (from 0° section to 90° section) and then decrease (from 90° section to 180° section), indicating that the channel bend can also improve turbulence along the water depth, especially near the bed regions. However, the k values of the big case (red line) increase more than the other two cases along the curved channel. The result suggests that the bed roughness and curved channel both improve turbulence along the water depth, especially near the bed regions, which is the same as the experimental results. Additionally, the thick sediment layer can promote *TKE* more than the curved channel does.

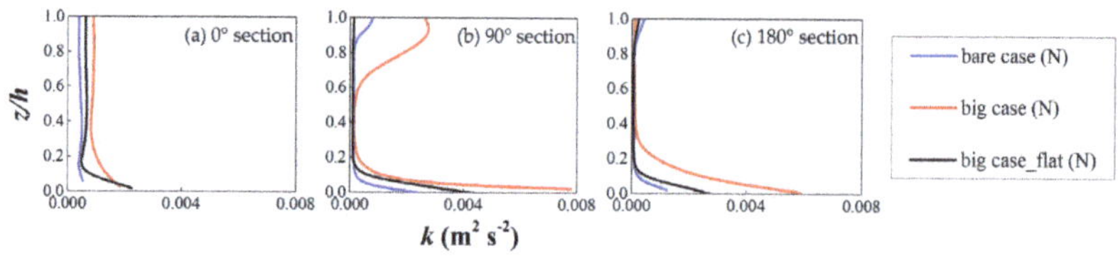

Figure 10. The profiles of *TKE* (k) at the 0°, 90° and 180° sections.

In order to explore the effect of distance from the convex bank on the distributions of k, three characteristic positions, i.e., $\frac{1}{4}$cm (=1/4 B), $\frac{1}{2}$cm (=1/2 B) and $\frac{3}{4}$cm (=3/4 B) from the convex bank along the transverse direction, are provided at the 90° section for three cases as shown in Figure 11. For the bare case (Figure 11a), the k values of three lines basically collapse together, meaning that the transverse positive has little effect to the k values for the smooth bed. For the big case (Figure 11b), the k values near convex bank is slightly larger when z/h = 0.1–0.6. The thick sediment layer with bed roughness strengthens the turbulence more near convex bank, which is the same as the experimental result. However, for the big case_flat (Figure 11c), the k values near concave bank at z/h > 0.1 is larger. The thick sediment layer increases the *TKE* (k) values, especially near convex bank.

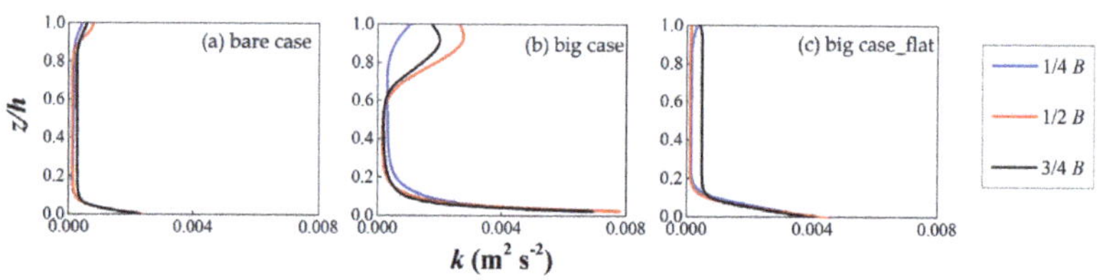

Figure 11. The distributions of *TKE* (k) at the 90° section for three cases.

The secondary flow that can interact dynamically with the primary flow is important for meandering rivers and useful for studies of diffusion and navigation in natural waterways, because it is partly responsible for the large-scale bed topography of natural alluvial channel bends [51]. The vector of the cross-stream motion (v, w) at the 90° section is shown in Figure 12. For the bare case (Figure 12a), there is a classical helical motion obviously, called the center-region cell. In addition, a weaker counterrotating cell of cross-stream circulation, called the outer-bank cell, exists in the corner formed by the water surface and the outer bank as Blanckaert and Graf [2] revealed. The center-region cell plays an important role in the advective momentum transport, and the outer-bank cell is believed to be crucial to bank erosion processes [52]. For the big case (Figure 12b), the center-region cell and outer-bank cell still exist. The center-region cell is confined to the near-bed area, and the outer-bank cell is stronger than that for the bare case, possibly accelerating the bank erosion. This is because the rough bed disturbs the flow, mainly reducing the longitudinal flow velocity near bed, and also affecting the distribution of the transverse velocity along the water depth. For the big case_flat (Figure 12c), the center-region cell is confined to a smaller area compared to the bare and big cases, which means that the thick sediment layer can influence more riverbed regions. Furthermore, the outer-bank cell is weaker than that for the big case.

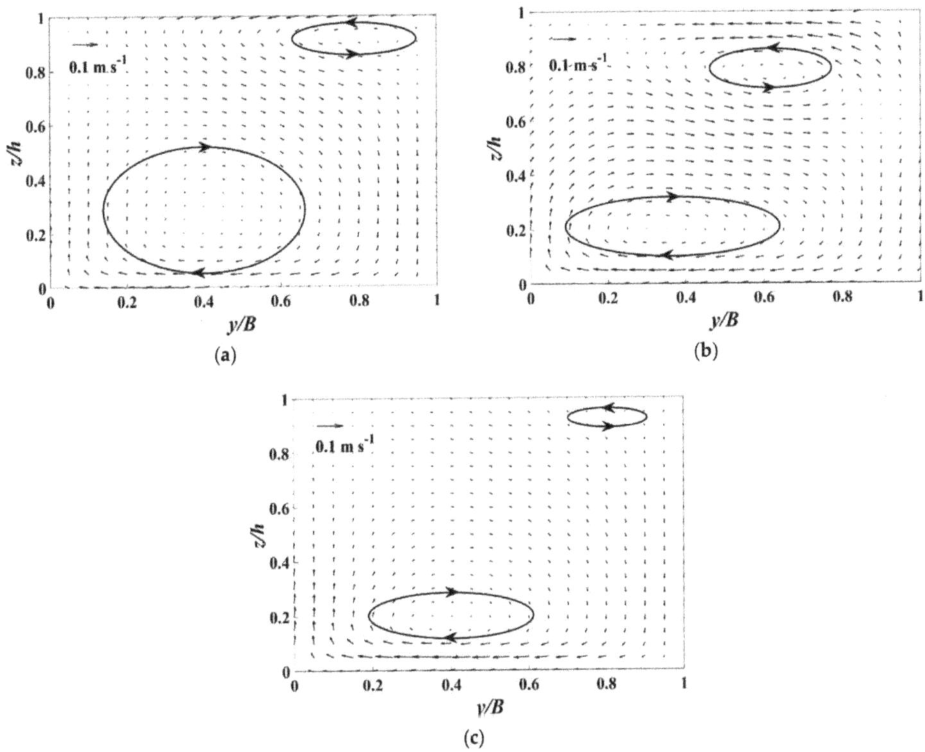

Figure 12. Numerical results of velocity vector for cross-stream motions: (**a**) the bare case, (**b**) the big case, and (**c**) the big case_flat. Note: the arrowed circle represents the secondary flow.

The transverse velocity distribution can be used to display the intensity of secondary flow. Figure 13 shows the secondary flow distributions of different cases at the mid-width of the flume (20 cm from inner bank) at the 90° section, where X-axis is $vr/(\overline{u}h)$, Y-axis is z/h, r is curvature radius of bend, and $\overline{u}$ is average longitudinal velocity. When the calculated values ($vr/(\overline{u}h)$) distribute on the same side of $vr/(\overline{u}h)$ = 0, i.e., all $vr/(\overline{u}h)$ values are larger or smaller than zero, it means that the direction of transverse velocity is the same at this position. In other words, there is no secondary flow. Once the calculated values ($vr/(\overline{u}h)$) fall on the two sides of $vr/(\overline{u}h) = 0$, there is secondary flow existing. Additionally, the secondary flow is stronger as the slope of transverse velocity distribution increases.

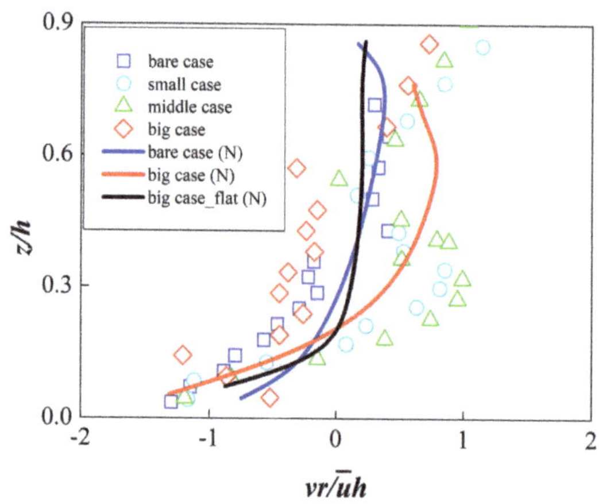

Figure 13. The transverse velocity distribution for experimental results (symbols) and numerical simulations (solid lines).

From Figure 13, it can be seen that numerical simulations (solid line) are close to experimental measurements (individual symbols). The slope of transverse velocity distribution for the big case is steeper than that for the bare case, which means that the rough bed would strengthen the secondary flow and possibly causes more riverbed erosion. Besides, the position of calculated values ($vr/(\overline{u}h) = 0$) for the big case is lower than that of the bare case, representing the secondary flow is confined to the near-bed area as the results discussed above. For the big case and big case_flat, the slope of transverse velocity distribution is comparable, i.e., similar secondary flow intensity.

4. Conclusions

Experiments and numerical simulation of curved open channel flows over rough beds under the condition of constant downstream water depth were carried out. Based on the ADV data, hydrodynamic characteristics of flow over rough bed such as longitudinal velocity distribution, turbulent characteristics, secondary flow, turbulent bursting and so on was analyzed. Additionally, the numerical results were used to show different flow patterns for the rough bed with and without thick sediment layers. The following conclusions are drawn:

1. In the study, the experiments cases were carried in a 180 degree U-shaped curved and rough bed flume under the condition of constant downstream water level. Different as previous studies, as the bed particles become larger, the water depth along the curved channel decreases, leading to greater velocity compared with that over smooth curved channels. The longitudinal velocity near bed significantly decreases due to the existence of rough bed, resulting in the larger gradient of longitudinal velocity profile for rough bed than that in the bare case. As the both effect of channel bend and rough bed, the longitudinal velocity profiles exhibit a trend of increasing in the bottom and decreasing near the water surface. The grain size of bed particles only changes the magnitude of velocity, making the phenomena more remarkable. Rough bed also enlarges the centrifugal effect of channel bend, and as a result the longitudinal velocity of concave bank is higher than that of convex bank.
2. Rough bed will enhance the mixing between the water bodies and rough bed, strengthening turbulence and increasing turbulent kinetic energy. From the power spectral density analysis, the structure of turbulent vortex along the vertical direction is resembled. However, the turbulent energy of near-bed position is bigger than that of top position. The power spectra of the longitudinal velocity components follow the

classic Kolmogorov -5/3 law in the inertial subrange, and the existence of rough bed shortens the inertial subrange and reaches to the viscous dissipation in advance.

3. The contributions of sweeps and ejections are more important than those of the outward and inward interactions over a rough bed. In addition, bed roughness is the dominant factor to the turbulent bursting in curved channels when the grain size of bed particles is intermediate (the small case and middle case), but the influence of rough bed for the big case to the turbulent bursting weakens due to the disturbances of bottom boundary layer.
4. Numerical simulations were used to discuss flow patterns over two different rough bed settings (the big and big_flat cases). Since the bed surface roughness are the same for the two cases, their skin drag to flow is identical. However, the form drag is greater for bed covered with a thick sediment layer (the big case) rather than the flat bed (the big_flat case). Simulation results show that for both rough bed cases the secondary flow is confined to the near-bed area and the intensity of secondary flow is improved, possibly causing more serious bed erosion along a curved channel. As a result, the advective momentum transport increases over rough and curved channels. Furthermore, the thick sediment layer (the big case), i.e., larger form drag, can delay the shifting of the core region of maximum longitudinal velocity towards the concave bank.
5. The effect of rough bed to turbulent kinetic energy is also provided using numerical methods. The numerical results show that the thick sediment layer can lead to larger k values than the curved channel and increase the *TKE* (k) values, especially near convex bank.

The findings of the research could improve the understanding of the interactive effects between the rough bed and the strongly curved channel flow. Based on the reported results, future researches can focus on studying the effect of flume curvature on the hydrodynamic characteristics of curved channel flow with gravel beds.

Author Contributions: Conceptualization, Y.-J.C., Y.-T.L. and X.J.; Methodology, Y.-J.C. and Y.-T.L.; Laboratory measurement, Y.Y.; Data analysis, Y.-T.L. and Y.Y.; Writing—Original Draft Preparation, Y.-T.L. and Y.Y.; Writing—Review and Editing, Y.-J.C.; Software, X.J.; Funding Acquisition, Y.-J.C. and X.J. All authors have read and agreed to the published version of the manuscript.

Funding: This research was funded by National Key Research and Development Project of China [grant number 2017YFC0405205]; Ministry of Science and Technology of Taiwan [grant number MOST 109-2625-M-002-016]; Zhejiang Provincial Natural Science Foundation of China [grant number LY20A020009); State Key Laboratory of Hydraulic Engineering Simulation and Safety, Tianjin University [grant number HESS-2014]; Fundamental Research Funds for the Central Universities [grant number 2020QNA4038].

Institutional Review Board Statement: Not applicable.

Informed Consent Statement: Not applicable.

Data Availability Statement: All data generated or analyzed during this study are included in this article.

Conflicts of Interest: The authors declare no conflict of interest.

References

1. Wei, M.; Blanckaert, K.; Heyman, J.; Li, D.; Schleiss, A.J. A parametrical study on secondary flow in sharp open-channel bends: Experiments and theoretical modelling. *J. Hydro-Environ. Res.* **2016**, *13*, 1–13. [CrossRef]
2. Blanckaert, K.; Graf, W.H. Momentum transport in sharp open-channel bends. *J. Hydraul. Eng.* **2004**, *130*, 186–198. [CrossRef]
3. Zeng, J.; Constantinescu, G.; Blanckaert, K.; Weber, L. Flow and bathymetry in sharp open-channel bends: Experiments and predictions. *Water Resour. Res.* **2008**, *44*, W09401. [CrossRef]
4. Blanckaert, K. Topographic steering, flow recirculation, velocity redistribution, and bed topography in sharp meander bends. *Water Resour. Res.* **2010**, *46*, W09506. [CrossRef]
5. Blanckaert, K.; De Vriend, H.J. Meander dynamics: A nonlinear model without curvature restrictions for flow in open-channel bends. *J. Geophys. Res.-Earth Surf.* **2010**, *115*, F04011. [CrossRef]

6. Vaghefi, M.; Akbari, M.; Fiouz, A.R. An experimental study of mean and turbulent flow in a 180 degree sharp open channel bend: Secondary flow and bed shear stress. *KSCE J. Civ. Eng.* **2016**, *20*, 1582–1593. [CrossRef]
7. Bomminayuni, S.; Stoesser, T. Turbulence statistics in an open-channel flow over a rough bed. *J. Hydraul. Eng.* **2011**, *137*, 1347–1358. [CrossRef]
8. Grass, A.J.; Stuart, R.J.; Mansour-Tehrani, M. Vortical structures and coherent motion in turbulent flow over smooth and rough boundaries. *Philos. Trans. R. Soc. Lond. A* **1991**, *336*, 35–65. [CrossRef]
9. Ferro, V. Friction factor for gravel-bed channel with high boulder concentration. *J. Hydraul. Eng.* **1999**, *125*, 771–778. [CrossRef]
10. Ferro, V. Flow resistance in gravel-bed channels with large-scale roughness. *Earth Surf. Process. Landf.* **2003**, *28*, 1325–1339. [CrossRef]
11. Nikora, V.; Goring, D.; McEwan, I.; Griffiths, G. Spatially averaged open-channel flow over rough bed. *J. Hydraul. Eng.* **2001**, *127*, 123–133. [CrossRef]
12. Nikora, V.; Koll, K.; McEwan, I.; McLean, S.; Dittrich, A. Velocity distribution in the roughness layer of rough-bed flows. *J. Hydraul. Eng.* **2004**, *130*, 1036–1042. [CrossRef]
13. Mignot, E.; Barthélemy, E.; Hurther, D. Double-averaging analysis and local flow characterization of near-bed turbulence in gravel-bed channel flows. *J. Fluid Mech.* **2009**, *618*, 279–303. [CrossRef]
14. Dey, S.; Das, R.; Gaudio, R.; Bose, S.K. Turbulence in mobile-bed streams. *Acta Geophys.* **2012**, *60*, 1547–1588. [CrossRef]
15. Qi, M.; Li, J.; Chen, Q.; Zhang, Q. Roughness effects on near-wall turbulence modelling for open-channel flows. *J. Hydraul. Res.* **2018**, *56*, 648–661. [CrossRef]
16. Jin, Y.; Steffler, P.M.; Hicks, F.E. Roughness effects on flow and shear stress near outside bank of curved channel. *J. Hydraul. Eng.* **1990**, *116*, 563–577. [CrossRef]
17. Blanckaert, K.; Duarte, A.; Chen, Q.; Schleiss, A.J. Flow processes near smooth and rough (concave) outer banks in curved open channels. *J. Geophys. Res.-Earth Surf.* **2012**, *117*, F04020. [CrossRef]
18. Hersberger, D.S.; Franca, M.J.; Schleiss, A.J. Wall-roughness effects on flow and scouring in curved channels with gravel beds. *J. Hydraul. Eng.* **2016**, *142*, 4015032. [CrossRef]
19. Jamieson, E.C.; Post, G.; Rennie, C.D. Spatial variability of three-dimensional Reynolds stresses in a developing channel bend. *Earth Surf. Process. Landf.* **2010**, *35*, 1029–1043. [CrossRef]
20. Pradhan, A.; Kumar Khatua, K.; Sankalp, S. Variation of velocity distribution in rough meandering channels. *Adv. Civ. Eng.* **2018**, *2018*, 1–12. [CrossRef]
21. Kraus, N.C.; Lohrmann, A.; Cabrera, R. New acoustic meter for measuring 3D laboratory flows. *J. Hydraul. Eng.* **1994**, *120*, 406–412. [CrossRef]
22. Goring, D.G.; Nikora, V.I. Despiking acoustic doppler velocimeter data. *J. Hydraul. Eng.* **2002**, *128*, 117–126. [CrossRef]
23. Wang, X.; Yang, Q.; Lu, W.; Wang, X. Experimental study of near-wall turbulent characteristics in an open-channel with gravel bed using an acoustic doppler velocimeter. *Exp. Fluids* **2012**, *52*, 85–94. [CrossRef]
24. Li, C.; Xue, W.; Huai, W. Effect of vegetation on flow structure and dispersion in strongly curved channels. *J. Hydrodyn.* **2015**, *27*, 286–291. [CrossRef]
25. Nogueira, H.I.S.; Adduce, C.; Alves, E.; Franca, M.J. Analysis of lock-exchange gravity currents over smooth and rough beds. *J. Hydraul. Res.* **2013**, *51*, 417–431. [CrossRef]
26. Yalin, M.S. *River Mechanics*; Elsevier Science & Technology: Kent, UK, 1992.
27. Schlichting, H.; Gersten, K. *Boundary-Layer Theory*, 8th ed.; Springer: Berlin, Germany, 2000.
28. Kironoto, B.A.; Graf, W.H. Turbulence characteristics in rough uniform open-channel flow. *Proc. Inst. Civ. Eng. Water Marit. Energy* **1994**, *106*, 333–344. [CrossRef]
29. Glatzel, T.; Litterst, C.; Cupelli, C.; Lindemann, T.; Moosmann, C.; Niekrawietz, R.; Streule, W.; Zengerle, R.; Koltay, P. Computational fluid dynamics (CFD) software tools for microfluidic applications—A case study. *Comput. Fluids* **2008**, *37*, 218–235. [CrossRef]
30. Al-Qadami, E.H.H.; Abdurrasheed, A.S.; Mustaffa, Z.; Yusof, K.W.; Malek, M.A.; Ghani, A.A. Numerical modelling of flow characteristics over sharp crested triangular hump. *Results Eng.* **2019**, *4*, 100052. [CrossRef]
31. Ghaderi, A.; Dasineh, M.; Aristodemo, F.; Aricò, C. Numerical simulations of the flow field of a submerged hydraulic jump over triangular macroroughnesses. *Water* **2021**, *13*, 674. [CrossRef]
32. Qi, H.; Zheng, J.; Zhang, C. Modeling excess shear stress around tandem piers of the longitudinal bridge by computational fluid dynamics. *J. Appl. Water Eng. Res.* **2021**. [CrossRef]
33. Huang, T.; Jan, C.; Hsu, Y. Numerical simulations of water surface profiles and vortex structure in a vortex settling basin by using flow-3d. *J. Mar. Sci. Technol.* **2017**, *25*, 531–542. [CrossRef]
34. Abbaspour, A.; Kia, S.H. Numerical investigation of turbulent open channel flow with semi-cylindrical rough beds. *KSCE J. Civ. Eng.* **2014**, *18*, 2252–2260. [CrossRef]
35. Bai, Y.; Song, X.; Gao, S. Efficient investigation on fully developed flow in a mildly curved 180° open-channel. *J. Hydroinform.* **2014**, *16*, 1250–1264. [CrossRef]
36. Gholami, A.; Akhtari, A.A.; Minatour, Y.; Bonakdari, H.; Javadi, A.A. Experimental and numerical study on velocity fields and water surface profile in a strongly-curved 90 degrees open channel bend. *Eng. Appl. Comp. Fluid Mech.* **2014**, *8*, 447–461. [CrossRef]

37. Blanckaert, K. Flow and Turbulence in Sharp Open Channel Bends. Ph.D. Thesis, Ecole Polytechnique Fédérale de Lausanne, Lausanne, Switzerland, 2003.
38. Barbhuiya, A.K.; Talukdar, S. Scour and three dimensional turbulent flow fields measured by ADV at a 90 ° horizontal forced bend in a rectangular channel. *Flow Meas. Instrum.* **2010**, *21*, 312–321. [CrossRef]
39. Singh, K.M.; Sandham, N.D.; Williams, J.J.R. Numerical simulation of flow over a rough bed. *J. Hydraul. Eng.* **2007**, *133*, 386–398. [CrossRef]
40. Reidenbach, M.A.; Limm, M.; Hondzo, M.; Stacey, M.T. Effects of bed roughness on boundary layer mixing and mass flux across the sediment-water interface. *Water Resour. Res.* **2010**, *46*, 58–72. [CrossRef]
41. Jiménez, J. Turbulent flows over rough walls. *Annu. Rev. Fluid Mech.* **2004**, *36*, 173–196. [CrossRef]
42. Huai, W.; Zhang, J.; Wang, W.; Katul, G.G. Turbulence structure in open channel flow with partially covered artificial emergent vegetation. *J. Hydrol.* **2019**, *573*, 180–193. [CrossRef]
43. Najafabadi, E.F.; Afzalimehr, H.; Sui, J. Turbulence characteristics of favorable pressure gradient flows in gravel-bed channel with vegetated walls. *J. Hydrol. Hydromech.* **2015**, *63*, 154–163. [CrossRef]
44. Mazumder, B.S.; Pal, D.K.; Ghoshal, K.; Ojha, S.P. Turbulence statistics of flow over isolated scalene and isosceles triangular-shaped bedforms. *J. Hydraul. Res.* **2009**, *47*, 626–637. [CrossRef]
45. Kassem, H.; Thompson, C.E.L.; Amos, C.L.; Townend, I.H. Wave-induced coherent turbulence structures and sediment resuspension in the nearshore of a prototype-scale sandy barrier beach. *Cont. Shelf Res.* **2015**, *109*, 78–94. [CrossRef]
46. Li, Y.; Wei, J.; Gao, X.; Chen, D.; Weng, S.; Du, W.; Wang, W.; Wang, J.; Tang, C.; Zhang, S. Turbulent bursting and sediment resuspension in hyper-eutrophic Lake Taihu, China. *J. Hydrol.* **2018**, *565*, 581–588. [CrossRef]
47. Lu, S.S.; Willmarth, W.W. Measurements of the structure of the Reynolds stress in a turbulent boundary layer. *J. Fluid Mech.* **1973**, *60*, 481–511. [CrossRef]
48. Duan, J.; He, L.; Wang, G.; Fu, X. Turbulent burst around experimental spur dike. *Int. J. Sediment Res.* **2011**, *26*, 471–486. [CrossRef]
49. Keshavarzy, A.; Ball, J.E. An analysis of the characteristics of rough bed turbulent shear stresses in an open channel. *Stoch. Hydrol. Hydraul.* **1997**, *11*, 193–210. [CrossRef]
50. Henderson, F.M. *Open Channel Flows*; Macmillan: New York, NY, USA, 1966.
51. Falcon, M. Secondary flow in curved open channels. *Annu. Rev. Fluid Mech.* **1984**, *16*, 179–193. [CrossRef]
52. Blanckaert, K.; De Vriend, H.J. Secondary flow in sharp open-channel bends. *J. Fluid Mech.* **2004**, *498*, 353–380. [CrossRef]

water MDPI

Article

Secondary Currents with Scour Hole at Grade Control Structures

Mouldi Ben Meftah *, Diana De Padova, Francesca De Serio and Michele Mossa

Department of Civil, Environmental, Land, Building Engineering and Chemistry, Polytechnic University of Bari, Via E. Orabona 4, 70125 Bari, Italy; diana.depadova@poliba.it (D.D.P.); francesca.deserio@poliba.it (F.D.S.); michele.mossa@poliba.it (M.M.)
* Correspondence: mouldi.benmeftah@poliba.it; Tel.: +39-080-5963508

Abstract: Most studies on local scouring at grade control structures have principally focused on the analysis of the primary flow field, predicting the equilibrium scour depth. Despite the numerous studies on scouring processes, secondary currents were not often considered. Based on comprehensive measurements of flow velocities in clear water scours downstream of a grade control structure in a channel with non-cohesive sediments, in this study, we attempted to investigate the generation and turbulence properties of secondary currents across a scour hole at equilibrium condition. The flow velocity distributions through the cross-sectional planes at the downstream location of the maximum equilibrium scour depth clearly show the development of secondary current cells. The secondary currents form a sort of helical-like motion, occurring in both halves of the cross-section in an axisymmetric fashion. A detailed analysis of the turbulence intensities and Reynolds shear stresses was carried out and compared with previous studies. The results highlight considerable spatial heterogeneities of flow turbulence. The anisotropy term of normal stresses dominates the secondary shear stress, giving the impression of its crucial role in generating secondary flow motion across the scour hole. The anisotropy term shows maximum values near both the scour mouth and the scour bed, caused, respectively, by the grade control structure and the sediment ridge formation, which play fundamental roles in maintaining and enhancing the secondary flow motion.

Keywords: scour; equilibrium condition; velocity field; secondary currents; turbulence

Citation: Ben Meftah, M.; De Padova, D.; De Serio, F.; Mossa, M. Secondary Currents with Scour Hole at Grade Control Structures. *Water* **2021**, *13*, 319. https://doi.org/10.3390/w13030319

Academic Editor: Roberto Gaudio
Received: 25 December 2020
Accepted: 23 January 2021
Published: 28 January 2021

1. Introduction

The presence of natural or man-made structures on riverbeds plays an important role in the evolution of river morphology and sediment entrainment. Flow turbulence properties and secondary currents play a crucial role in sediment transport, and, in turn, suspended particle motion influences turbulence, such as Reynolds shear stress and velocity [1]. In addition to the fluid–particle interactions, which definitely influence the flow velocity distribution, the fluid–structure interactions, i.e., with natural vegetations, riverbed debris, bridge piers and abutments, sills, sluice gates, spillways, weirs, spur dikes, off-shore platforms, wind turbines, energy converters, etc., cause additional complex effects on the flow hydrodynamic characteristics. Local scouring is produced due to these complex flow patterns occurring in the surroundings of such structures. The local scouring process has attracted the attention and interest of many scientists for decades [2–18].

Experimental studies on the scouring process at grade control structures (GCSs) in riverbeds [3,5,9,17,18] showed that the scour often developed downstream of the structure. The extension of the scour hole is strongly influenced by the properties of the incoming flow, which is usually a two-dimensional jet-like flow. Owing to the high velocity of the entering jet flow, a large amount of sediment erosion locally occurs downstream of the GCS, forming the scour hole. Because of the large velocity gradients among the jet flow and that in the scour pool, the jet diffuses near the bed, and is redirected at a reduced bed velocity. The equilibrium state occurs when the path of the impinging jet becomes long enough and its diffused velocity is reduced to values below the minimum value required to move the sediments [17,18]. The entering jet produces a very complex flow field in the scour hole. For structures around which the flow passes (bypassed structures), i.e., bridge piers, abutments, offshore platforms, and wind turbines, the scour profile is considerably affected by the flow turbulence structures, generated in the form of a horseshoe vortex, wake vortex, and surface roller [14].

Most studies on local scouring downstream of GCSs have essentially focused on the prediction of the maximum scour dimensions at equilibrium, especially the maximum scour depth and the maximum length. For the sake of simplicity, in these studies, flows have been analyzed in the plane of flow symmetry (along the channel axis), neglecting the effect of the wall-normal velocity component. However, flows within scour holes, regardless of the GCS's geometry, are three-dimensional, and therefore, secondary currents can occur in some cross-sections of the scour hole. Secondary currents represent circulation of fluids around the axis of the primary flow [19].

In open channel flows, secondary currents affect the distribution of bed shear stress, Reynolds stress, and turbulence intensities across the channel [20]. In large channels, the secondary currents consist of a series of counter-rotating cells through the sectional planes. Between the cells, upwelling and downwelling alternating movement zones may occur, which extend over the entire water column. Secondary current cells often generate a sort of undulating bottom shear stress distribution in the transverse direction, affecting the whole depth-flow field and the free-surface flow pattern [20,21]. According to Albayrak and Lemmin [20], the dynamics of secondary currents are strongly affected by the channel aspect ratio. The authors [20] observed that the number of secondary current cells changes proportionally with the aspect ratio. Within the range of their experimental conditions, Albayrak and Lemmin [20] also argued that, for a given aspect ratio, the number of secondary current cells is not affected by changes in the Reynolds number or the Froude number. Papanicolaou et al. [22] proved that the presence of secondary currents increases the sidewall shear stress and affects the turbulent production within a channel flow.

Secondary flow circulation and scouring processes were also observed in stream confluences [23–25]. The increase of flow velocity due to tributary junction generates a zone of maximum scour located near the center of the confluence. This zone is characterized by dominant flow convergence and a consistent pattern of secondary circulation.

Despite the numerous studies conducted on local scouring, a lack of information regarding the structures and generation of a secondary current in scour holes has been noted in literature. The main factors generating secondary currents in straight non-circular channel flows have been and remains a topic of much discussion and conjecture. In this study, we essentially focus on the hydrodynamic characteristics of flow across scour holes developed downstream of a GCS in sand riverbeds. Thanks to several measurements of the flow velocities performed through the scour cross-section at the position of maximum equilibrium scour depth, the secondary current patterns in the scour hole were achieved. This paper aims to: (i) check the development of secondary currents in scour hole downstream of a GCS, (ii) analyze the evolution of the turbulence structure in the scour hole at the equilibrium condition, and (iii) try to understand the physical origin of secondary current cells across the equilibrium scour hole.

2. Theoretical Considerations

Sand ridges, also termed sediment strips, which are usually aligned parallel to the direction of the mean flow, have been widely observed in nature, i.e., in rivers with a bed composed of non-cohesive material, continental shelves, estuaries, and deserts [26]. These phenomena are strongly related to the development of secondary currents. In the literature (e.g., [20,21,26]), steady secondary currents have been classified into two categories: (i) secondary currents of the first kind (skew-induced stream-wise vorticity), taking origin from mean flow, but driven by curvature effect, and (ii) secondary currents that are generated by anisotropy of flow turbulence.

For steady flow and incompressible fluid, the time-averaged continuity and Navier–Stokes equations are (using index notation):

$$\frac{\partial U_i}{\partial x_i} = 0 \tag{1}$$

$$\frac{\partial U_j U_i}{\partial x_j} = -\frac{1}{\rho}\frac{\partial p}{\partial x_i} + \nu\frac{\partial^2 U_i}{\partial x_j^2} + \frac{\partial \tau_{ij}}{\partial x_j} + g_i \tag{2}$$

where x_i and x_j = (x, y, z) is the direction tensor, U_i and U_j = (U, V, W) is the time-averaged velocity tensor, in which U, V, and W are the velocity along x, y, and z, respectively, ρ is the water density, p is the pressure force, ν is the water kinematic viscosity, g_i is the gravitational acceleration tensor, $\tau_{ij} = -U'_iU'_j$ is the shear stress tensor, and $U'_iU'_j$ is the time-averaged shear stress of $U'_iU'_j(t)$ over the length of the time series. The instantaneous velocity is defined as $u_i(t) = U_i + U'_i(t)$, where $U'_i = (U', V', W')$ is the velocity fluctuation tensor of u, v, and w, respectively.

From Equation (2), one can obtain the equation of the transport of stream-wise momentum as:

$$\underbrace{-\frac{1}{\rho}\frac{\partial p}{\partial x}}_{I} - \underbrace{\left(U\frac{\partial U}{\partial x} + V\frac{\partial U}{\partial y} + W\frac{\partial U}{\partial z}\right)}_{II} + \underbrace{\nu\left(\frac{\partial^2 U}{\partial x^2} + \frac{\partial^2 U}{\partial y^2} + \frac{\partial^2 U}{\partial z^2}\right)}_{III} - \underbrace{\left(\frac{\partial U'U'}{\partial x} + \frac{\partial U'V'}{\partial y} + \frac{\partial U'W'}{\partial z}\right)}_{IV} = 0 \tag{3}$$

In Equation (3), *I* is the pressure gradient driving the flow, *II* indicates the convective transport of stream-wise momentum, *III* represents the transport due to viscous shear stresses, and *IV* is the transport due to turbulent stresses. The term *III* is generally negligible, except close to boundaries (bed and banks/walls). In the case of uniformly developed turbulent flow, $\partial/\partial x = 0$ in the terms *II*, *III*, and *IV*.

Since, in this study, the channel is straight, and the anisotropy of the flow turbulence could be the source of the secondary current formation. In this case, the secondary currents (steady and incompressible flow) can be described by the stream-wise vorticity equation as:

$$\begin{aligned}\underbrace{U\frac{\partial \omega_x}{\partial x} + V\frac{\partial \omega_x}{\partial y} + W\frac{\partial \omega_x}{\partial z}}_{A1} &= \underbrace{\omega_x\frac{\partial U}{\partial x} + \omega_y\frac{\partial U}{\partial y} + \omega_z\frac{\partial U}{\partial z}}_{A2} + \underbrace{\nu\left(\frac{\partial^2 \omega_x}{\partial x^2} + \frac{\partial^2 \omega_x}{\partial y^2} + \frac{\partial^2 \omega_x}{\partial z^2}\right)}_{A3} \\ &\quad \underbrace{+\frac{\partial}{\partial x}\left(\frac{\partial(U'V')}{\partial z} - \frac{\partial(U'W')}{\partial y}\right)}_{A4} + \underbrace{\frac{\partial^2}{\partial y \partial z}\left(V'^2 - W'^2\right)}_{A5} + \underbrace{\left(\frac{\partial^2}{\partial z^2} - \frac{\partial^2}{\partial y^2}\right)V'W'}_{A6}\end{aligned} \tag{4}$$

where $\omega_x = \frac{\partial W}{\partial y} - \frac{\partial V}{\partial z}$, $\omega_y = \frac{\partial U}{\partial z} - \frac{\partial W}{\partial x}$, and $\omega_z = \frac{\partial V}{\partial x} - \frac{\partial U}{\partial y}$ are the stream-wise, the span-wise, and the vertical vorticities, respectively. The term A1 in Equation (4) represents the rate of change of ω_x due to the convection of fluid by the mean flow velocities: primary flow U and secondary motions V and W. The term A2 represents the tilting and stretching of the three vorticities by the gradients of the mean flow velocities. The term A3 represents the viscous diffusion of ω_x. The terms A4, A5, and A6 represent the rates of change of the normal and shear stresses in the cross-section plane, which are theoretically responsible for developing and maintaining the second kind of secondary currents [27,28].

3. Experimental Set-Up

The experiments on the scour processes were carried out in a rectangular flume of closed-circuit flow at the Hydraulic Laboratory of the Mediterranean Agronomic Institute of Bari (Bari, Italy). The flume had glass sidewalls and a Plexiglas floor, allowing a good side view of the flow. It was 7.72 m long, 0.30 m wide, and 0.40 m deep. A pump with a maximum discharge of 24 L/s was used to deliver water from the laboratory sump to an upstream tank equipped with a baffle and lateral weir, maintaining a constant head upstream of a movable slide gate constructed at the inlet of the flume. The slide gate regulates channel flow discharge. To create a smooth flow transition from the upstream reservoir to the flume, a rectangular wooden ramp, playing the role of a first GCS, was installed at the inlet of the flume; it was 1.55 m long and 0.15 m thick, and of the same channel width (Figure 1). At the outlet of the flume, water was intercepted by a stilling tank, equipped with three vertical grids to stabilize water, and a triangular weir (V-notch sharp crested weir) to measure the discharge with a relative uncertainty of $\pm$ 8%. At the downstream end of the flume, a movable gate made of Plexiglas and hinged at the channel bottom was used to regulate the flow depth.

To simulate grade control structures to protect riverbeds against erosion, in this study, we used a series of sills that consisted of PVC plates 0.30 m wide and 0.01 m thick. The sills were installed on an experimental area extended 6 m along the channel, downstream of the wooden ramp. The sill height decreased progressively as it went downstream from the wooden ramp, respecting a determined initial slope S_0 of 0.0086. Different configurations were investigated, the difference between them being the distance between GCSs (sills).

The flume bottom between the GCSs was covered with an erodible bed material layer, consisting of almost uniform sand particles with a median size, d_{50}, of 1.8 mm and a density of 2650 kg/m^3. Along the experimental area, the sand layer was leveled respecting the maximum GCS heights, forming the original bed of the channel with a slope S_0 (Figure 1).

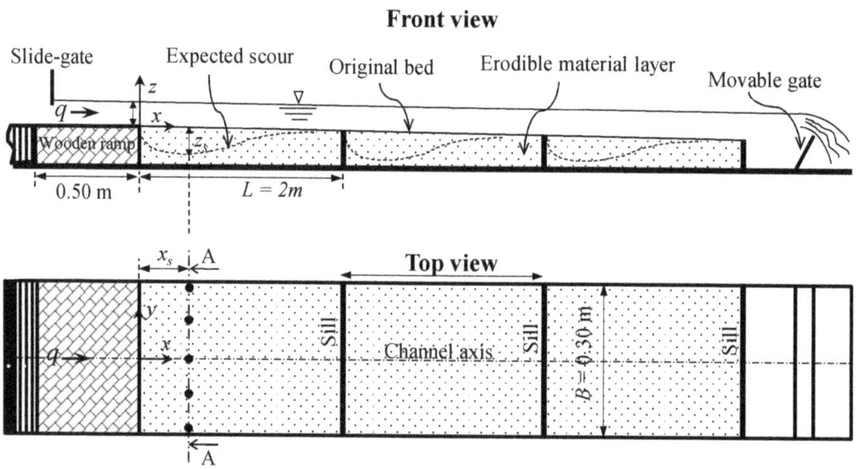

Figure 1. General sketch of the laboratory flume with the initial condition and expected scour hole (dashed profile) downstream of GCSs: B = channel width, L = distance among GCSs, q = discharge per unit width of channel, x_s = position of maximum scour depth at equilibrium, z_s = maximum scour depth at equilibrium, A-A = cross-section at x_s. This image qualitatively indicates the transversal locations ($y/B = -0.43, -0.25, 0, 0.25, 0.43$) where the vertical profiles of flow velocities were measured.

To check the development of secondary current cells within the scour hole, detailed measurements of the flow velocity field were carried out through cross-sectional planes of the scour at equilibrium condition. The velocity data were collected using a 3-D Acoustic Doppler Velocimeter (ADV) system, developed by Nortek, with a sampling rate of 25 Hz and for a time windows of 3 min. The sampling volume of the ADV was located 5 cm below the transmitter probe. The ADV was used with a velocity range equal to $\pm$0.30 m/s, a measured velocity accuracy of $\pm$1%, and a sampling volume of less than 0.25 cm^3. For high-resolution measurements, the manufacturer recommended a 15 dB signal-to-noise ratio (SNR) and a correlation coefficient larger than 70%. The acquired data were filtered based on the Tukey method, and bad samples (SNR < 15 dB and correlation coefficient < 70%) were also removed. Additional details concerning the ADV system operations can be found in [29–33]. Flow velocity measurements through the scour hole were carried out for different configurations in both the longitudinal plane of symmetry and in some transversal planes, such as the cross-section A-A in Figure 1.

Table 1 reports the hydraulic parameters of the experiments, where U_s is the mean flow velocity at the downstream location, x_s, of maximum equilibrium scour depth (in correspondence of h_s), h_s is the axial ($y = 0$) flow depth in the equilibrium scour hole at x_s, u^* is the mean bottom friction velocity, $A_r = B/hs$ is the channel aspect ratio at x_s, $F_{ds} = U_s/(gh_s)^{0.5}$ is the Froude number at x_s, $Re_s = U_sR_s/\nu$ is the Reynolds number at x_s, and $R_s = Bh_s/(B + 2h_s)$ is the hydraulic radius at x_s. For the sake of brevity, in this study, we focus in detail on the data of tests T10, T12, and T13 for the analysis of the turbulent parameters. Further information on the experiments can be found in Ben Meftah et al. [17].

Table 1. Hydraulic parameters of the investigated tests.

Tests	q (m²/s)	U_s (m/s)	x_s (m)	z_s (m)	h_s (m)	u^* (m/s)	A_r (-)	F_{ds} (-)	Re_s (-)
T10	0.027	0.227	0.54	0.057	0.123	0.022	2.44	0.20	14,270
T12	0.040	0.233	0.70	0.081	0.171	0.024	1.76	0.17	18,577
T13	0.046	0.236	0.73	0.090	0.194	0.025	1.55	0.18	20,466

4. Results and Discussion

4.1. Flow Velocities in Scour Cross-Sections

Figure 2 shows examples of the flow velocity distribution, V_{yz}, across the equilibrium scour hole at x_s for tests T10, T12, and T13. The V_{yz} velocity is the resultant of the span-wise V and vertical W time-averaged velocity components. The profiles of the free surface flow (solid line), the initial bed (dash-dotted line), and the bed at the equilibrium state (bullet) are also reported in Figure 2. The rest of the sediment at the channel bottom is indicated by a random point cloud. The transversal position y is normalized by the channel width B, and the vertical position z is normalized by the maximum scour depth at equilibrium z_s. In Figure 2, the coordinates y and z originate from the channel axis ($y/B = 0$) and the initial bed profile (dash-dotted line), respectively. The maximum scour depth z_s is determined based on the longitudinal profile of the equilibrium scour at the channel axis ($y/B = 0$), downstream of the wooden ramp (Figure 1). The absence of velocity measurements in the upper flow region is due to the limitation of the ADV down-looking probe, being that the uppermost 7 cm of the flow could not be sampled.

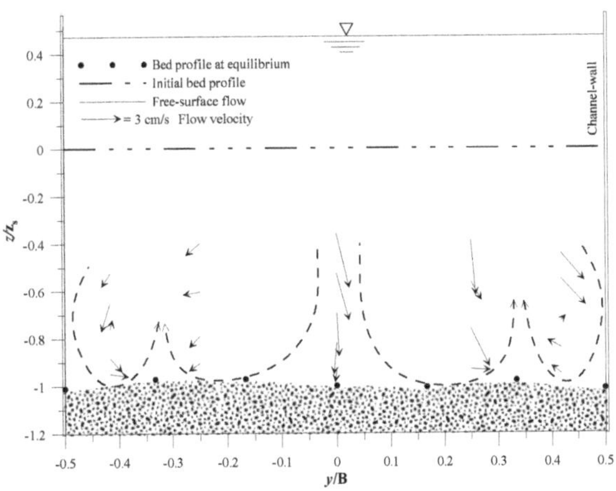

(a)

Figure 2. *Cont.*

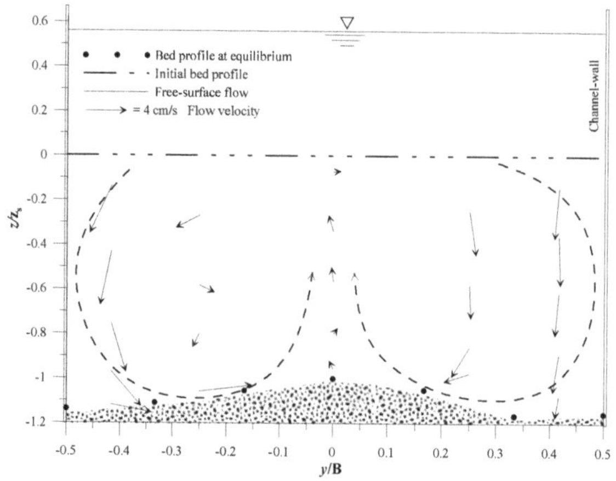

(b)

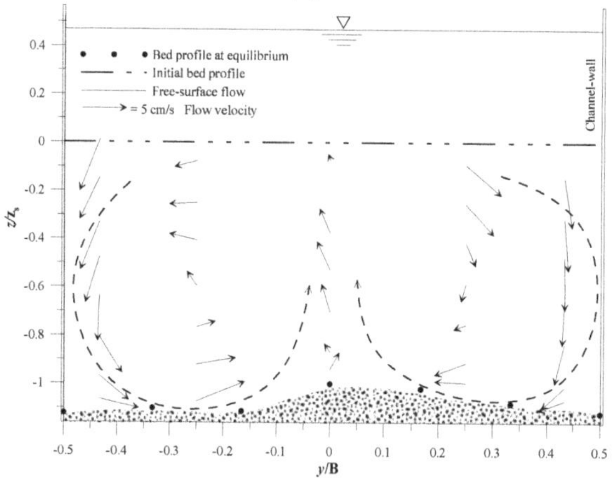

(c)

Figure 2. Vector map of the flow velocity, V_{yz}, across the equilibrium scour hole at x_s (location of maximum scour depth downstream of the GCS): (**a**) test T10 of z_s = 0.057 m; (**b**) test T12 of z_s = 0.081 m; (**c**) test T13 of z_s = 0.090 m. The random point cloud represents the remaining amount of sediment from the channel bottom at scour equilibrium.

To investigate the hydrodynamic structure across the scour hole well, a high-resolution time series of the flow velocities with high quality is required. The magnitudes of the flow velocity vectors reported in Figure 2 are the result of over 4500 measurements of the instantaneous velocity components. Figure 3, as an example, shows the velocity skewness and kurtosis distribution of all measured velocities across the scour hole at x_s for tests T12 and T13. Figure 3a shows that the skewness of the three velocity components u, v, and w was between −0.5 and 0.5 in most measuring positions. This implies that the measured velocity data were symmetrical. Figure 3b shows that u, v, and w kurtosis saturated to a value around 3. These results are similar to the Gaussian case (skewness = 0 and kurtosis = 3), indicating that most of the values of each measured velocity component clustered around the mean value, which means that the ADV instrument was performing satisfactorily.

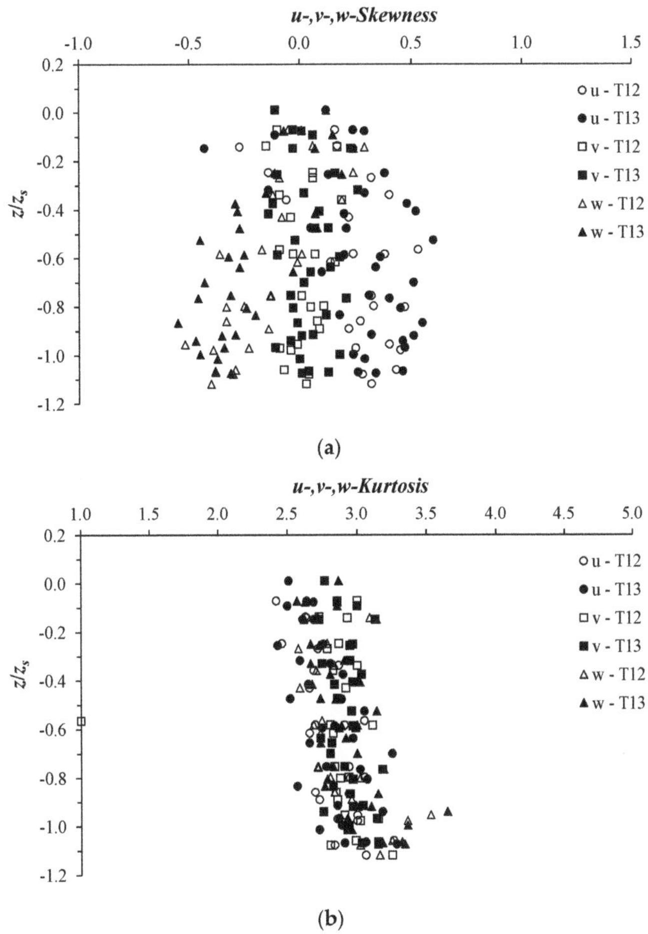

Figure 3. Velocity skewness and kurtosis: (**a**) u, v, w skewness; (**b**) u, v, w kurtosis.

In Figure 2, the flow velocity distributions in the scour cross-sections (looking downstream) clearly show the development of secondary current cells. The secondary current cells represent a circulation of fluids in the plane orthogonal to the axis of the primary flow. Figure 2 shows the formation of a helical motion rotating in opposite directions, as indicated by the curved arrow made of dashed lines in the figure. For test T10, Figure 2a shows the formation of a series of secondary current cells across the scour hole, similarly to what happens in large straight channels [20]. The four helical cells shown in the figure are symmetric with respect to the (x, z) plane at $y/B = 0$ (plane of flow symmetry). Figure 2a also shows an undulating sediment profile (manifested by two sediment ridges at $y/B = -0.25$ and 0.25) in the transverse direction. According to previous studies [20,21], the undulating sediment profile, generated by secondary current cells, may affect both the whole depth-flow field and the free surface flow pattern. Figure 2b,c indicates the formation of only two large counter-rotating vortices and a single sediment ridge, located around the channel axis ($y/B = 0$). For tests T12 and T13, the vortices pushed the water outward (towards the channel wall) in the upper part of the scour cross-section, while, near the bottom, the water was driven inward [19], forming the sediment ridge at the center of the channel. According to previous studies (e.g., [20,21]), the hydrodynamics of secondary currents is strongly influenced by the aspect ratio, A_r (Table 1), and the number of secondary current cells varies in proportion to the latter. This could explain the increase in the number of secondary current cells with test T10, of A_r = 2.44, compared to tests T12 and T13, of A_r = 1.76 and 1.55, respectively. Figure 2 shows that the width of the secondary current cells is $0.6h_s$, $0.9h_s$, and $0.8h_s$ with tests T10, T12, and T13, respectively. This finding agrees with that found in previous studies [20,34], claiming that the width of secondary current cells scales with the flow depth.

Figure 4 shows the vertical profiles of the normalized velocity U/u^* at the transversal positions $y/B = -0.43, -0.25, 0, 0.25$, and 0.43 across the scour hole at x_s. Here, the friction velocity u^* was determined based on the measured primary Reynolds shear stresses $U'W'$ near the scour bed, and $Z/h_{sy} < 0.2$, $u^* = (-U'W')^{0.5}$, Z, and h_{sy} are defined below. The vertical position z is normalized by z_s, and originates at the initial bed profile, as shown in Figure 2. The data refer to test T12 (Figure 4a) and test T13 (Figure 4b). At the different transversal positions y/B, U/u^* shows the same trend along the vertical. It is considerably reduced approaching the scour bed. For both tests, the profiles of U/u^* are shifted from each other, indicating an anisotropic transversal distribution. At the downwelling flow regions, at $y/B \leq -0.25$ and $y/B \geq 0.25$ (see Figure 2), the stream-wise velocity shows the largest values, while it is reduced at the upwelling flow region, around $y/B = 0$. This result is in good agreement with that found by Albayrak and Lemmin [20] and Stoesser et al. [27], confirming the significant influence on the primary flow by the secondary current cells.

Figure 5a shows an example of the normalized span-wise velocity distribution V/u^* at the transversal locations $y/B = -0.43, -0.25, 0, 0.25$, and 0.43, across the scour hole at x_s. For the sake of brevity, due to the similar behavior of V/u^* observed with the different tests, only the profiles of test T13 are included in Figure 5a. At $y/B = 0$, V/u^* shows small values tending to zero. This is consistent with the fact that, in the plane of flow symmetry ($y/B = 0$), the transversal velocity component is theoretically null. Moving right or left from the plane of flow symmetry, V/u^* shows considerable magnitudes. The vertical profiles clearly indicate that V/u^* changes sign from negative to positive, going towards the scour bed, in the right half of the channel ($y/B < 0$) and vice versa in the left half, forming an x-shape with the different vertical profiles. This behavior of V/u^* is a result of the development of secondary current cells across the scour hole. The x-shape of the different profiles is due to the symmetry of the vortices with respect to the (x, z) plane at $y/B = 0$, as shown in Figure 2c. To highlight the important role of secondary currents of the second kind in flow dynamics, a comparison of the data in the present study to field data of secondary currents of the first kind, which is usually generated in natural watercourses, is of high interest. Figure 5b shows a comparison between the vertical V/u^* profiles of T13, as shown in Figure 5a, and those obtained by Papanicolaou et al. [22] on a natural river, characterized by a sequence of channel expansions and contractions along its length. It is worth mentioning that the data of Papanicolaou et al. [22] were collected along a cross-section of a natural channel of arbitrary geometry (see Figure 3 in this cited study) without GCS. Seven vertical velocity profiles were taken along the cross-section at different transversal locations A-G, using a SonTek ADV. Analysis of the vector field of the flow velocity, V_{yz}, in this cross-section indicates the formation of a single counterclockwise flow circulation, occupying the right half ($y/B > 0$) of the river (see Figure 7 in Papanicolaou et al. [22]). For the sake of comparison, in Figure 5b, the vertical position Z (capital) originates from the channel/scour bed, as in Papanicolaou et al. [22], and is normalized by h_{sy}, the local flow depth at a given transversal position y. Applying the same procedure in normalizing the transverse direction by the width, B, of the cross-section, as in the present study, where $y = 0$ at the center of the cross-section, the profiles Pr. A, Pr. D, Pr. E, and Pr. G, obtained by Papanicolaou et al. [22], correspond to $y/B = -0.34$, 0, 0.16, and 0.40, respectively. Figure 5b shows that V/u^* in the present study behaves quite similarly to that obtained by Papanicolaou et al. [22]. Moreover, the order of magnitude of V/u^* within the secondary current cells for both studies seems comparable, as shown by the profiles at the counterclockwise vortices ($y/B < 0$ for the present study and Pr. E and Pr. G in Papanicolaou et al. [22]). The secondary current in Papanicolaou et al. [22]

(of the first kind) was strongly influenced by the distortion of the primary flow, caused by an upstream bend (convex to the right riverbank) of the channel. The distortion of the primary flow underlies the formation of a single counterclockwise flow vortex, located near the right side of the channel. This could also explain the considerable increase of V/u^* in the left half of the cross-section at Pr. A and Pr. D ($y/B = 0$), contrary to what happens in the present study conducted in a straight channel, where $V/u^* \approx 0$ at $y/B = 0$, due to the flow symmetry.

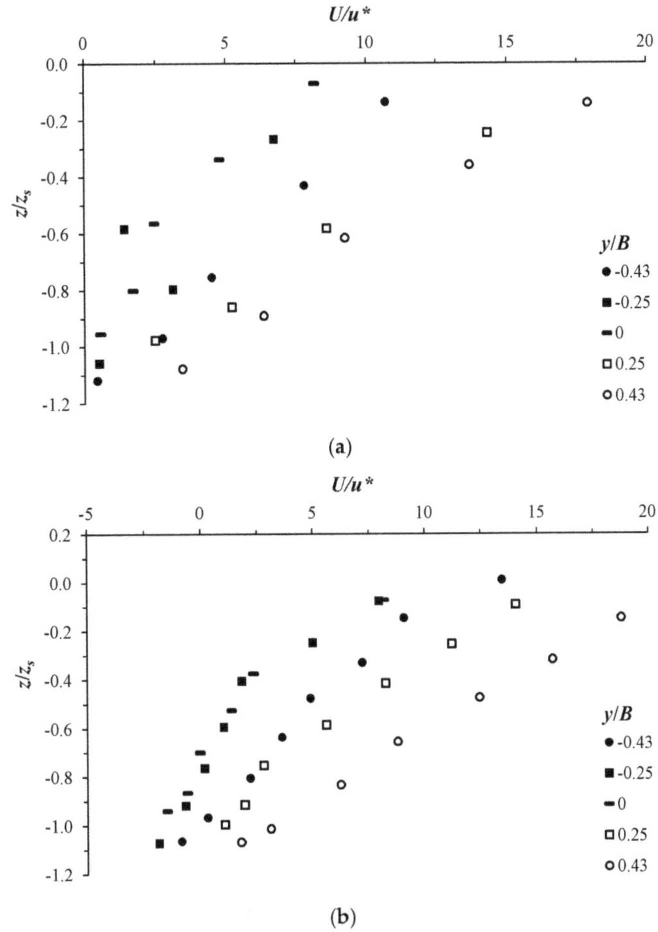

Figure 4. Vertical profiles of the normalized stream wise velocity U/u^* at different transversal position y/B across the equilibrium scour hole at x_S: (**a**) test T12; (**b**) test T13.

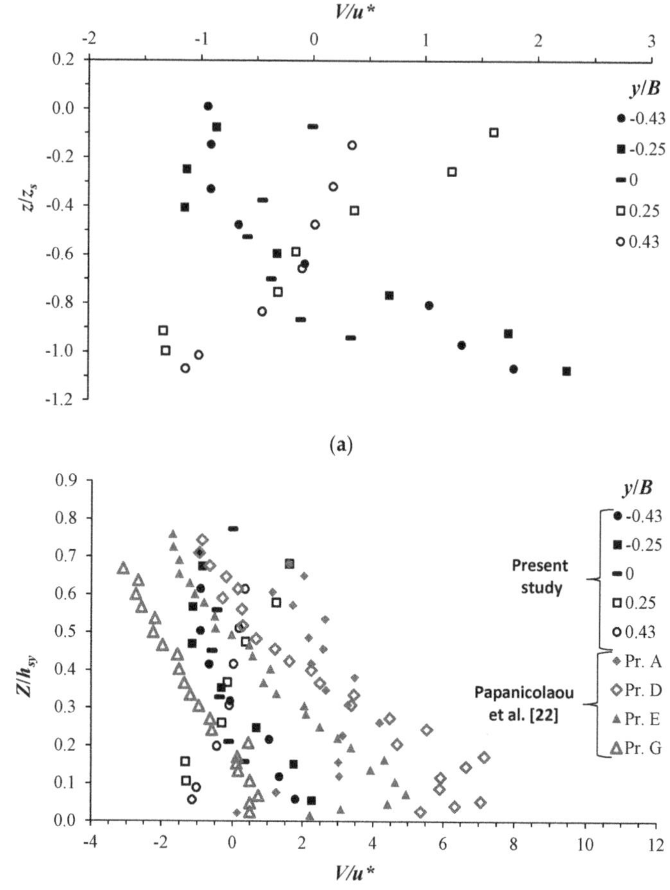

(a)

(b)

Figure 5. Normalized span-wise velocity V/u^* at different transversal positions y/B across the equilibrium scour hole at x_s: (**a**) data refer to test T13, and the vertical position z (small) originates from the initial bed profile and is normalized by z_s, as in Figure 2; (**b**) comparison between the data of the present study of test T13 and the data obtained by Papanicolaou et al. [22] on a natural river. Here, the vertical coordinate Z (capital) originates at the scour/channel bed and is normalized by the flow depth h_{sy}. The profiles Pr. A, Pr. D, Pr. E, and Pr. G were taken at $y/B = -0.34$, 0, 0.16, and 0.40, respectively.

Figure 6 shows the profiles of the normalized vertical velocity W/u^* for the present study and that by Papanicolaou et al. [22]. Figure 6a illustrates the profiles of W/u^* of test T13 of the present study in the scour hole at the downstream position x_s. To clearly show the flow features in the scour hole, we also plotted W/u^* versus z/z_s (as explained above). In the downwelling flow regions ($y/B < -0.25$ and $y/B < 0.25$) (see Figure 2), W/u^* undergoes negative values, varying in an arched way (of concavity to the right) along the vertical: it increases in magnitude with increasing depth, reaches extreme values of O(−3) at $z/z_s \approx -0.5$, and then begins to gradually decrease (in magnitude), tending toward zero close to the scour bed. In the upwelling flow region (around $y/B = 0$), W/u^* shows positive values, behaving in a similar fashion (in an arcuate way, but with concavity to the left) along the vertical as in the downwelling flow regions. In the intermediate regions ($y/B = -0.25$ or 0.25, as an example), W/u^* can change sign from negative to positive and vice versa along the vertical. This behavior of W/u^* through the scour hole is a result of the flow circulation caused by the secondary current cells. Figure 6b represents a comparison between the W/u^* profiles of test T13 of the present study and those by Papanicolaou et al. [22]. Here, W/u^* is plotted versus Z/h_{sy} (as explained above). The profiles shown in Figure 6b almost indicate the same order of magnitude of W/u^* for both studies, ranging between about −3 and 3. At relative transversal positions within a secondary current cell, Figure 6b also reveals that W/u^* follows a similar trend along the vertical direction for both studies. The different vertical profiles of W/u^* relative to a single secondary current cell show a sort of φ-shaped distribution.

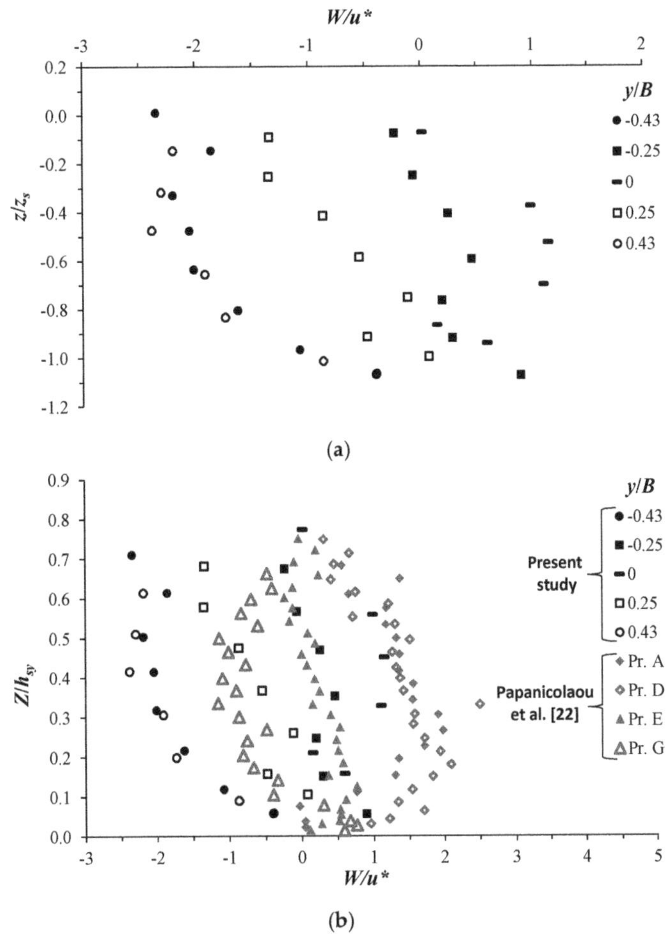

Figure 6. Normalized vertical velocity W/u^* at different transversal positions y/B across the equilibrium scour hole at x_s: (**a**) W/u^* of test T13 plotted versus z/z_s; (**b**) comparison between W/u^* of test T13 of the present study and that of Papanicolaou et al. [22], considering the vertical coordinate system Z/h_{sy}.

4.2. Turbulence Intensities and Shear Stress Across Equilibrium Scour Hole

The profiles of the flow turbulence intensities at the different transversal locations $y/B = -0.43, -0.25, 0, 0.25$, and 0.43 across the equilibrium scour hole at the downstream position x_s are shown in Figure 7. The normalized turbulence intensities in the stream wise direction, U'^*, span-wise direction, V'^*, and vertical direction, W'^* for tests T12 and T13 are plotted in Figure 7a–f, respectively. Here, U'^*, V'^*, and W'^* are defined as the ratio of the standard deviation of the stream-wise, span-wise, and vertical flow velocity component fluctuations in the friction velocity, u^*, respectively. To show the profiles inside of the scour hole, in Figure 7, we adapt the normalized vertical coordinate z/z_s (as explained in Section 4.1).

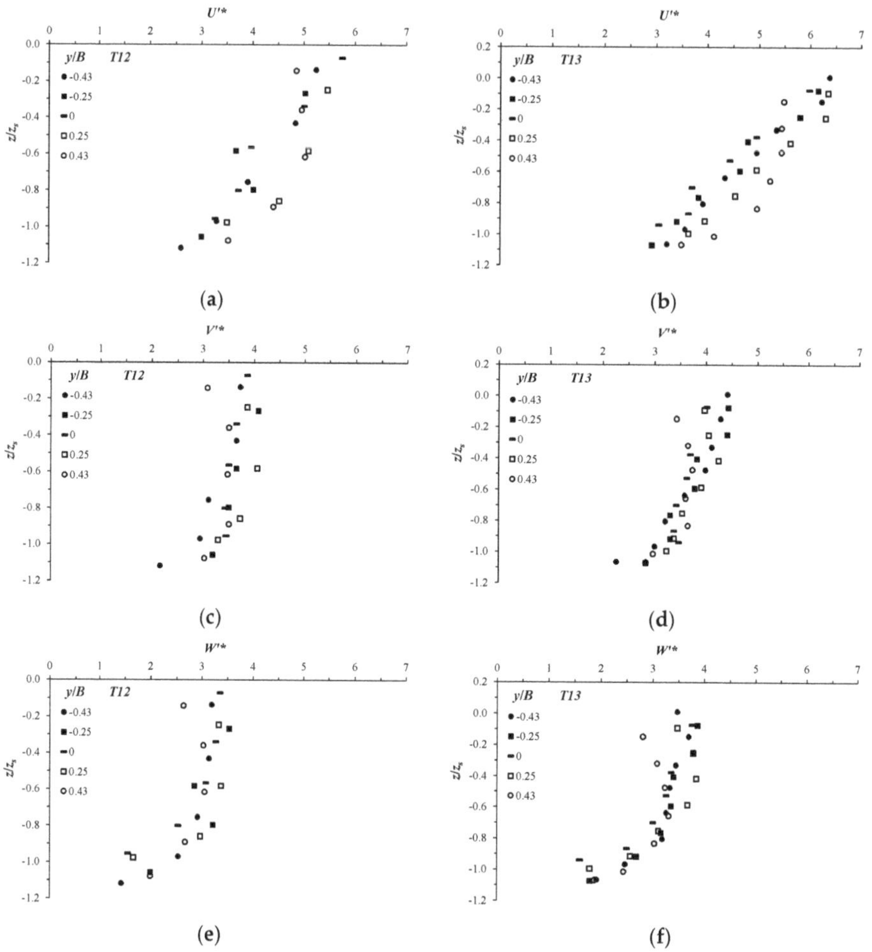

Figure 7. Vertical profiles of the flow turbulence intensities at different transversal locations across the equilibrium scour hole at x_S: (**a**,**b**) stream-wise turbulence intensity U'^* for tests T12 and T13, respectively; (**c**,**d**) span-wise turbulence intensity V'^* for tests T12 and T13, respectively; (**e**,**f**) vertical turbulence intensity W'^* for tests T12 and T13, respectively.

Figure 7a,b points out that the stream-wise turbulence intensity generally behaves in the same way at the different transversal locations along the water column in the scour hole. U'^* decreases going down towards the scour from values of O(6) and O(7) at $z/z_s \approx 0$ to values of O(2) and O(3) near the scour bed in tests T12 and T13, respectively. In the upwelling flow region, U'^* decreases almost linearly as z/z_s decreases. In the downwelling flow region, U'^* decreases in a curved way, which is more pronounced at y/B = 0.43. The profiles illustrated in Figure 7c–f also show, at first glance, a decreasing trend of V'^* and W'^* going to the scour bed. At $z/z_s \approx 0$, V'^* and W'^* have an order of magnitude almost half that of U'^*. Contrary to what happens with U'^*, V'^* and W'^* seem to change more gradually with the distance from the initial bed profile at $z/z_s = 0$. Looking closely at Figure 7c–f, it can be noted that V'^* and W'^* change their magnitudes in an arched way along the vertical; it slightly increases starting from $z/z_s \approx 0$, peaks at almost $z/z_s = -0.5$, and then begins to decrease reaching the scour bed. This is more pronounced at the downwelling flow regions, especially with W'^*. The trend of variation of V'^* and W'^* is physically consistent with the behaviors of V/u^* and W/u^*, as shown in Figures 5 and 6. Similar results of the flow turbulence intensity behaviors were achieved in Chang et al. [35], studying the scour process downstream of a groundsill.

In Figure 8, we plotted the vertical profiles of U'^*, V'^*, and W'^* of test T13, together with those conducted by Papanicolaou et al. [22]. Note that, for the sake of consistency between both studies, we adopted in these figures the normalized vertical coordinate system Z/h_{sy}. In Figure 8, U'^*, V'^*, and W'^* show similar trends along the water column. Furthermore, U'^*, V'^*, and W'^* experience quite comparable values, despite the difference in hydraulic and geometric conditions among both studies. A shifting between the profiles of turbulence intensities occurred for both studies, which is pronounced with the data of Papanicolaou et al. [22], possibly due to the strong bending effect. Contrary to conventional findings in open channels, Figures 7 and 8 show that the three turbulence intensities U'^*, V'^*, and W'^* decrease as the channel bottom approaches. This means that the momentum exchange occurs from the bottom towards the surface and not the opposite, as usually confirmed by conventional findings, which may be attributed to the presence of secondary flows [22].

The analysis of the normalized shear stresses across the equilibrium scour hole at x_s was also performed. Figure 9 reports the vertical profiles of the normalized Reynolds shear stresses at different transversal locations. The measured Reynolds shear stresses, $U'_iU'_j$ (see Section 2) were normalized by the square friction velocity, u^{*2}, and for simplicity, they are denoted as $(U'_iU'_j)^*$. Figure 9a,b shows the profiles of $(U'V')^*$ for tests T12 and T13, respectively. Figure 9c,d shows the profiles of $(U'W')^*$ for tests T12 and T13, respectively. Figure 9e,f shows the profiles of $(V'W')^*$ for tests T12 and T13, respectively. The shear stresses in Figure 9 indicate similar behavior to the turbulence intensities. They all show a tendency of decreasing going further towards the scour bed. Figure 9a,b shows that the span-wise shear stress $(U'V')^*$ is of considerable magnitudes through the scour hole; it is of magnitudes almost half those of the primary shear stress $(U'W')^*$, shown in Figure 9c,d. The presence of positive and negative values of $(U'V')^*$ indicates the flow symmetry with respect to the vertical (x, z) plane at $y/B = 0$. Theoretically, at $y/B = 0$, $(U'V')^*$ must be null, which is difficult to reach experimentally. The development of secondary currents in the scour hole contributes to a significant momentum transfer in the transverse direction [27]. The secondary (cross-plane) shear stress $(V'W')^*$, as illustrated in Figure 9e,f, shows the smallest strengths through the scour cross-section, which are two to four times lower than those of $(U'V')^*$ and $(U'W')^*$. By looking at each profile separately in Figure 9, one can note that the shear stresses show one or more peaks along the vertical, showing in general a curved profile of left or right concave distribution, which agrees well with previous studies (e.g., [20,21]). Figure 9 highlights that the three shear stress components slightly increase when comparing tests T13 with T12, which reflects a proportionality to the Reynolds number Re_s (Table 1). Figure 9 also points out that the shear stresses $(U'V')^*$, $(U'W')^*$, and $(V'W')^*$ exhibit a slight increase in the downwelling flow regions compared to the upwelling regions. This behavior agrees with that observed by Albayrak and Lemmin [20] in the wall region. Albayrak and Lemmin [20] noted an opposite behavior, away from the wall region, where $(U'W')^*$ is larger in the upwelling flow regions than that in the downwelling regions. This may imply, due to the channel's narrowness in the present study, that the flow in the equilibrium scour hole behaves similarly to that in the wall region in straight open channels with uniform flows (without scours).

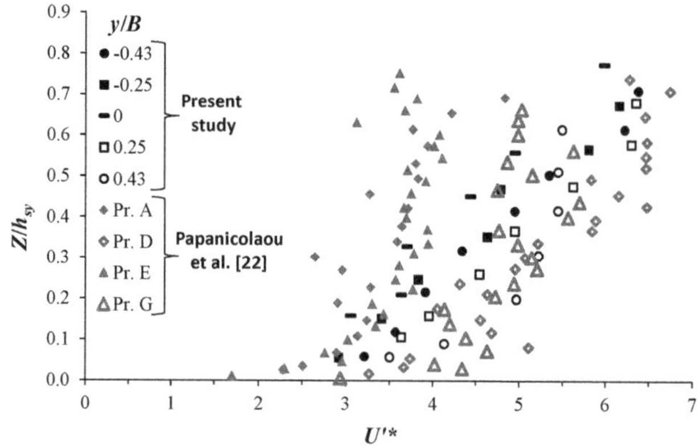

(a)

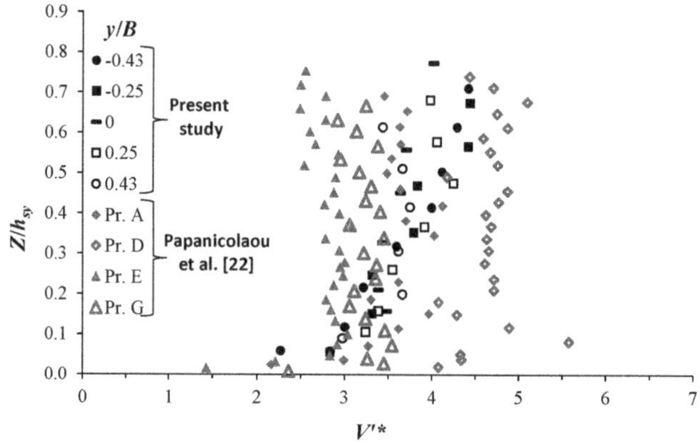

(b)

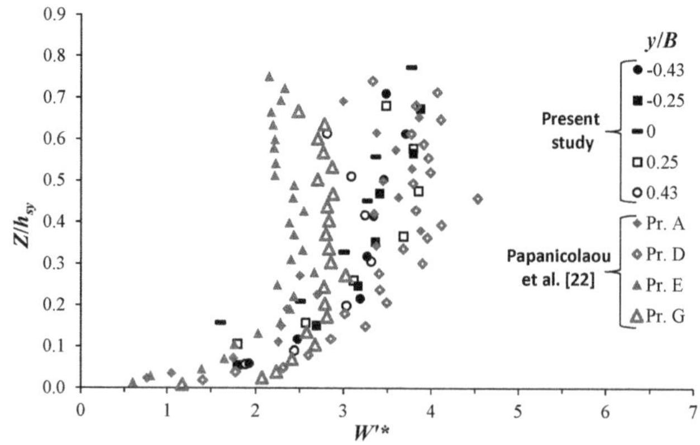

(c)

Figure 8. Comparison of the flow turbulence intensities of test T13 in the present study and those in Papanicolaou et al. [22]: (**a**) the stream-wise turbulence intensity U'^*; (**b**) the span-wise turbulence intensity V'^*; (**c**) the vertical turbulence intensity W'.

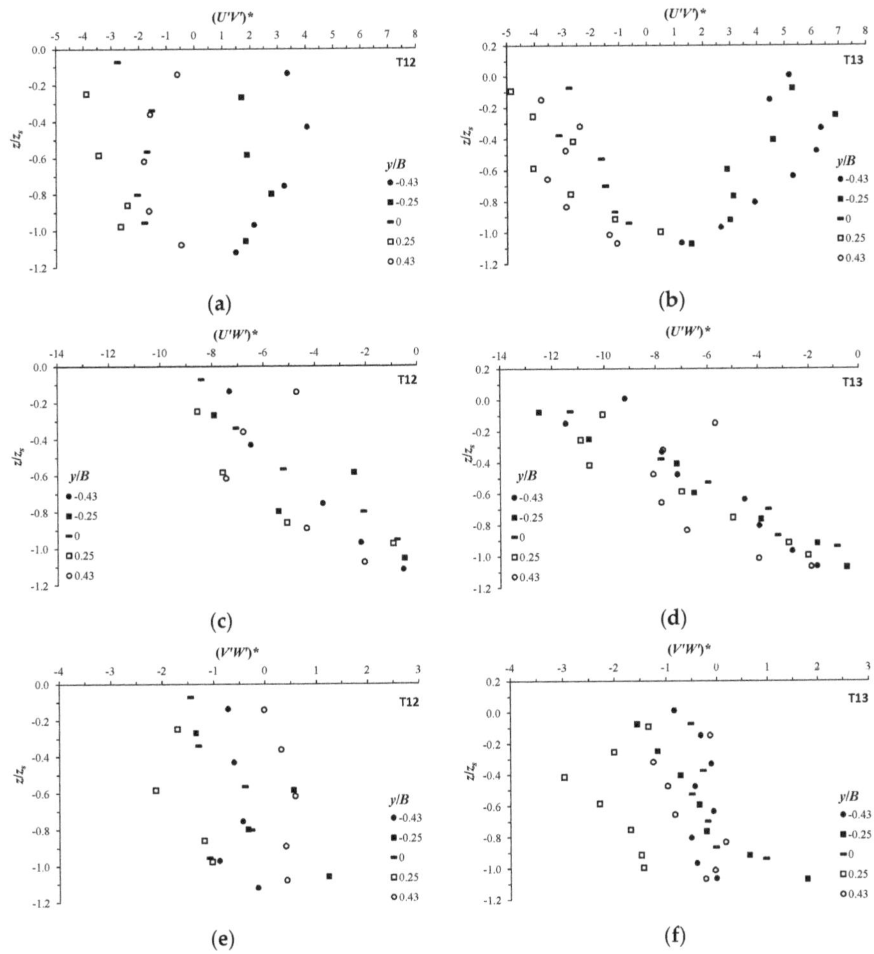

Figure 9. Reynolds shear stresses normalized by the square friction velocity at different transversal positions y/B across the scour hole at x_s: (a,b) span-wise shear stress $(U'V')^*$ for tests T12 and T13, respectively; (c,d) primary shear stress $(U'W')^*$ for tests T12 and T13, respectively; (e,f) secondary shear stress $(V'W')^*$ for tests T12 and T13, respectively.

Figure 10 presents a comparison between the shear stress $(U'W')^*$ of test T13 of the present study and that obtained by Papanicolaou et al. [22]. Here, we adopt the normalized vertical coordinate system Z/h_{sy} (as explained above). Figure 10 indicates that, for both studies, $(U'W')^*$ profiles have a similar tendency, especially over $Z/h_{sy} < 0.3$, where the data of both studies tend to collapse together. Furthermore, at $Z/h_{sy} < 0.3$, $(U'W')^*$ in Papanicolaou et al. [22] shows the largest strengths in the downwelling regions (Pr. G) compared to that in the upwelling regions (Pr. D and Pr. E). The variation of the Reynolds shear stress values along the transverse direction, between downwelling and upwelling flow regions, reflects the oscillatory nature of the flow induced by the secondary currents, as also confirmed by Albayrak and Lemmin [20].

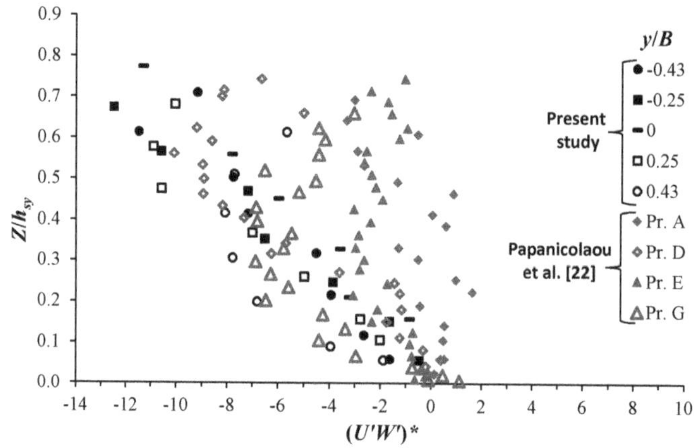

Figure 10. Comparison of the primary Reynolds shear stress $(U'W')^*$ of test T13 in the present study with those in Papanicolaou et al. [22]. Here, we adopt the normalized vertical coordinate Z/h_{sy}.

4.3. Contribution of Normal Stress Spatial Anisotropy to Stream-Wise Vorticity

Since the generation of the second kind of secondary currents in the cross-section plane of a straight channel is induced by the inhomogeneity of the Reynolds normal and shear stresses (terms A4, A5, and A6 in Equation (4)), as cited in several previous studies (e.g., [20,26]), in Figure 11, we illustrate the vertical profiles of the anisotropy term of the normal stresses $(V'^2 - W'^2)^*$ at different transversal locations y/B. Here, by anisotropy, we are addressing the spatial heterogeneities of $(V'^2 - W'^2)^*$. This term is normalized by the square friction velocity as: $(V'^2 - W'^2)^* = (V'^2 - W'^2)/u^{*2}$. It is worth mentioning that, according to the literature (e.g., [28]), the driving mechanism of stress-induced secondary flows only induces secondary currents in straight non-circular turbulent channel flows, not in laminar channel flows. Figure 11 shows considerable values of the anisotropy term of the normal stresses for both T12 and T13 tests. The profiles of $(V'^2 - W'^2)^*$ indicate that the latter varies considerably, both transversely and vertically. This implies the significant contribution of the term A5, as shown in Equation (4), to stream-wise vorticity generation, confirming previous findings in rough-bed, open-channel flows [27]. By treating each profile separately, it can be noted that $(V'^2 - W'^2)^*$ shows a maximum value near the scour mouth, and at $z/z_s = 0$, it decreases, reaching a minimum at a vertical position of $-0.8 < z/z_s < -0.6$, and then increases, attaining another maximum near the scour bed. This tendency is more pronounced with test T13, and is in good agreement with that shown in a previous study by Stoesser et al. [27].

For the sake of simplicity, if we neglect all gradients with respect to the stream-wise direction, i.e., $\partial/\partial x = 0$ (which is physically valid only along a small strip dx around the downstream location of maxim equilibrium scour depth x_s), the first addends in terms A1 and A3 disappear, while the term A4 vanishes completely. After some additional mathematical operations, the term A2 reduces to zero. After these assumptions, Equation (4) of the mean stream-wise vorticity for turbulent straight non-circular channel flows becomes:

$$\underbrace{V\frac{\partial \omega_x}{\partial y} + W\frac{\partial \omega_x}{\partial z}}_{A1} = \underbrace{\nu\left(\frac{\partial^2 \omega_x}{\partial y^2} + \frac{\partial^2 \omega_x}{\partial z^2}\right)}_{A3} + \underbrace{\frac{\partial^2}{\partial y \partial z}\left(V'^2 - W'^2\right)}_{A5} - \underbrace{\left(\frac{\partial^2}{\partial y^2} - \frac{\partial^2}{\partial z^2}\right)V'W'}_{A6} \quad (5)$$

The remaining terms in Equation (5) represent a balance between the convection process (term A1), the diffusion process (term A3), production process (term A5), and suppression process (term A6). The convection by secondary flow serves the transport of vorticity from production regions to the regions of diffusion by viscosity (destruction of vorticities) [28]. Previous studies [20,26] stated that the secondary current is enhanced by the production term A5 and suppressed by the term A6, which are of opposite signs, and are dominant with respect to the terms A1 and A3. Equation (5) reflects the highest importance of the production/suppression processes of vorticity, revealing the critical consideration and analysis of terms A5 and A6.

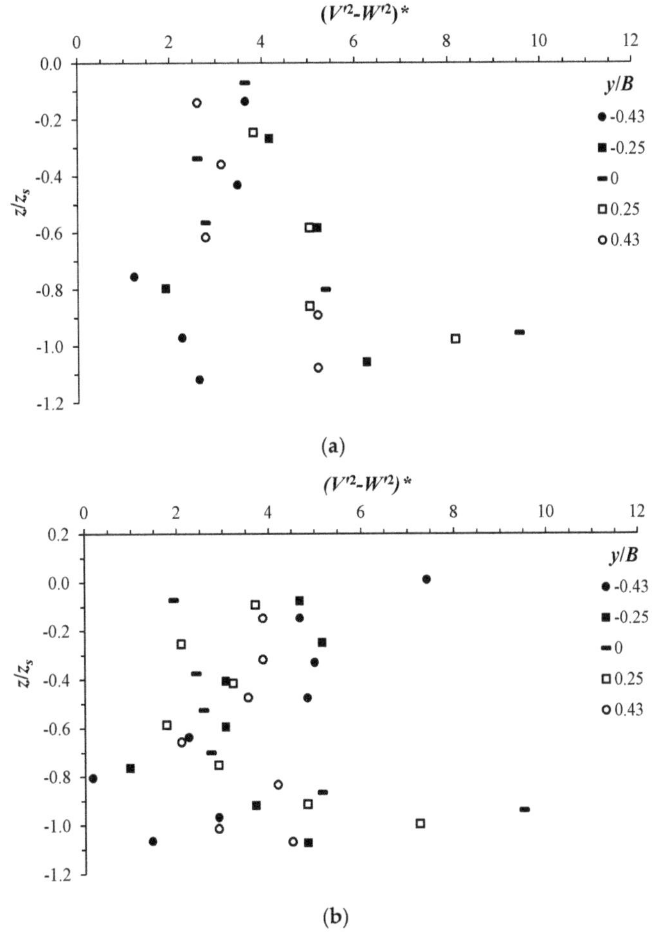

Figure 11. Vertical profiles of the normalized anisotropy term of normal stresses at different transversal locations y/B across the scour hole at x_S: (**a**) data of test T12; (**b**) data of test T13.

The comparison between the anisotropy term of the normal stresses $(V'^2 - W'^2)^*$, shown in Figure 11, and the secondary shear stress $(V'W')^*$, shown in Figure 9, indicates that they have opposite signs, and $(V'^2 - W'^2)^*$ has an order of magnitude larger than $(V'W')^*$, agreeing with most of the findings in the literature. Furthermore, $(V'^2 - W'^2)^*$ shows greater (y, z) spatial variation compared to $(V'W')^*$, obviously leading to higher gradients. In the literature (e.g., [26,28,34–36]), the condition for the generation of a secondary current in straight non-circular turbulent channel flows was considerably discussed, giving rise to two opinion groups. The first group suggested that the anisotropy term of normal stresses, A5, is dominant, providing an essential condition to induce the secondary currents. The second group argued that this term does not play a main role in the generation of secondary currents, but that the latter are the results of the variation of bed roughness or morphology. As indicated above, Figure 11 shows maximum values of $(V'^2 - W'^2)^*$ near both the scour mouth and scour bed, which is more pronounced with test T13 in Figure 11b. The large values of $(V'^2 - W'^2)^*$ in the vicinity of the scour mouth may be the results of high turbulence intensities induced by the incoming jet flow from the grade control structure [17], while those close to the bed are perhaps enhanced by the deformation of the transversal scour bed and development of sediment ridges [1,20,22]. A significant increase of $(V'^2 - W'^2)^*$ near the bed channel is also clearly distinguishable with the data of Papanicolaou et al. [22], as shown in Figure 12. Figure 12 reports a comparison among the data of test T13 in the present study and those of Papanicolaou et al. [22], plotted with the normalized vertical coordinate Z/h_{sy}. Since the secondary currents (of the first kind) in Papanicolaou et al. [22] were driven by a vortex tilting mechanism due to the channel curvature effect, the presence of high turbulence in the cross-sectional planes, generating an imbalance of normal Reynolds stresses, provides additional driving forces that maintain and enhance secondary flow motions.

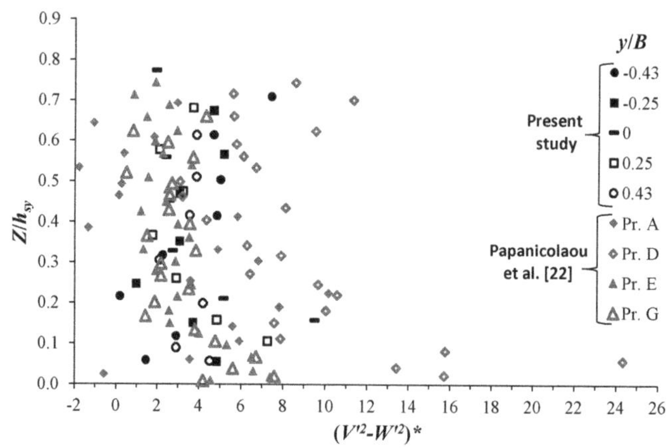

Figure 12. Comparison of the normalized anisotropy term of normal stresses of test T13 in the present study with those in Papanicolaou et al. [22]. Here, the data are plotted versus the normalized vertical coordinate Z/h_{sy}.

5. Conclusions

Secondary current structures across an equilibrium scour hole downstream of GCSs in erodible bed channels were investigated. The measurement of the flow velocities across the equilibrium scour hole, at the downstream position of its maximum depth x_s, clearly shows the formation of secondary current cells. The number of secondary current cells changes proportionally to the channel aspect ratio A_r at x_s. For $A_r < 2$, the secondary currents across the scour hole are manifested by two large counter-rotating vortices. For $A_r > 2$, there has been observed a development of multi-cellular secondary currents along the scour cross-section. The secondary current structure is accompanied by the appearance of upwelling and downwelling flow regions. These flow regions are responsible for the channel bed deformation across the scour hole. The upwelling regions are characterized by a slight reduction in the primary current velocity U, giving rise to the development of sediment ridges. The downwelling regions, on the other hand, show the greatest magnitude of primary current velocity, giving the space for trough region formation across the bed scour. The alternation of several regions of upwelling and downwelling flows, in the case of a high aspect ratio, gives rise to an undulating sediment profile in the transverse direction.

The distribution of the time-averaged, span-wise, V, and vertical, W, velocity components in the cross-sectional plane of the equilibrium scour hole at the location of maximum scour depth was analyzed, confirming the development of secondary current cells of the second kind (for straight and non-circular channels). The comparison of the normalized components V/u^* and W/u^* with those obtained in a previous study by Papanicolaou et al. [22], for secondary currents of the first kind (driven by the curvature effect) occurred in natural channels without GCS, shows comparable orders of magnitudes for both kinds, especially for W/u^*. For secondary currents of the first kind, V undergoes a slight increase, due to the considerable curvature effect, compared to that of the present study. The vertical profiles of V/u^* and W/u^*, taken at different transversal locations through a second current cell, indicate similar behavior of both components in the water column for both secondary kinds of currents.

The analysis of the turbulence intensities in the three flow directions across the equilibrium scour hole shows an anisotropic spatial distribution. The flow turbulence intensity in the stream-wise direction, U'^*, exhibits the largest values, twice greater than those of V'^* and W'^*, at the mouth of the scour depth ($z/z_s = 0$). Near the scour bed, U'^*, V'^*, and W'^* have minimum values of O (2). In the present study, U'^*, V'^*, and W'^* manifest similar behavior along the water column, and exhibit an order of magnitude through a secondary current cell relatively comparable to that in Papanicolaou et al. [22]. The decrease of U'^*, V'^*, and W'^* going towards the scour bed indicates that, contrary to conventional findings, the turbulence momentum transfer occurs from the bed towards the mouth of the scour depth, at $z/z_s = 0$.

The Reynolds shear stress distribution shows similar tendency to turbulence intensity. As compared with previous studies (e.g., [20–22]), the Reynolds shear stresses across the equilibrium scour hole at x_s show a similar distribution to the wall layer in normal (without scours) open channel flows. This implies that the secondary currents significantly affect the flow patterns and, therefore, the morphology of the scour bed. Stronger Reynolds shear stresses were observed in the downwelling flow regions, while the weakest were found within the upwelling flow regions. The transversal undulation of the Reynolds stresses strongly influences the mass transport, and consequently the distribution of the sediment particles across the scour hole. The decrease of the Reynolds stresses going towards the scour bed indicates the flow momentum transfer from the bed towards the mouth of the scour, which may explain the generation of vertical lifting forces, giving rise to particles moving upwards.

The physical origin of the secondary currents in straight non-circular channel flows was and remains a challenging task. In the present study, we tried to understand the development of secondary current cells across the equilibrium scour hole at x_s. A detailed analysis of turbulence properties was conducted. The turbulence distribution in the scour hole has shown considerable inhomogeneity. The anisotropy term of normal stresses has exhibited significant magnitudes, dominating the other terms in the mean stream-wise vorticity equation, which may reflect its potential effect in generating the secondary current motion. Furthermore, the anisotropy term of normal stresses has shown a substantial increase at the level of the scour mouth and especially near the scour bed. The interaction between the secondary currents and these boundary conditions (grade control structure and scour bed topography) may play a fundamental role in maintaining and enhancing secondary flow motion.

Author Contributions: M.B.M. performed the experiments, analyzed the data, designed the study, and wrote the paper; D.D.P., F.D.S., and M.M. contributed suggestions and discussions and reviewed the manuscript. All authors have read and agreed to the published version of the manuscript.

Funding: This research received no external funding.

Institutional Review Board Statement: Not applicable.

Informed Consent Statement: Not applicable.

Data Availability Statement: The data presented in this study are available on request from the corresponding author.

Acknowledgments: The experiments were carried out at the Hydraulic Laboratory of the Mediterranean Agronomic Institute of Bari (Italy).

Conflicts of Interest: The authors declare that they have no conflict of interest.

References

1. Yang, S.-Q. Influence of sediment and secondary currents on velocity. *Water Manag.* **2009**, *162*, 299–307. [CrossRef]
2. Carstens, M.R. Similarity laws for localized scour. *J. Hydraul. Div.* **1966**, *92*, 13–34.
3. Bormann, N.E.; Julien, P.Y. Scour Downstream of Grade-Control Structures. *J. Hydraul. Eng.* **1991**, *117*, 579–594. [CrossRef]
4. Balachandar, R.; Kells, J.A. Local channel in scour in uniformly graded sediments: The time-scale problem. *Can. J. Civ. Eng.* **1997**, *24*, 799–807. [CrossRef]
5. Gaudio, R.; Marion, A.; Bovolin, V. Morphological effects of bed sills in degrading rivers. *J. Hydraul. Res.* **2000**, *38*, 89–96. [CrossRef]
6. Lenzi, M.A.; Marion, A.; Comiti, F. Local scouring at grade-control structures in alluvial mountain rivers. *Water Resour. Res.* **2003**, *39*, 1176. [CrossRef]
7. D'Agostino, V.; Ferro, V. Scour on Alluvial Bed Downstream of Grade-Control Structures. *J. Hydraul. Eng.* **2004**, *130*, 24–37. [CrossRef]
8. Adduce, C.; Sciortino, G. Scour due to a horizontal turbulent jet: Numerical and experimental investigation. *J. Hydraul. Res.* **2006**, *44*, 663–673. [CrossRef]
9. Meftah, M.B.; Mossa, M. Scour holes downstream of bed sills in low-gradient channels. *J. Hydraul. Res.* **2006**, *44*, 497–509. [CrossRef]
10. Tregnaghi, M.; Marion, A.; Gaudio, R. Affinity and similarity of local scour holes at bed sills. *Water Resour. Res.* **2007**, *43*, W11417. [CrossRef]
11. Pagliara, S.; Amidei, M.; Hager, W.H. Hydraulics of 3D Plunge Pool Scour. *J. Hydraul. Eng.* **2008**, *134*, 1275–1284. [CrossRef]
12. Scurlock, S.M.; Thornton, C.I.; Abt, S.R. Equilibrium Scour Downstream of Three-Dimensional Grade-Control Structures. *J. Hydraul. Eng.* **2012**, *138*, 167–176. [CrossRef]
13. Lu, J.Y.; Hong, J.-H.; Chang, K.-P.; Lu, T.-F. Evolution of scouring process downstream of grade-control structures under steady and unsteady flows. *Hydrol. Process.* **2013**, *27*, 2699–2709. [CrossRef]
14. Manes, C.; Brocchini, M. Local scour around structures and the phenomenology of turbulence. *J. Fluid Mech.* **2015**, *779*, 309–324. [CrossRef]
15. Papanicolaou, A.N.T.; Bressan, F.; Fox, J.; Kramer, C.; Kjos, L. Role of Structure Submergence on Scour Evolution in Gravel Bed Rivers: Application to Slope-Crested Structures. *J. Hydraul. Eng.* **2018**, *144*, 03117008. [CrossRef]
16. Wang, L.; Melville, B.W.; Guan, D.; Whittaker, C.N. Local Scour at Downstream Sloped Submerged Weirs. *J. Hydraul. Eng.* **2018**, *144*, 04018044. [CrossRef]
17. Ben Meftah, M.; De Serio, F.; De Padova, D.; Mossa, M. Hydrodynamic Structure with Scour Hole Downstream of Bed Sills. *Water* **2020**, *12*, 186. [CrossRef]
18. Ben Meftah, M.; Mossa, M. New Approach to Predicting Local Scour Downstream of Grade-Control Structure. *J. Hydraul. Eng.* **2020**, *146*, 04019058. [CrossRef]
19. Njenga, K.J.; Kioko, K.J.; Wanjiru, G.P. Secondary Current and Classification of River Channels. *Appl. Math.* **2013**, *4*, 70–78. [CrossRef]
20. Albayrak, I.; Lemmin, U. Secondary Currents and Corresponding Surface Velocity Patterns in a Turbulent Open-Channel Flow over a Rough Bed. *J. Hydraul. Eng.* **2011**, *137*, 1318–1334. [CrossRef]
21. Nezu, I.; Nakagawa, H. *Turbulence in Open-Channel Flows*; IAHR monograph series; Balkema: Rotterdam, The Netherlands, 1993.
22. Papanicolaou, A.N.; Elhakeem, M.; Hilldale, R. Secondary current effects on cohesive river bank erosion. *Water Resour. Res.* **2007**, *43*, W12418. [CrossRef]
23. Rhoads, B.L.; Kenworthy, S.T. Time-averaged flow structure in the central region of a stream confluence. *Earth Surf. Process. Landf.* **1998**, *23*, 171–191. [CrossRef]
24. Lane, S.N.; Bradbrook, K.F.; Richards, K.S.; Biron, P.M.; Roy, A.G. Secondary circulation cells in river channel confluences: Measurement artefacts or coherent flow structures? *Hydrol. Process.* **2000**, *14*, 2047–2071. [CrossRef]
25. Ianniruberto, M.; Trevethan, M.; Pinheiro, A.; Andrade, J.F.; Dantas, E.; Filizola, N.; Santos, A.; Gualtieri, C. A field study of the confluence between Negro and Solimões Rivers. Part 2: Bed morphology and stratigraphy. *C. R. Geosci.* **2018**, *350*, 43–54. [CrossRef]
26. Yang, S.-Q.; Tan, S.K.; Wang, X.K. Mechanism of secondary currents in open channel flows. *J. Geophys. Res.* **2012**, *117*, 1–13. [CrossRef]
27. Stoesser, T.; McSherry, R.; Fraga, B. Secondary Currents and Turbulence over a Non-Uniformly Roughened Open-Channel Bed. *Water* **2015**, *7*, 4896–4913. [CrossRef]
28. Stanković, B.D.; Belošević, S.V.; Crnomarkovic, N.D.; Stojanovic, A.D.; Tomanovic, I.D.; Milicevic, A.R. Specific aspects of turbulent flow in rectangular ducts. *Therm. Sci.* **2016**, *20* (Suppl. 6), S1–S16. [CrossRef]
29. Meftah, M.B.; Mossa, M. Turbulence Measurement of Vertical Dense Jets in Crossflow. *Water* **2018**, *10*, 286. [CrossRef]

30. Meftah, M.B.; Malcangio, D.; De Serio, F.; Mossa, M. Vertical dense jet in flowing current. *Environ. Fluid Mech.* **2018**, *18*, 75–96. [CrossRef]
31. Meftah, M.B.; De Serio, F.; Mossa, M.; Pollio, A. Experimental study of recirculating flows generated by lateral shock waves in very large channels. *Environ. Fluid Mech.* **2008**, *8*, 215–238. [CrossRef]
32. Meftah, M.B.; Mossa, M. Partially obstructed channel: Contraction ratio effect on the flow hydrodynamic structure and prediction of the transversal mean velocity profile. *J. Hydrol.* **2016**, *542*, 87–100. [CrossRef]
33. Meftah, M.B.; Mossa, M. A modified log-law of flow velocity distribution in partly obstructed open channels. *Environ. Fluid Mech.* **2016**, *16*, 453–479. [CrossRef]
34. Kotsovinos, N.E. Secondary currents in straight wide channels. *Appl. Math. Model.* **1988**, *12*, 22–24. [CrossRef]
35. Chang, C.-K.; Lu, J.-Y.; Lu, S.-Y.; Wang, Z.-X.; Shih, D.-S. Experimental and Numerical Investigations of Turbulent Open Channel Flow over a Rough Scour Hole Downstream of a Groundsill. *Water* **2020**, *12*, 1488. [CrossRef]
36. Yang, S.-Q. Mechanism for initiating secondary currents in channel flows. *Can. J. Civ. Eng.* **2009**, *36*, 1506–1516. [CrossRef]

water MDPI

Article

Experimental Analysis of Velocity Distribution in a Coarse-Grained Debris Flow: A Modified Bagnold's Equation

Donatella Termini [1,*] and Antonio Fichera [2]

[1] Department of Engineering, University of Palermo, 90128 Palermo, Italy
[2] Faculty of Engineering, University Enna Kore, 94100 Enna, Italy; antonio.fichera@unikore.it
* Correspondence: donatella.termini@unipa.it; Tel.: +39-09123896522

Received: 26 February 2020; Accepted: 11 May 2020; Published: 15 May 2020

Abstract: Today, Bagnold's theory is still applied to gravity-driven flows under the assumption of uniform sediment concentration. This study presents findings of flume experiments conducted to investigate the velocity and concentration distributions within the debris body by using high-resolution images. The analysis has shown that the concentration and mobility of grains vary along the depth. A linear law to interpret the grains concentration distribution, starting from the knowledge of the packing concentration and of the surface concentration, Cs, has been identified. By considering such a law, modified expressions of the Bagnold's number and the velocity in stony-type debris flows are also presented. By using these expressions, three regimes of motion have been identified along the depth, and the velocity profile within the debris body is determined as a function of the parameter Cs. It has been verified that the velocity profiles estimated by using the modified equation compare well (mean square error less than 0.1) with the literature's measured profiles when Cs is correctly measured or estimated. Results of cutting tests, conducted for a sample of the used material, have also allowed us to verify that Cs could be determined as a function of the static friction angle of the material.

Keywords: debris flows; flow velocity; sediment concentration; prevision

1. Introduction

Coarse-grained debris flows represent the gravity-driven motion of granular sediment, which especially occurs in steep valleys and rivers [1,2]. These flows consist of rapid movements of mass that might determine forces of high-impact and, consequently, serious hazards, loss of human life, and structural damage.

The identification of laws governing hydrodynamics of flows in steep river reaches is less advanced than in lowland river systems [3]. This is especially related to the fact that different flow behaviors can be obtained, depending on the concentration and composition of sediments. Consequently, a wide range of coarse-grained debris flows, characterized by different internal stress distributions, can be observed [4–6]. The sediment transport processes in flows characterized by small sediment concentrations have been largely analyzed, and different approaches to treat the bed load and the suspended load can be found in the literature, among others [7–10]. In flows characterized by high sediment concentrations, such as in debris flows, these approaches are generally unsuitable [6]. It should also be considered that, depending on the composition of the mixture determining the dominant effect (i.e., either the inertial stress due to the interstitial fluid or the collision stress due to coarser sediments) in flow dynamics, different regimes of motion can be obtained [11–14]. According to Takahashi's [14] classification, bedload or suspended load and small values of collision stress can occur when the sediment concentration is less than about 0.02. When sediment concentration ranges from 0.02 to 0.2, the collision stress dominates only in the lower mixing layer of the debris body, and the so-called immature debris flow develops. For values of sediment concentrations larger than 0.2, collision stress becomes important in the entire depth, and the dynamic debris flow develops. In this last case, the grain mobility changes according to the quantity of finer sediment in the mixture. In particular, the stony-type debris flows, characterized by a low portion of finer sediment and a high percentage of coarse sediment (40%–75%), have a large resistance to motion and are dominated by collision stresses that are responsible for the dispersion of grains in the entire flow depth. In muddy-type debris flows and in viscous debris flows, which contain higher percentages of fines and a low portion of coarse sediment (around 10%–30%), the turbulent mixing stresses and the viscous stresses dominate, respectively [14,15].

The first conceptual model for interpreting the dynamics of water/sediment mixtures was formulated by Bagnold [16], who conducted experiments in a concentric cylinder rheometer to examine the effects determined by the dispersion of grains on flow by simultaneously measuring the acting shear and normal forces. Although Bagnold's [16] theory was based on simplified and restrictive experimental conditions (see as an example the exhaustive critical review of [17]), it introduced important and still actual concepts, such as the existence of two different regimes of flow motion (i.e., the macro-viscous and grain-inertia regimes). To identify the two regimes of motion, Bagnold determined a no-dimensional parameter, N_{Ba}, that is the ratio between the inertial grain stress and the viscous shear stress. The results of Bagnold's measurements were then used to analyze different processes such as the grains sorting, the flows in gravel beds, and the critical conditions under which a particle remains suspended [18,19]. Takahashi [14] considered a uniform layer of granular loose material and applied Bagnold's [16] constitutive equations to stony-type debris flow by integrating them under the assumption of constant grain concentration in the flow depth.

Especially in recent years, researchers have increased their interest in coarse-grained flows, attempting to identify more convincing conceptual models.

Many theoretical analyses to interpret the dynamics of coarse-grained mixtures can be found in the literature. Different aspects influencing the stress distribution and the motion mechanisms have been investigated, devoting particular attention either to debris flow generated along steep valleys by high run-offs (see, as examples, in [4,15,20–25]), or to the additional bedload transport occurring when runoff-generated debris flows entrain in erodible-bed channels (see, for example, [3,26]), or to the transition processes from the intense bedload transport due to turbulence (as it could occur in coarse beds) to dilute suspension, which also produce the bed and suspended loads in granular flows [27,28]. Some researchers [29–32] have also proposed to apply the kinetic theory to granular flows. Such an approach shows limited applicability because it requires large dimensions of the control volume, compared either with the sediment particle size or with their mean distance and is based on the equation of state, which is not directly related to the shear rate [6,32].

Due to the complexity of the fluid/sediment interactions in debris flow, the elaborated models generally include several empirical parameters and are not easily generalizable.

However, it is clear that the identification of the internal velocity distribution is a key feature in analyzing the dynamics of coarse-grained flow and in estimating the consequent impact forces.

To this aim, several experimental investigations have been conducted, especially in laboratory flumes, by using different typologies of course material, generally artificial, under different hydrodynamic conditions. Lanzoni [33] conducted a series of flume experiments to analyze the velocity profiles of a debris flow formed by a loose sediment layer either of crusher gravels or of glass spheres. Armanini et al. [34] reproduced the granular–liquid mixture by using artificial particles and analyzed the propagation phenomenon of mature debris flow under temporally and spatially uniform conditions. Sanvitale et al. [35] measured velocity profiles of a propagating fluid mixture, consisting of hydrocarbon oil and borosilicate glass of non-uniform size, over a fixed-bed laboratory flume. Iverson et al. [36] conducted experiments in a large-scale flume, releasing a saturated heterogeneous mixture of sand–gravel and sand–gravel–mud over a rough-fixed bed. Sarno et al. [37] measured the velocity profiles of a steady mixture consisting either of well-sorted round-shaped silica sand or of artificial plastic beads of constant diameter. Hsu et al. [38] and Kaitna et al. [13] considered different gravel mixtures saturated by water and mud. More recently, Fichera et al. [39] analyzed the velocity profiles within a debris flow consisting of crushed gravels with assorted diameters.

Although the aforementioned studies have allowed significant progress in understanding the velocity distribution and the different rheological behaviors of the coarse-grained mixtures, their dynamics have still not been completely defined today. This could be due both to the difficulty of reproducing the real propagation conditions of these flows and to the difficulty in identifying the characteristics of the fluid/sediment mixture accurately [40,41].

In such a context, although Bagnold's theory was based on simplified experimental conditions, inducing high uncertainty in the flow velocity estimation, it is still widely applied to gravity-driven flows, especially in numerical models. The point is that, while Bagnold's results were obtained with fixed concentrations and shear rates [14,16], the studies conducted in this field highlight that the velocity distribution, which depends on the variation of the shear rate within the debris body, can be in turn affected by the variation of the grain concentration in the flow depth.

This suggests that the major problem in the application of Bagnold's theory to gravity-driven flows could be related to the assumption of uniform sediment distribution in the entire depth of flow [42,43].

To the authors' knowledge, no systematic research has been conducted to explore the variation of the grain concentration within the debris body and its effect on the velocity profile. This information could be especially used for interpreting the flow dynamics in numerical models.

The present work is inspired by the aforementioned consideration. In particular, focusing the attention on stony-type debris flows, which are frequent in the Dolomites (Italian Alps) and in alpine catchments, especially during the summer period when high-intensity rainfall storms occur [44–46], the present study aims (1) to gain some insights on the variation of grains concentration within the debris body and its effect on the velocity profile; (2) to explore the applicability of Bagnold's theory to debris flows, according to Takahashi [14,42,43], by removing the hypothesis of uniform sediment concentration. This would allow us to obtain more accurate results in the application of Bagnold's theory to stony-type debris flows.

The analysis is conducted with the aid of data both appositely collected in a laboratory flume and available in the literature. The paper is organized as follows: Section 2 summarizes peculiar aspects of the application of Bagnold's theory to debris flows and describes the experimental apparatus; Section 3 presents the experimental results; discussion is reported in Section 4; finally, conclusions are drawn in Section 5.

2. Material and Methods

2.1. Pertinent Aspects of Bagnold's Theory and Its Application to Debris Flow

Before proceeding further, it seems important to briefly summarize here some peculiar aspects of Bagnold's theory and its application to debris flows.

As mentioned in the introduction, Bagnold, on the basis of experimental measurements and physical arguments, introduced the concept of two different regimes of motion: the "macro-viscous" regime, which occurs for small shear rates and is dominated by the viscosity, and the "grain-inertia" regime, which occurs for larger shear rates and is dominated by the inertial forces. To distinguish the two aforementioned regimes, Bagnold defined the dimensionless number (i.e., the so-called Bagnold's N_{Ba}) given by the ratio between the stresses due to the inertial forces and those due to the viscosity:

$$N_{Ba} = (\rho_s \gamma {d_p}^2 \lambda^{1/2})/\mu_f \quad (1)$$

where d_p indicates the particle diameter, ρ_s is the density of grains, μ_f is the fluid's dynamic viscosity, γ is the shear rate, and λ is the linear concentration of grains given by the following relation:

$$\lambda = \frac{1}{[C_*/C]^{1/3} - 1} \tag{2}$$

In Equation (2), C is the grain concentration, and C_* is the maximum possible concentration (i.e., the so-called packing concentration), which can be obtained in static conditions where the friction between the sediment particles is of primary importance [47,48].

Bagnold established that the macro-viscous regime occurs for very low Bagnold's numbers (i.e., for $N_{Ba} < 40$), while the grain-inertia regime occurs for high values of the Bagnold number (i.e., for $N_{Ba} > 450$). The intermediate-range of the Bagnold numbers indicates a transitional regime between the aforementioned ones. While in the macro-viscous regime, the stresses are determined by the interaction of the granular phase with the interstitial fluid, in the inertia regime, the inertia associated with the individual grains is more important.

The point is that Bagnold's results were obtained with fixed concentrations and shear rates [6,14,16]. In accordance with Iverson [4], this would imply that the grain concentration should be specified rather than determined by the grain interaction mechanisms.

According to Takahashi [14,42,49], by considering a stony-type debris flow, for which the effect of the interstitial fluid would be negligibly small, and by assuming uniform grain concentration and shear stress proportional to the square of the vertical gradient of the longitudinal velocity u, the integration of Bagnold's constitutive equations (under the boundary condition $u = 0$ at $z = 0$, with u = longitudinal velocity and z = direction orthogonal to the bed) gives

$$u = \frac{1}{d_p}\left[\frac{g\cos\theta}{a_i\cos\alpha_i}\right]C\left(1 - \frac{\rho}{\rho_s}\right)^{1/2}\frac{1}{\lambda}\left[h^{3/2} - (h - z)^{3/2}\right] \tag{3}$$

with

$$C = \frac{\rho\tan\beta'}{(\rho_s - \rho)(\tan\alpha - \tan\beta')} \tag{4}$$

where g is the gravity acceleration, β' is the inclination, h is the local flow depth, ρ is the fluid density, a_i is the so-called "friction coefficient" [16,50], and α is the collision angle. According to Bagnold's experimental tests, it can be assumed $a_i = 0.04$ and $\tan\alpha = 0.75$; Takahashi [14,49] suggested to assume $a_i = 0.42$ for loose beds.

Equation (3) (with λ estimated by Equation (2)) depends both on grain concentration C and on the maximum possible concentration C_*. While C_* could be estimated by using either experimental data appositely collected or physically-based literature data, more investigations should be conducted to evaluate grain concentration C and its variation within the debris body.

2.2. Experimental Apparatus and Procedure

In order to analyze the velocity and grains concentration distributions within the debris body, data collected in a straight reach of a laboratory flume, constructed at the Department of Engineering, University of Palermo (Italy), have been used. The channel was 0.17 m wide and with Perspex transparent side-walls (see Figure 1). For the experimental run considered in the present work, the bed slope was 15° and consisted of assorted gravels with mean diameter D_{50} = 3 mm. These experimental conditions were selected after preliminary setting runs and in accordance with the indications of previous literature works [33,39,51,52].

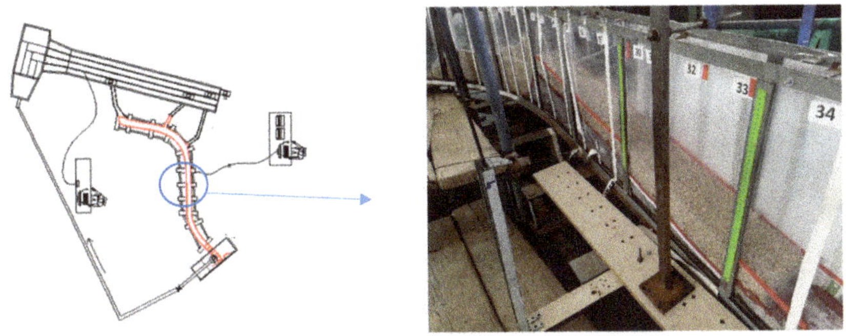

(a) (b)

Figure 1. (**a**) Plan view of the experimental apparatus; (**b**) particular of the channel reach considered for the analysis (bed slope = 15°; assorted gravels with mean diameter D_{50} = 3 mm; the numbers indicate the sections in which the channel was discretized).

Figure 2 reports the grain-size distribution of the bed material, which was determined by using an electric oscillating sieve. The debris flow was generated by releasing the water discharge $Q = 3.5 \times 10^{-3}$ m^3/s over a loose sediment layer of a thickness of around 0.10 m, which was previously saturated by slowly releasing a low water discharge $Q_0 = 0.8 \times 10^{-3}$ m^3/s. A permeable ground sill was positioned at the downstream end of the examined reach to avoid the degradation of the sediment layer.

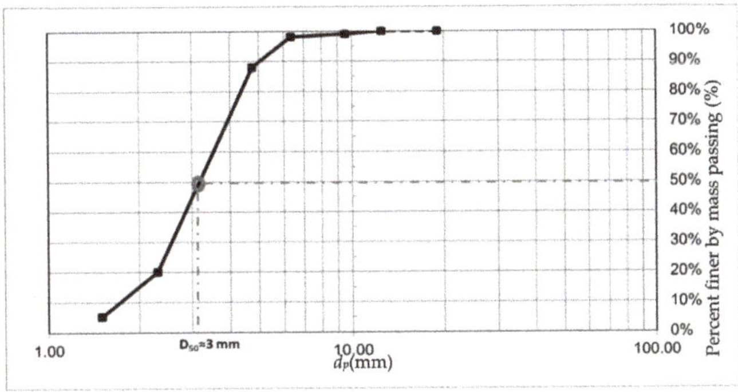

Figure 2. Grain size distribution of the bed material (the x-axis indicates the grain diameter; the y-axis indicates the percent of finer by mass passing).

During the experiment, the water discharge was measured through the ultrasonic instrument Mainstream EH7000 (by Endress+Hauser S.p.a.), which is based on the Doppler effect, located in the upstream inflow channel. A high-resolution digital camera (AOS Technology AG) was used to record the flow motion through the transparent right side-wall. The camera was set at a frame rate of 300 fps, and the investigated area was around 20 cm wide (see Figure 3). The rate of the acquisition frequency was defined, during the preliminary runs, according to the camera's characteristics and the extension of the covered area. During the passage of the debris body, around 600 frames were recorded. Furthermore, three samples of the mixture were taken at the free surface to estimate the free surface grains concentration Cs. Each sample was weighed, after having removed the water, with the help of a bake in order to determine the corresponding grain concentration. Finally, the "measured" free surface grain concentration was assumed as the mean value of the grain concentration of the three samples, which was equal to $Cs_{,m}$ = 0.305.

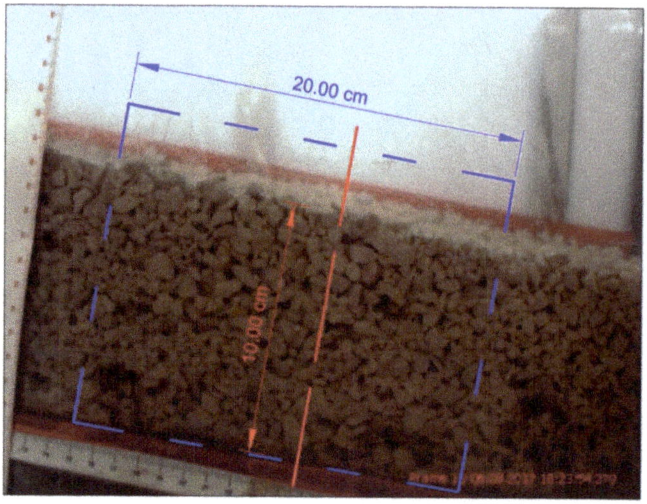

Figure 3. Scheme of the investigated area (in blue).

The frames recorded by the high-resolution camera during the passage of the debris body were used to estimate the instantaneous values both of the longitudinal velocity and the grain concentration at different distances, z, from the channel bed. To this aim, the sediment layer was divided into sub-layers, parallel to the channel bed, of thickness $St = D_{50}$. Within each sub-layer of thickness St, the grain's instantaneous longitudinal velocity, $u(t)$, was estimated as the ratio between the grain's displacement for two consecutive frames (i.e., two consecutive times) and the time interval. In particular, by considering the orthogonal local reference system with an origin at the southwest corner of the frame, it was estimated $u(t) = (s_{i+1} - s_i)/(t_{i+1} - t_i)$ (where t_{i+1} and t_i are two consecutive times and s_{i+1} and s_i are the corresponding grain's positions). Then, the time series of $u(t)$, obtained for the grains within each sub-layer St by considering all the recorded frames, were used to determine both the corresponding time-averaged value u and the mean, $\overline{u}$, of the time-averaged velocities of the considered sub-layer St.

The recorded frames were also used to estimate the instantaneous values of the grain concentration within each sub-layer of thickness St. To this aim, a four-step procedure, appositely implemented in a Matlab environment and allowing us to count the grains in the considered sub-layer, was applied to each frame. Then, the time series of the grain concentration obtained for each sub-layer St were used to determine the corresponding time-averaged value C. Thus, the grain concentration distribution along the depth was analyzed by using the values of C obtained for all the sub-layers. The sensitivity analysis of the estimated grain concentration distribution with the value of the thickness St was also performed, as explained in the following Section 3.1.

3. Results

3.1. Grains Concentration Distribution

Both by the direct analysis during the experimental run and as a result of the aforementioned four-steps procedure, it was verified that the grain concentration varies along the direction orthogonal to the bed. In particular, as Figure 4 clearly shows, as one passes from the free surface to the bed, the movement of the grains decreases and the grain concentration increases. Close to the bed, the frictional stress is so high that no motion is possible (static configuration), and the grain concentration assumes its maximum value (i.e., the maximum packing concentration) C_*. The thickness of the static bed, z_o, was estimated from the analysis of both of the recorded frames and of the measured velocity profiles.

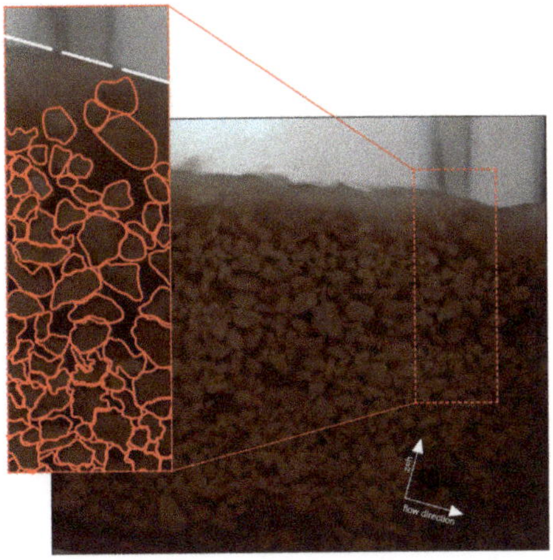

Figure 4. Variation of grains concentration in the depth of flow and particular of the grain configuration.

The values of grain concentration C obtained for all the sub-layers were plotted against the distance from the upper level of the static bed, as reported in Figure 5.

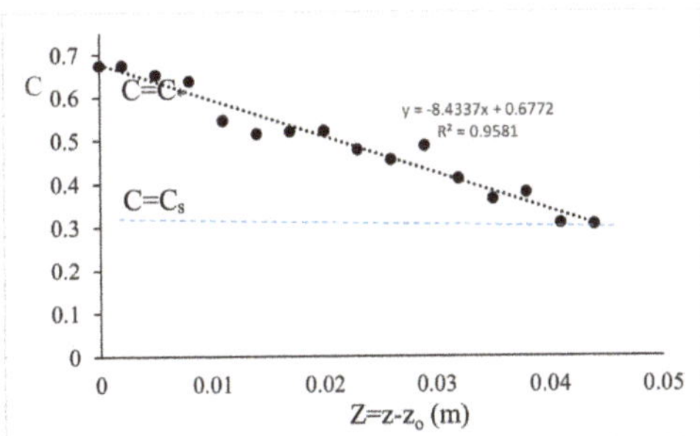

Figure 5. Measured grain concentrations C and interpolating law (Z indicates the direction orthogonal to the bed with an origin at the upper level of the static bed).

As Figure 5 shows, the points can be interpolated by a linear law (with regression coefficient $R^2 = 0.96$). The values of the grain concentration at the extreme points of the interpolating line are very close, respectively, to those of the surface concentration, Cs, and of the maximum possible concentration C_* at the bed. Based on this, the following linear law is considered to approximate the distribution of the grain concentration

$$\frac{C(z) - C*}{C_s - C*} = \frac{z - z_0}{h - z_0} \tag{5}$$

where $C(z)$ indicates the value of the grain concentration at level z. Based on Equation (5), the variation of the grain concentration within the debris body can be estimated as a function of the two parameters $C*$ and Cs.

For the present application, it is $C* = 0.676$ and $Cs = 0.305$, and thus Equation (5) can be written as $C(z) = -8.793Z + 0.68$.

In order to investigate the sensitivity of the estimated grains concentration distribution to the thickness of the sub-layer, other two values of St have been considered (i.e., $St = 2D_{50} = 6$ mm; $St = 3D_{50} = 9$ mm) and the corresponding values of grain concentration C have also been estimated by applying the aforementioned four-step procedure. Then, the values of the grain concentration determined for each value of St were compared with the theoretical distribution given by Equation (5), as reported in Figure 6.

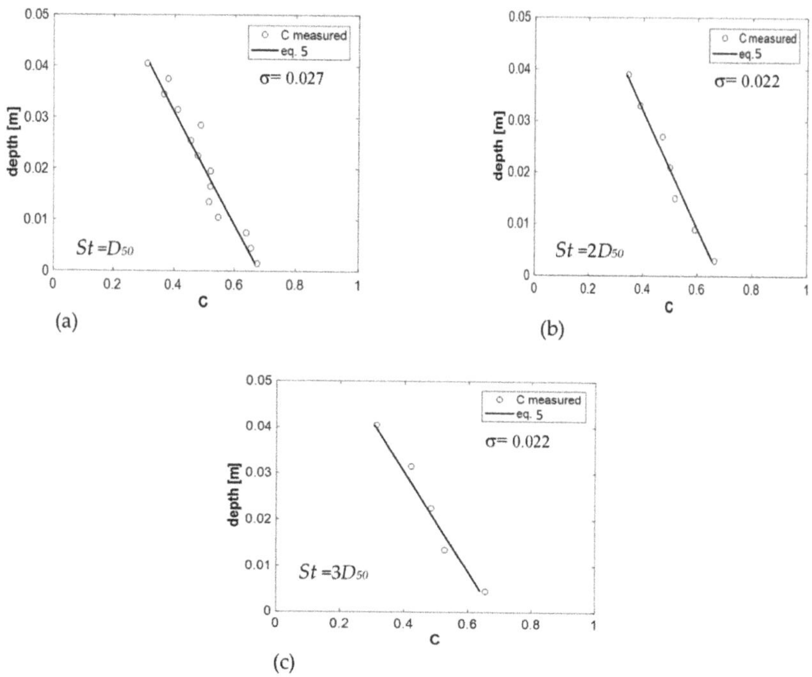

Figure 6. Comparison between the theoretical distribution of the grain concentration (Equation (5)) and the estimated values (C measured) by assuming different sub-layer thickness St: (**a**) for $St = D_{50}$; (**b**) for $St = 2D_{50}$; (**c**) for $St = 3D_{50}$. The x-axis indicates the grain concentration; the y-axis indicates the distance from the upper level of the static bed.

Figure 6 also reports the values of the root mean square error (σ) between the values of the grain concentration calculated in the i-th sub-layer by using Equation (5), $C_{i,c}$, and those measured, $C_{i,m}$:

$$\sigma = \sqrt{\frac{\sum_{i=1,N_{ts}}(C_{i,m} - C_{i,c})}{N_{ts}}} \tag{6}$$

where N_{ts} is the number of the sub-layers of thickness St. It can be observed that σ assumes small and almost equal values for all the considered values of St. This suggests that the estimated grains concentration distribution does not depend on the selected value of St. Finally, a value of $St = 2D_{50}$ (6 mm) has been assumed for the subsequent analysis.

3.2. *Modified Bagnold's Equation Applied to Debris Flow*

Based on the results obtained in Section 3.1, Equation (3) is rewritten as follows:

$$u = \frac{1}{d_p}\left[\frac{g\cos\theta}{a_i\cos\alpha_i}\right]C(z)\left(1-\frac{\rho}{\rho_s}\right)^{1/2}\frac{1}{\lambda(z)}\left[h^{3/2}-(h-z)^{3/2}\right] \tag{7}$$

with

$$\lambda(z) = \frac{1}{[C(z)/C_*]^{1/3}-1} \tag{8}$$

where $C(z)$ is given by Equation (5).

In Figure 7, the velocity profile estimated by applying Equation (7) is compared to that determined by applying Equation (3). This figure also reports both the profile of the mean measured velocity $\overline{u}$ (indicated as "experimental profile") and the velocity u of selected grains. From Figure 7, it is clear that while the velocity profile estimated by applying Equation (3) (i.e., with uniform grain concentration within the debris body) deviates from the experimental one, the velocity profile estimated by using Equation (7) (i.e., by considering the linear variation of the grain concentration within the debris body) well approximates the experimental profile.

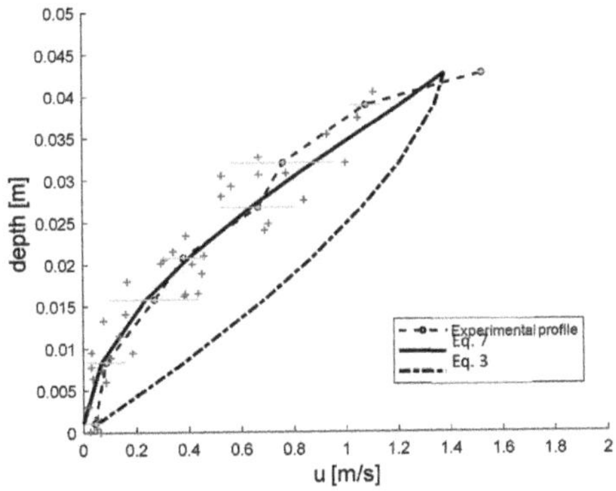

Figure 7. Comparison between the $\overline{u}$ profile (experimental profile) and the estimated ones by applying Equation (3) and by applying Equation (7). The dots represent the velocity u of selected grains (the x-axis indicates the longitudinal velocity; the y-axis indicates the distance from the upper level of the static bed).

This result indicates that Equation (5) well interprets the variation of the grain concentration within the debris body, also confirming that the assumption of uniform grain concentration in the entire flow depth is not realistic.

Thus, the Bagnold number N_{Ba} has been rewritten as follows:

$$N_{Ba}(z) = [\rho_s \gamma d_p{}^2 \lambda(z)^{1/2}]/\mu_f \tag{9}$$

By applying Equation (9), the regime of flow motion has been identified in the entire depth of flow, as reported in Figure 8.

In particular, Figure 8 reports both the N_{Ba}-values obtained by using Equation (9), with the velocity and the grain concentration distributions respectively given by Equations (7) and (5), and the N_{Ba}-values (hereon indicated as "experimental N_{Ba}-values") obtained by using the experimental velocity profile. Figure 8 also reports the N_{Ba}-values obtained by using Equations (1) and (3). Figure 8 shows that while the distribution of the N_{Ba}-values estimated by using Equation (9) is in agreement with the profile of the experimental N_{Ba}-values, the distribution of the N_{Ba}-values estimated by using Equation (1) deviates from it. In particular, the last one has N_{Ba}-values greater than 1000, close to the bed, and a decreasing trend as one passes from the bed to the free surface. On the contrary, Equation (9) determines N_{Ba}-values less than 1000 close to the bed and greater than 1000 close to the free surface. The latter trend of the N_{Ba}-values is consistent with that observed in previous literature works [34,48,51], indicating the development of the frictional flow regime for N_{Ba}-values <1000 and of the collisional-inertial flow regime for N_{Ba}-values >1000.

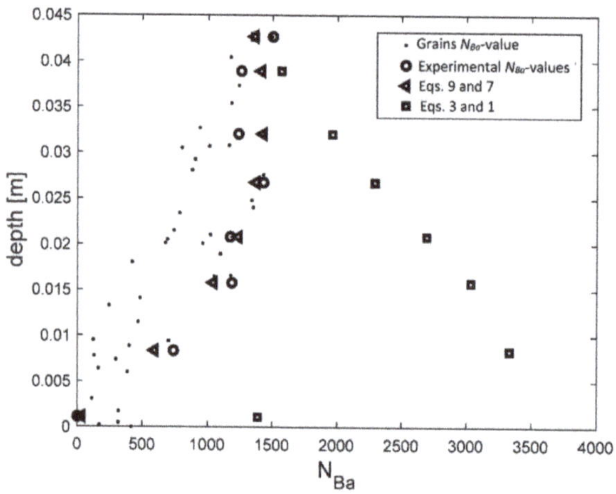

Figure 8. Comparison between the N_{Ba}-values by using the experimental velocity profile (experimental N_{Ba}-values), the N_{Ba}-values estimated by applying Equations (9) and (7) and the N_{Ba}-values estimated by applying Equations (3) and (1). The dots represent the N_{Ba}-values of selected grains; (the x-axis indicates the N_{Ba}-values; the y-axis indicates the distance from the upper level of the static bed).

Based on the obtained results, it can be concluded that, in the present application, three different regimes of motion can be identified within the debris body (see Figure 9): the collisional regime close to the free surface with N_{Ba}-values >1000 and C ≅ *Cs*, the frictional regime in the intermediate zone with N_{Ba}-values <1000 and C = *C(z)*, and the static zone close to the bed with N_{Ba}-values <<1000 and C ≅ *C*∗.

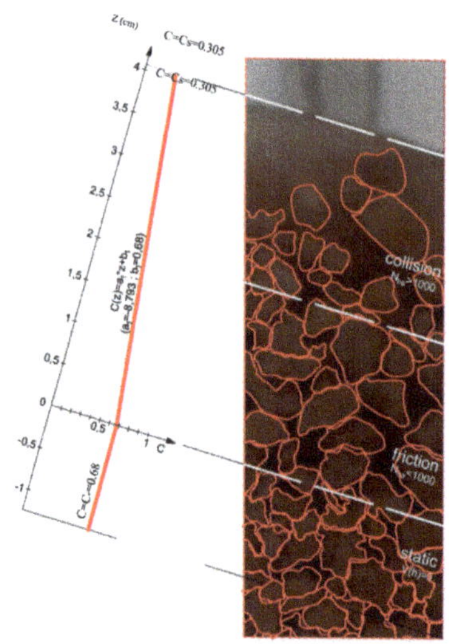

Figure 9. Regimes of motion within the debris body.

3.3. Free Surface Grains Concentration and Procedure for Its Estimation

Based on Equation (7), the longitudinal velocity profile should be determined as a function of the variation of the grain concentration defined by Equation (5) and, thus, as a function of the parameters *C*∗ and *Cs*. While the parameter *C*∗ could be easily identified, as mentioned in Section 2.1, it is difficult to determine the free surface grain concentration, *Cs*.

Thus, in this part of the work, experimental velocity profiles taken from the literature, i.e., data from [33,35,37,39] were considered and the parameter Cs was determined by minimizing the mean square error between the velocity values determined by using Equation (7) and the experimental ones:

$$\min[\sigma_v] = \min\sqrt{\frac{\sum_{i=1,n}(u_{m,i} - u_{c,i})^2}{n-1}} \tag{10}$$

where n is the number of the measurement points, $u_{m,i}$ and $u_{c,i}$ indicate, respectively, the measured and the calculated velocity values in the i-th sub-layer.

Figure 10 shows the comparison between the values of the parameter Cs obtained as a result of the minimization process (indicated as "optimal" values of Cs) and those taken from the literature [33,35,37,39]. The line of the perfect agreement is also reported in Figure 10. From this figure, it can be observed that, generally, the "optimal" values of Cs are slightly greater than those measured, and a very good agreement between the "optimal" values of Cs and those measured by Fichera et al. [39] can be observed. The different behavior observed in the latter case could be due to the fact that while Fichera et al. [39] measured the free surface concentration during the passage of the debris body in the examined reach, as well as in the present work, the other values of Cs were measured either not during the passage of the debris body or in sections different from those examined or not at the free surface. This further indicates the importance of the correct definition of the parameter Cs.

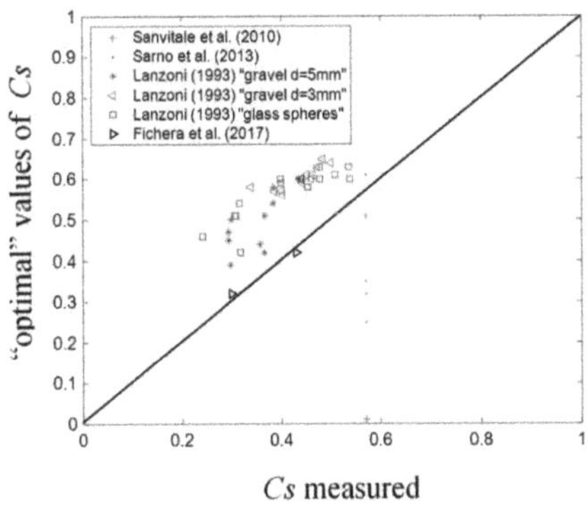

Figure 10. Comparison between the estimated "optimal" values of the parameter Cs and those measured and taken from the literature.

In order to identify a more generalizable procedure to estimate Cs, it has been taken into account that, according to Armanini et al. [34] (see also in [51]), in a flowing layer of grains over a loose bed of inclination β, fully saturated with water, and, with grain concentration assumed coincident with the transport concentration Cs, it is

$$\tan\beta = \left[\frac{(\rho_s - \rho)C_s}{\rho + (\rho_s - \rho)C_s}\right]\tan\varphi_c \tag{11}$$

where φ_c represents the critical friction angle.

For negligible interstitial overpressures (as it occurs at the free surface), in Equation (11), φ_c could be assumed equal to the static friction angle of the material. This means that Equation (11) would allow us to obtain information on the grain concentration Cs, starting from the knowledge of the bed slope and of the static friction angle, which can be determined by simple shear tests.

Thus, the existence of the relationship (Equation (11)) between the static friction angle and the concentration Cs is explored for the present application. To this aim, the static friction angle φ_c was determined by using the direct cutting apparatus available at the Geotechnical Laboratory of the Engineering Department, University of Palermo. Such an apparatus is specially adapted to the present application because it is equipped by a large cutting box ($30 \times 30 \times 20$ cm^3—see Figure 11), allowing us to test a sample, of adequate volume, of the material used for the experimental run.

Figure 11. Box used for the cutting tests.

The cutting tests were performed, at a constant volume and in saturated conditions, for three samples of the debris body by inducing a constant horizontal force. As a result, the relation between the cutting shear stress, τ, and the corresponding cutting force, σ_n, was obtained for each test. More details of the testing procedure can be found in Fichera [53]. From the cutting tests, it was also verified that all the examined samples assumed a very similar behavior. By considering the maximum value of the shear stress, τ_{max}, of each test, the couples of values [τ_{max}, σ_n] were reported on a Cartesian plane. As Figure 12 shows, the coupled values [τ_{max}, σ_n] are arranged around an interpolating line, having an angular coefficient equal to $\varphi_{c,e} = 38°$. Such an angular coefficient represents the "experimental" value of the static friction angle of the used granular material.

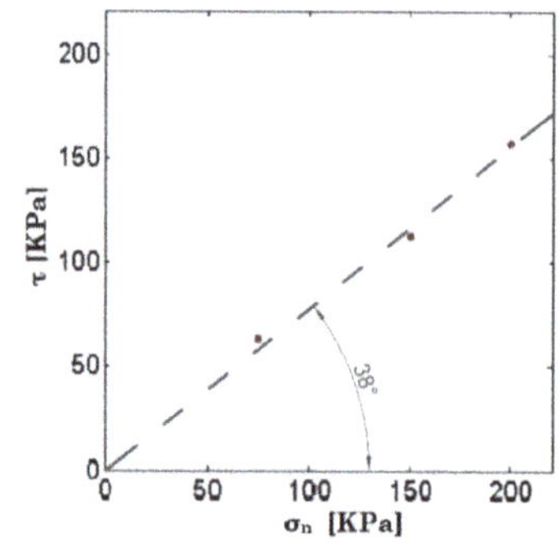

Figure 12. Coupled values [τ_{max}, σ_n] and interpolating line.

Then, the static friction angle was also determined from Equation (11) by considering that, for the examined case, the measured surface grain concentration is $Cs_{,m} = 0.305$ and the inclination is $\beta = 15°$. Finally, a value $\varphi_c = 38°$ was obtained from Equation (11). It is clear that this value of φ_c is equal to that obtained from the aforementioned direct cutting tests.

Thus, it can be concluded that, in the first approximation, the free surface grain concentration, Cs, can be estimated by applying Equation (11), starting from the knowledge of the bed slope and of the static friction angle of the material.

4. Discussion

4.1. Velocity Profiles and Comparisons with Literature Data

The applicability of the Equations (5) and (7) was also assessed by comparing the estimated velocity profiles with measured velocity profiles taken from the literature. In particular, measured profiles taken from Lanzoni [33] (see also in [53,54]), Sanvitale et al. [35], Sarno et al. [37], and Fichera et al. [39] were considered. In total, measured profiles from 46 pieces of literature were used for the comparison.

In Figure 13, the velocity profiles estimated by using Equations (5) and (7) are compared with those measured. In particular, this figure reports the velocity profiles estimated by assuming for the parameter *Cs* both the values taken from the literature and the calculated "optimal" values.

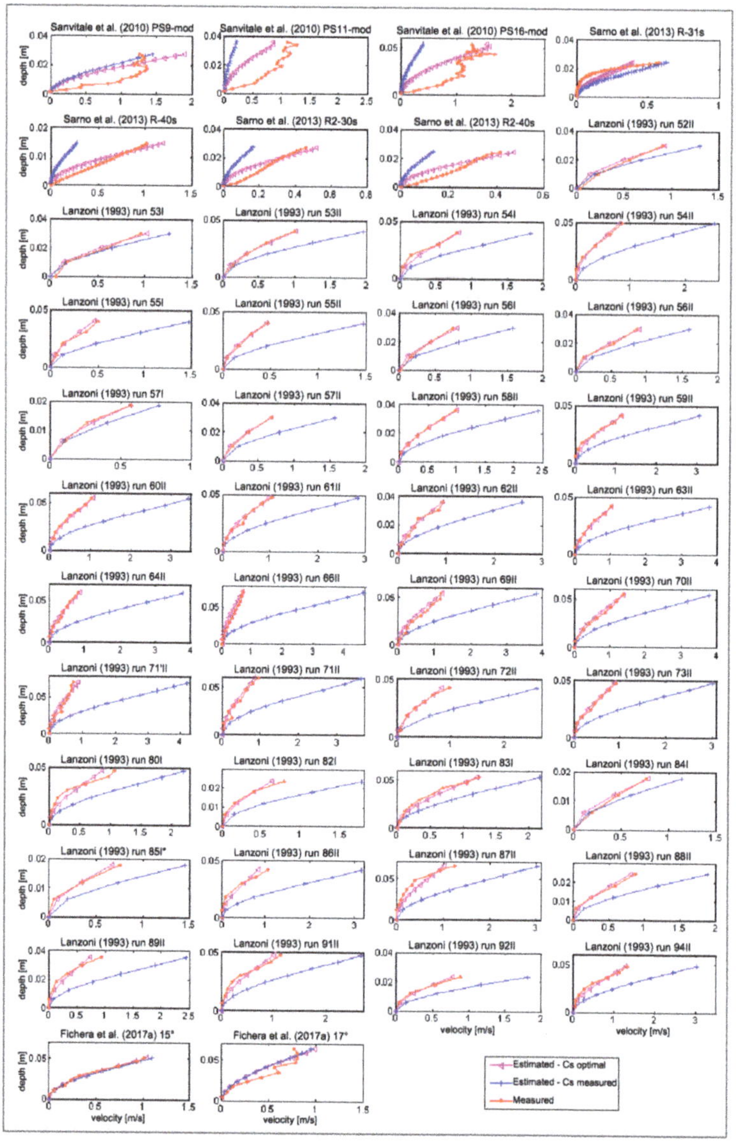

Figure 13. Comparison between the estimated velocity profiles and those taken from the literature.

It can be seen that, while in the first case the estimated velocity profiles differ from those measured when the "optimal" values of *Cs* are used, the estimated velocity profiles compare well with the measured ones and, with the exception of the comparisons with SanVitale et al.'s [35] data, a value of the mean square error less than 0.1 is obtained. The differences obtained within the debris body between the estimated velocity profiles and those measured by SanVitale et al. [35] could be related to the fact that SanVitale et al. [35] analyzed a fluid/sediment mixture (consisting of borosilicate glass and hydrocarbon oil) of chemical/physical characteristics different from those examined by the other authors. This confirms previous literature findings indicating that the chemical/physical characteristics of the material could affect the stress distribution and, thus, the behavior of the fluid/sediment mixture (see, as examples, [55,56]).

Thus, on the one hand, Figure 13 demonstrates that Equation (7) allows us to correctly interpret the velocity distribution within the debris body by taking into account the grain concentration variation through Equation (5); on the other hand, it confirms the importance of the correct definition of the parameter *Cs*. Based on the results presented in Section 3.3, *Cs* could be determined either by direct measurements during the passage of the debris body or by applying Equation (11) as a function of the static friction angle of the material.

4.2. Main Aspects Derived by the Proposed Modified Expressions and Procedure

As mentioned in the introduction, in natural environments, stony-type debris flows can be generated in rapid valleys and in high slope streams [1,2]. In these types of debris flows, in which finer sediments do not affect the overall behavior of the mixture, the assumption of the grain concentration being uniform everywhere is generally accepted in analyzing flow dynamics [11,12,14].

The analysis conducted in the present work has substantially highlighted that the identification of the grain concentration distribution within the debris body as an important aspect in estimating the velocity profile. In particular, the presented experiment has provided useful information on the distribution of the grain concentration within the debris body. It has been observed that as one passes from the free surface towards the bed, the grain concentration increases, and the movement of the grains decreases. At the bed, the frictional stress is so high that no motion is possible, the static configuration is obtained, and the grain concentration assumes its maximum value (i.e., the so-called maximum packing concentration), C_*. This confirms that the assumption of uniform sediment distribution in the entire depth, which is generally used in analyzing the hydrodynamics of stony-debris flow, is not realistic.

The analysis of the measured values of the grain concentration has shown that the grain concentration distribution within the debris body can be approximated by a linear law, which can be identified starting from the knowledge of the maximum packing concentration at the bed, C_*, and the value of the grain concentration at the free surface, *Cs*.

Furthermore, the obtained results have shown that the variation of grain concentration along the depth strongly affects the grains' mobility, thus confirming previous literature findings [13,51] suggesting the development of a spatially variable behavior and stress regime within the debris flow.

By considering the obtained linear law of the grain concentration distribution, modified expressions of Bagnold's number N_{Ba} and of the velocity for stony-type debris flows have been obtained.

The analysis of the values of Bagnold's number (N_{Ba}), estimated by using the aforementioned modified expression, has highlighted that three different flow regimes can be identified within the debris body. In particular, the collisional regime is obtained close to the free surface, where the grain concentration is $C = Cs$ and the N_{Ba}-values are greater than 1000; the frictional regime is obtained in the intermediate zone, where $C = C(z)$ and the N_{Ba}-values are lower than 1000; close to the bed, where the concentration is $C = C_*$ and the N_{Ba}-values are strongly lower than 1000, the static zone is established. The observed behavior is consistent with results obtained in other literature works (see in [34,51,53]), indicating the development of the frictional flow regime for N_{Ba}-values <1000 and of the collisional-inertial flow regime for N_{Ba}-values >1000.

The proposed modified expression to estimate the velocity in stony-type debris flows allows us to evaluate the flow velocity profile within the debris body as a function of two parameters, that are the maximum packing concentration, C_*, and free surface concentration, *Cs*. While parameter C_* can be easily identified by using physically-based literature data, it is difficult to determine parameter *Cs*. Thus, first, the "optimal" values of parameter *Cs* were calculated by minimizing the mean square error between the velocity values estimated by using the proposed expression and the experimental ones taken from the literature. As is clear from the previous subsection, this analysis has allowed us to highlight the important role of parameter *Cs* in defining the velocity distribution within the debris body. The comparison between the velocity profiles estimated by considering the "optimal" values of *Cs* and the literature's measured profiles (from [33,35,37,39]) has demonstrated that the proposed modified expression, which takes into account the variation of the grain concentration within the debris body, correctly interprets the measured velocities.

A useful general procedure for estimating free surface concentration Cs is also provided by using the results of cutting tests, also performed in the ambit of the present study, of a sample of the granular material used for the experimental run. Based on the relationship between the grain concentration and the bed slope, according to Armanini et al. [34], it has been verified that, in first approximation, surface grain concentration *Cs* could be determined, starting from the knowledge of the bed slope and of the static friction angle of the material, which could be easily estimated by simple direct cutting tests.

In summary, based on the aforementioned considerations, the use of the proposed expressions allows us to easily estimate the velocity profile and the stress regime variation within the debris body according to the procedure shown in the flow chart of Figure 14.

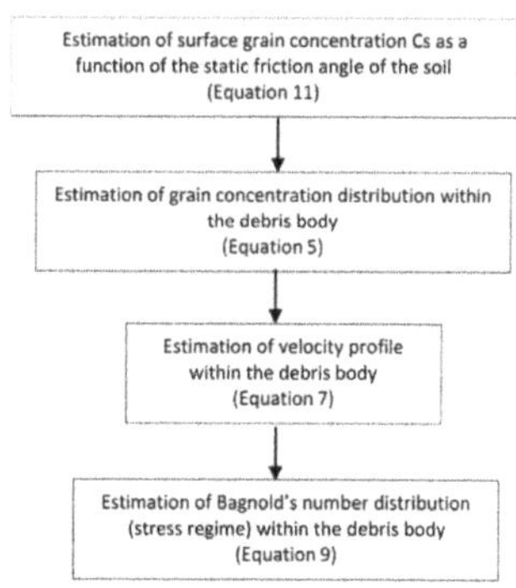

Figure 14. Flow chart of the procedure by using the proposed modified equations.

5. Conclusions

The present work deals with the dynamics of stony-type debris flows, focusing attention on the velocity and grain concentration distributions. The results obtained can be summarized as follows:

(1) the distribution of the grain concentration can be interpreted by a linear law obtained between the value of the maximum package value, C_*, at the bed and the value of the free surface concentration, *Cs*;

(2) by removing the hypothesis of uniform grain concentration along the entire depth, modified expressions of Bagnold's number and of the longitudinal velocity, which take into account the variation of the grain concentration in the entire depth, are presented. The expression of the velocity profile includes two parameters: the maximum package value, C_*, which could be determined by using either experimental data appositely collected or physically-based literature data, and the value of the free surface concentration, *Cs*;

(3) by using the modified expression of Bagnold's number, it has been verified that a varying stress regime can develop within the debris flow. The N_{Ba}-values are strongly lower than 1000 when close to the bed (frictional regime) and are greater than 1000 (collisional-inertial regime) when close to the free surface;

(4) it has been verified that, in the first approximation, surface concentration *Cs* can be estimated as a function of the static friction angle of the material, which can be determined by simple shear tests.

In summary, based on the results presented in this work, the velocity and the grain concentration distributions within the debris body can be estimated, starting from the knowledge of parameters C_* and *Cs* that can be easily identified through physically-based literature data and by simple shear tests, respectively. This result is of great importance, especially in numerical modeling of stony-type debris flows, which in nature especially occur in rapid valleys or in high-slope streams.

Author Contributions: Conceptualization, D.T. and A.F.; methodology, D.T. and A.F.; validation, D.T. and A.F.; formal analysis, D.T.; investigation, A.F.; resources, D.T. and A.F.; data curation, A.F.; writing—original draft preparation, D.T.; writing—review and editing, D.T.; visualization, A.F.; supervision, D.T. and A.F. All authors have read and agreed to the published version of the manuscript.

Funding: This research received no external funding.

Acknowledgments: This work was partially supported by Italian National Research Programme PRIN 2017, with the project "IntEractions between hydrodyNamics flows and bioTic communities in fluvial Ecosystems: advancement in dischaRge monitoring and understanding of Processes Relevant for ecosystem sustaInability by the development of novel technologieS with fIeld observatioNs and laboratory testinG (ENTERPRISING)".

Conflicts of Interest: The authors declare no conflict of interest.

References

1. Takahashi, T. Debris flow. *Annu. Rev. Fluid Mech.* **1981**, *13*, 57–77. [CrossRef]
2. Cannon, S.; Gartner, J.; Wilson, R.C.; Bowers, J.C.; Laber, J. Storm rainfall conditions for floods and debris flows from recently burned areas in south western Colorado and southern California. *Geomorphology* **2008**, *96*, 250–269. [CrossRef]
3. Theule, J.I.; Liébault, F.; Laigle, D.; Loye, A.; Jaboyedoff, M. Channel scour and fill by debris flows and bedload transport. *Geomorphology* **2015**, *243*, 92–105. [CrossRef]
4. Iverson, R.M. The physics of debris flows. *Rev. Geophys.* **1997**, *35*, 245–296. [CrossRef]
5. Kaitna, R.; Rickenmann, D.; Schatzmann, M. Experimental study on rheologic behaviour of debris-flow material. *Acta Geotech.* **2007**, *2*, 71–85. [CrossRef]
6. Armanini, A. Granular flows driven by gravity. *J. Hydraul. Res.* **2013**, *51*, 111–120. [CrossRef]
7. Van Rijn, L.C. Sediment transport, part I: Bed load transport. *J. Hydraul. Eng.* **1984**, *110*, 1431–1456. [CrossRef]
8. Van Rijn, L.C. Sediment transport, part II: Suspended load transport. *J. Hydraul. Eng.* **1984**, *110*, 1613–1641. [CrossRef]
9. Joshi, S.; Xu, Y.J. Assessment of suspended sand availability under different flow conditions of the Lowermost Mississippi River at Tarbert Landing during 1973–2013. *Water* **2015**, *7*, 7022–7044. [CrossRef]
10. Joshi, S.; Xu, Y.J. Bedload and suspended load transport in the 140-km reach downstream of the Mississippi River avulsion to the Atchafalaya River. *Water* **2017**, *9*, 716. [CrossRef]
11. Campbell, C.S. Granular shear flows at the elastic limit. *J. Fluid Mech.* **2002**, *465*, 261–291. [CrossRef]
12. Campbell, C.S. Stress-controlled elastic granular shear flows. *J. Fluid Mech.* **2005**, *539*, 273–297. [CrossRef]
13. Kaitna, R.; Dietrich, W.E.; Hsu, L. Surface slopes, velocity profiles and fluid pressure in coarse-grained debris flows saturated with water and mud. *J. Fluid Mech.* **2014**, *7*, 377–403. [CrossRef]
14. Takahashi, T. Debris Flows: Mechanics, prediction and countermeasures. Balkema-Proc and Monogr. In *Eng., Water and Earth Sc.*; Taylor and Francis: New York, NY, USA, 2007.
15. Bonnet-Staub, I. Definition d'une typologie des deposits de laves torrentielles et identification de critres granulomtriques et geotechniques concernant les zones sources. *J. Bull. Eng. Geol. Environ.* **1999**, *57*, 359–367. [CrossRef]
16. Bagnold, R.A. Experiments on a gravity-free dispersion of large solid spheres in a Newtonian fluid under shear. *Proc. R. Soc. Lond. A* **1954**, *225*, 49–63.
17. Hunt, M.; Zenit, R.; Campbell, C.S.; Brennen, C.E. Revisiting the 1954 suspension experiments of RA Bagnold. *J. Fluid Mech.* **2002**, *452*, 1–24. [CrossRef]
18. Bagnold, R.A. The shearing and dilation of dry sand and the 'singing' mechanism. *Proc. R. Soc. Lond. A* **1966**, *295*, 219–232.
19. Bagnold, R.A. An empirical correlation of bedload transport rates in flumes and natural rivers. *Proc. R. Soc. Lond. A* **1980**, *372*, 453–473.
20. Egashira, S.; Miyamoto, K.; Itoh, T. Constitutive equations of debris flow and their applicability. In Proceedings of the 1st International Conference on Debris-Flow Hazards Mitigation: Mechanics, Prediction, and Assessment, San Francisco, CA, USA, 7–9 August 1997; pp. 340–349.

21. Iverson, R.M.; Denlinger, R.P. Flow of variably fluidized granular masses across three-dimensional terrain: Coulomb mixture theory. *J. Geophys. Res.* **2001**, *106*, 537–552. [CrossRef]
22. Iverson, R.M. Elementary theory of bed-sediment entrainment by debris flows and avalanches. *J. Geophys. Res.* **2012**, *117*, F03006. [CrossRef]
23. Iverson, R.M.; Ouyang, C. Entrainment of bed material by Earth-surface mass flows: Review and reformulation of depth-integratedtheory. *Rev. Geophys.* **2015**, *53*, 27–58. [CrossRef]
24. Hungr, O.; McDougall, S.; Bovis, M. Entrainment of material by Debris flows. In *Debris-Flow Hazards and Related Phenomena*; Jakob, M., Hungr, O., Eds.; Springer: Berlin/Heidelberg, Germany, 2005; pp. 135–158.
25. Scheidl, C.; Rickenmann, D. Empirical prediction of debris-flow mobility and deposition on fans. *Earth Surf. Process. Landf.* **2009**, *35*, 157–173. [CrossRef]
26. Berger, C.; McArdell, B.W.; Schlunegger, F. Direct measurement of channel erosion by debris flows, Illgraben, Switzerland. *J. Geophys. Res.* **2011**, *116*, F01002. [CrossRef]
27. Chiodi, F.; Claudin, P.; Andreotti, B. A two-phase flow model of sediment transport: Transition from bedload to suspended load. *J. Fluid Mech.* **2014**, *755*, 561–581. [CrossRef]
28. Thibaud, R.; Chauchat, J. A two-phase model for sheet flow regime based on dense granular flow rheology. *J. Geophys. Res. Am. Geophys. Union* **2013**, *118*, 619–634.
29. Chapman, S.; Cowling, T.G. *The Mathematical Theory of Non-Uniform Gases*, 3rd ed.; Cambridge University Press: Cambridge, UK, 1971.
30. Jenkins, J.T.; Savage, S.B. A theory for rapid flow of identical, smooth, nearly elastic, spherical particles. *J. Fluid Mech.* **1983**, *130*, 186–202. [CrossRef]
31. Lun, C.K.K.; Savage, S.B.; Jeffrey, D.J.; Chepurniy, N. Kinetic theories for granular flow: Inelastic particles in Couette flow and slightly in elastic particles in a general flow field. *J. Fluid Mech.* **1984**, *140*, 223–256. [CrossRef]
32. Jenkins, J.T.; Richman, M.W. Kinetic theory for plane flows of a dense gas of identical, rough, inelastic, circular disks. *Phys. Fluids* **1985**, *28*, 3485–3494. [CrossRef]
33. Lanzoni, S. Meccanica di miscugli solio-liquido in regime granulo-inerziale. Ph.D. Thesis, University of Padova, Padua, Italy, 1993.
34. Armanini, A.; Capart, H.; Fraccarollo, L.; Larcher, M. Rheological stratification in experimental free-surface flows of granular-liquid mixtures. *J. Fluid Mech.* **2005**, *532*, 269–319. [CrossRef]
35. Sanvitale, N.; Bowman, E.T.; Genevois, R. Experimental measurements of velocity through granular-liquid flows. *Ital. J. Eng. Geol. Environ. Book Casa Editrice Università La Sapienza* **2010**. [CrossRef]
36. Iverson, R.M.; Logan, M.; Lahusen, R.G.; Berti, M. The perfect debris flow? *Aggregated results from 28 large-scale experiments. J. Geophys. Res.* **2010**, *115*, F03005.
37. Sarno, L.; Papa, M.N.; Tai, Y.C.; Carravetta, A.; Martino, R. A reliable PIV approach for measuring velocity profiles of highly sheared granular flows. In Latest Trends in Engineering Mechanics, Structures, Engineering Geology. 2013. ISBN: 978-960-474-376-6. Available online: https://www.researchgate.net/publication/278961892_A_reliable_PIV_approach_for_measuring_velocity_profiles_of_highly_sheared_granular_flows (accessed on 13 May 2020).
38. Hsu, L.; Dietrich, W.E.; Sklar, L.S. Experimental study of bedrock erosion by granular flows. *J. Geophys. Res.* **2008**, *113*, F02001. [CrossRef]
39. Fichera, A.; Stancanelli, L.M.; Lanzoni, S.; Foti, E. Stony Debris Flow Debouching in a River Reach: Energy Dissipative Mechanisms and Deposit Morphology. In Proceedings of the Workshop on World Landslide Forum, Ljubljana, Slovenia, 9 May–2 June 2017; Springer: Cham, Switzerland, 2017; pp. 377–383.
40. Marchi, L.; Dalla Fontana, G. GIS morphometric indicators for the analysis of sediment dynamics in mountain basins. *Environ. Geol.* **2005**, *48*, 218–228. [CrossRef]
41. Gruber, S.; Huggel, C.; Pike, R. Modelling mass movements and landslide susceptibility. In *Geomorphometry;* Hengl, T., Reuter, H.I., Eds.; Elsevier: Amsterdam, The Netherlands, 2008; pp. 527–550. Available online: http://www.zora.uzh.ch/id/eprint/6475/ (accessed on 13 May 2020).
42. Takahashi, T. Mechanical characteristics of debris flow. *J. Hydraulics Div. ASCE* **1978**, *104*, 1153–1169.
43. Takahashi, T. Debris flow on prismatic open channel. *J. Hydraulics Div. ASCE* **1980**, *106*, 1153–1169.
44. Gregoretti, C.; Dalla Fontana, G. The triggering of debris flows due to channel-bed failure in some alpine headwater basins of Dolomites: Analyses of critical runoff. *Hydrological Process.* **2008**, *22*, 2248–2263. [CrossRef]

45. Berti, M.; Genevois, R.; Simoni, A.; Tecca, P.R. Field observations of a debris flow event in the Dolomites. *Geomorphology* **1999**, *29*, 256–274. [CrossRef]
46. Moscariello, A.; Marchi, L.; Maraga, F.; Mortara, G. Alluvial fans in the Italian Alps: Sedimentary facies and processes. In *Flood & Megaflood Processes and Deposits—Recent and Ancient Example*; Martini, P., Baker, V.R., Garzon, G., Eds.; Blackwell Science: Oxford, UK, 2002; pp. 141–166.
47. Savage, S.B.; Lun, C.K.K. Particle size segregation in inclined chute flow of dry cohesionless granular solids. *J. Fluid Mech.* **1988**, *189*, 311–335. [CrossRef]
48. Aharonov, E.; Sparks, D. Rigidity phase transition in granular packings. *Phys. Rev. E* **1999**, *60*, 6890–6896. [CrossRef]
49. Takahashi, T. An occurrence mechanism of mud-debris flows, and their characteristics in motion. *Annu. DPRI* **1977**, *20*, 405–435.
50. Tsai, Y.F.; Tsai, H.K.; Ch, Y.L. Study on the configurations of debris-flow fans. *IJEGE* **2011**, *2011*, 03.B-032.
51. Lanzoni, S.; Gregoretti, C.; Stancanelli, L.M. Coarse-grained debris flow dynamics on erodible beds. *J. Geophys. Res. Earth Surf.* **2017**, *122*, 596–614. [CrossRef]
52. Chen, S.C.; An, S. Flume experiment of debris flow confluence formed alluvial fan in the main channel. In *River, Coastal and Estuarine Morphodynamics: RCEM 2007*; Dohmen-Janssen, C.M., Hulscher, S.J.M.H., Eds.; Taylor and Francis: London, UK, 2007; pp. 829–835.
53. Fichera, A. Analisi e caratterizzazione del comportamento di una colata detritica. Ph.D. Thesis, University of Enna "KORE", Italy, Kore, 2017.
54. Lanzoni, S.; Tubino, M. Rheometric experiments on mature debris flows. In Proceedings of the XXV International Association for Hydro-Environment, Tokyo, Japan, 30 August–3 September 1993. Editor: Japan Society of Civil Engineers, Local Organizing Committee of the XXV Congress, IAHR, 1993.
55. ASM International Handbook Committee. *Engineered Materials Handbook*; Desk Edition 1995; ASM International: Novelty, OH, USA, 1995; ISBN 0-87170-283-5.
56. Zhao, Y.; Liu, Z. Study of Material Composition Effects on the Mechanical Properties of Soil-Rock Mixtures. Hindawi. *Adv. Civil Eng.* **2018**, *2018*, 3854727. [CrossRef]

water

MDPI

Article

Deciphering Morphological Changes in a Sinuous River System by Higher-Order Velocity Moments

Jyotismita Taye [1], Jyotirmoy Barman [1], Bimlesh Kumar [1,*] and Giuseppe Oliveto [2]

1 Department of Civil Engineering, Indian Institute of Technology, Guwahati 781039, India; taye176104022@iitg.ac.in (J.T.); jyoti174104121@iitg.ac.in (J.B.)

2 School of Engineering, University of Basilicata, 85100 Potenza, Italy; giuseppe.oliveto@unibas.it

* Correspondence: bimk@iitg.ac.in; Tel.: +91-361-2582420

Received: 29 January 2020; Accepted: 9 March 2020; Published: 11 March 2020

Abstract: Bank erosion in a sinuous alluvial channel is a continuous phenomenon resulting in bank instability and migration of sediment. In this study, flume experiments were conducted in a sinuous channel to investigate its morphological changes and hydrodynamics. High-order velocity fluctuation moments are analyzed at outer and inner banks to explain the morphological variation in a sinuous river channel. The variance of streamwise velocity fluctuations on both banks of the sinuous channel follows a logarithmic law from a particular depth. In the outer bend region, the magnitude of velocity fluctuation moment is significant, indicating erosion. The trend of velocity fluctuation at higher even-order moments is similar to the variance of streamwise velocity fluctuations where the outer bend magnitude is greater than the inner bend. The premultiplied probability density functions (PDFs) and the flatness factor show greater magnitude in the outer bend of the channel as compared to the inner bend.

Keywords: sinuous channel; velocity fluctuations; river bend erosion; structure function

1. Introduction

The study of turbulence in a sinuous channel is a complex topic as compared to a straight channel. There is uniformity in a straight channel in both banks; however, in a sinuous channel, there are continuous erosion and deposition processes at outer and inner banks, respectively. Turbulence studies in a sinuous channel have been discussed quite elaborately over the past 2–3 decades. Researchers such as Rozovskiĭ [1], Anwar [2], de Vriend and Geldof [3], Blanckaert and Graf [4], Booij [5], Sukhodolov and Kaschtschejewa [6], and Engel and Rhoads [7] have studied the turbulent characteristics of both infield and laboratory and forwarded theories and articles regarding the uniqueness of flow behaviour. Several researchers (e.g., [8,9]) have also explained the relations between turbulence characteristics and erosion at the outer bank. Blanckaert [10] stated that the Reynolds shear stress (RSS) in a sinuous bend indicates the presence of helical flow, which contributes to the erosion and deposition processes. Esfahani and Keshavarzi [11] studied the bursting process by octant analysis. Their study was based on two models of 17° and 30° bend angle. They found that the effect of river bends on flow characteristics and bursting events is inversely proportional to the curvature of the bend. The velocity distribution at channel bends is conflicting. In sinuous bends, the maximum velocity is not found always towards the outer bend because of the velocity redistribution [3,12]. de Vriend and Geldof [3] mentioned that there is a shift of maximum velocity in the inner bend before the flow enters the bend apex. This paper aims to address that the mean velocity may not be always higher towards the outer bend. Thus, the paper highlights the erosion and deposition patterns in a sinuous channel by analyzing the high-order streamwise velocity fluctuations.

Previous studies have testified the morphological changes in a sinuous channel [13–15]. In sinuous channel, the shifting of centerline and migration of sediment is a continuous process. Experiments conducted with erodible beds in a meandering channel have reported bed formations as point bars, mid-channel bars, free bars, and sand bars [16–19]. Xu and Bai [12] in their experiments with erodible bed and fixed walls, observed depositional bars at the inner bank and pools at the outer bank. Zhang et al. [20] observed the flood events in a step-pool channel. Scour appeared in the pool at threshold condition, and after which the scour depth increased linearly until the step collapsed. Riverbanks are vulnerable to erosion, and recent studies have been done in riverbanks of Langat River in Malaysia [21] and Parlung Tsangpo River in China [22]. Under natural conditions in rivers, the flow turbulence governs the bed deformation and channel migration [23]. Small-scale physical models of sinuous channels were prepared successfully in the laboratory [24–27], although it is very challenging to simulate a dynamic realistic sinuous channel. However, to find a solution to a specific problem to understand the fundamentals of river processes, experimental channels are quite useful to explore the mechanisms associated. In this study, scale effects were neglected, and a laboratory study was conducted to assess flow turbulence effects.

Many researchers have worked towards high-order velocity moments in the turbulent boundary layer to acquire the information on flow behaviour. Meneveau and Marusic [28] studied the higher-order moments of velocity fluctuations at different Reynolds numbers and presented a generalized logarithmic law. de Silva et al. [29] studied the statistical properties of wall turbulence using higher-order moments of streamwise velocity fluctuations. They found that at higher Reynolds numbers, the even-order moments follow the logarithmic behaviour. Sharma and Kumar [30] studied the higher-order structure functions for seepage-affected channels. They found that higher-order moments of velocity fluctuations increase with the application of downward seepage. Another parameter called the flatness factor or Kurtosis helps in observing the distribution of the fluctuating velocity. Meneveau and Marusic [28] in their study estimated the Reynolds number dependence on sub-Gaussian statistics. The flatness factor evaluated in the inertial region is below the Gaussian value 3 and found no noticeable dependence of the Reynolds number on sub-Gaussianity.

The previous studies present the significance of the high-order velocity moments and the correlation of the turbulent flow with sediment transport. Previously, outer bend erosion was explained by using various turbulent parameters such as Reynolds shear stress, turbulent intensities, and turbulent kinetic energy. However, the previous studies had not addressed the scope to understand the fluvial morphology in association with the higher-order velocity fluctuations. Here, we have considered only the streamwise velocity to analyze the higher-order velocity fluctuations. Though the velocity field is three-dimensional in our channel and has an impact on the morphology, the study has been limited to acknowledge specifically the scour and deposition at the outer and inner bend, respectively. In this paper, firstly, we have analyzed the mean streamwise velocity and the morphology of the sinuous channel. The morphological behaviour around the bend apex (i.e., scour and deposition) needed more insights, for which the mean velocity distribution is not sufficient to make a statement on the bed deformation. Concerning this, we have applied the high-order structure function to understand the scour and deposition in a sinuous bend.

2. Experimental Methods and Program

The experiments were conducted in a glass-sided recirculating flume of length 17.2 m and width 1 m (Figure 1). In the main channel, we constructed a rectangular sinuous channel with rigid sides of length 5.64 m and width 0.3 m. The centerline of the sinuous channel follows the sine-generated function forwarded by Leopold and Langbein [31]. The function is expressed as

$$\theta = \theta_0 \cos\left(2\pi \frac{m}{L}\right) \tag{1}$$

where θ_o is the maximum angle the channel makes with the downvalley axis, θ is the angle at distance m measured along the centerline of the channel, and L is the arc length (length along the centerline of the channel between two repeating points). The channel centerline at the crossover made an angle of 65° with the horizontal, known as the deflection angle. The channel sinuosity was calculated as $\sigma = L/\lambda$ [32], where L and λ are the arc length and wavelength of the meander, respectively (Figure 1b). The wavelength was calculated as $\lambda \approx 2\pi B$ [33], where $B = 0.3$ m is the width of the sinuous channel. The calculated sinuosity and wavelength were equal to 1.25 and 1.88 m, respectively. At the entry and exit of the sinuous channel, guide vanes were provided to direct the flow into and out from the sinuous channel smoothly. The flow was administered into the main channel from an overhead tank with the help of a control valve. The water first falls into an inlet tank (2.8 m length, 1.5 m width, and 1.5 m deep) at the upstream of the flume and it gradually enters the main channel. A tailgate at the downstream of the flume manages the flow depth during experiments. The main channel discharge was measured using a rectangular notch located at the downstream end of the channel. The experiments were conducted on a uniform river sand bed of median diameter $d_{50} = 1.1$ mm. The flow discharge attained was equal to 0.0156 m^3/s with a flow depth of 0.117 m. The flow depth was measured along the centerline of the channel. In our study, the channel achieved the Reynolds number and Froude number as 57,731 and 0.40, respectively. Therefore, we have maintained a subcritical flow condition (Froude number < 1) and turbulent flow condition (Reynolds number > 10,000). We have physically modeled a sinuous channel in a laboratory flume. Scale effects may arise in physical modeling of sediment transport processes when all the forces in the model and the real field river flows are nonidentical. As gravity is the primary driving force in open channel flows, we tried to achieve a Froude number similar to field conditions. In the field study by Engel and Rhoads [23], the onsite Froude number is about 0.3 (subcritical flow), and the Reynolds number ranges from 292,698 to 397,254 (turbulent flow). The Froude number observed in our tests is very close to the field conditions. Therefore, scale effects due to this may be negligible. However, the Reynolds number of the laboratory flow is greater than 50,000. Most of the river flows are turbulent and in the hydraulic rough regime, where losses are independent of the Reynolds number. Therefore, the Reynolds number of the laboratory flows was greater than 50,000 and the shear Reynolds number ($R_* = 75$) was achieved greater than 70 so that the laboratory flow was in the fully turbulent hydraulic rough regime to better account for the losses [24].

The instantaneous flow velocities u, v, and w in three directions X (streamwise), Y (transverse), and Z (vertical) were measured by a velocimeter. The instantaneous velocities are equal to the sum of mean velocities ($\overline{u}, \overline{v}$, and $\overline{w}$) and fluctuating velocities (u', v', and w') in the form $u = \overline{u} + u'$, $v = \overline{v} + v'$, and $w = \overline{w} + w'$. The experimental investigation of recording the velocity and morphological changes were performed along the second bend of the sinuous channel. We used an acoustic doppler velocimeter (ADV) by Nortek®(vectrino+ 4-beam down-looking probe, Nortek AS, 1351 Rud, Norway) to record the velocity data at five locations (1, 2, 3, 4, and 5) of the bend at section $r-r$, $s-s$, and $t-t$ (Figure 1c). The velocimeter works on the principle of Doppler effect and has four downward-looking probes with the sampling volume located 0.05 m beneath the central transmitter. Readings were taken at the five locations throughout the flow depth. At each point of measurement, the velocimeter recorded 12,000 samples for 120 s (sampling rate 100 Hz). For higher-order statistical analysis, Schwarz et al. [34] recommended that at least 10,000 samples should be collected.

The uncertainty associated with the ADV measurements was tested by taking 17 pulses for a duration of 120 s recorded at near-bed depth $z \sim 3$ mm, where z is upward and is positive in the vertical direction (Table 1). The data collected from the ADV contains spikes, and therefore it should be filtered. The data were filtered using the acceleration threshold method [35]. During the measurement, the signal-to-noise ratio (SNR) and correlation were greater than 15 dB and 60%–70%, respectively. The correlations were reduced by ± 5% approximately near the channel bed [36]. The spikes were filtered such that the despiked data satisfies the Kolmogorov's 5/3 law in the inertial subrange (Figure 2), where the acceleration threshold value ranged from 1 to 1.5 by trial and error [37].

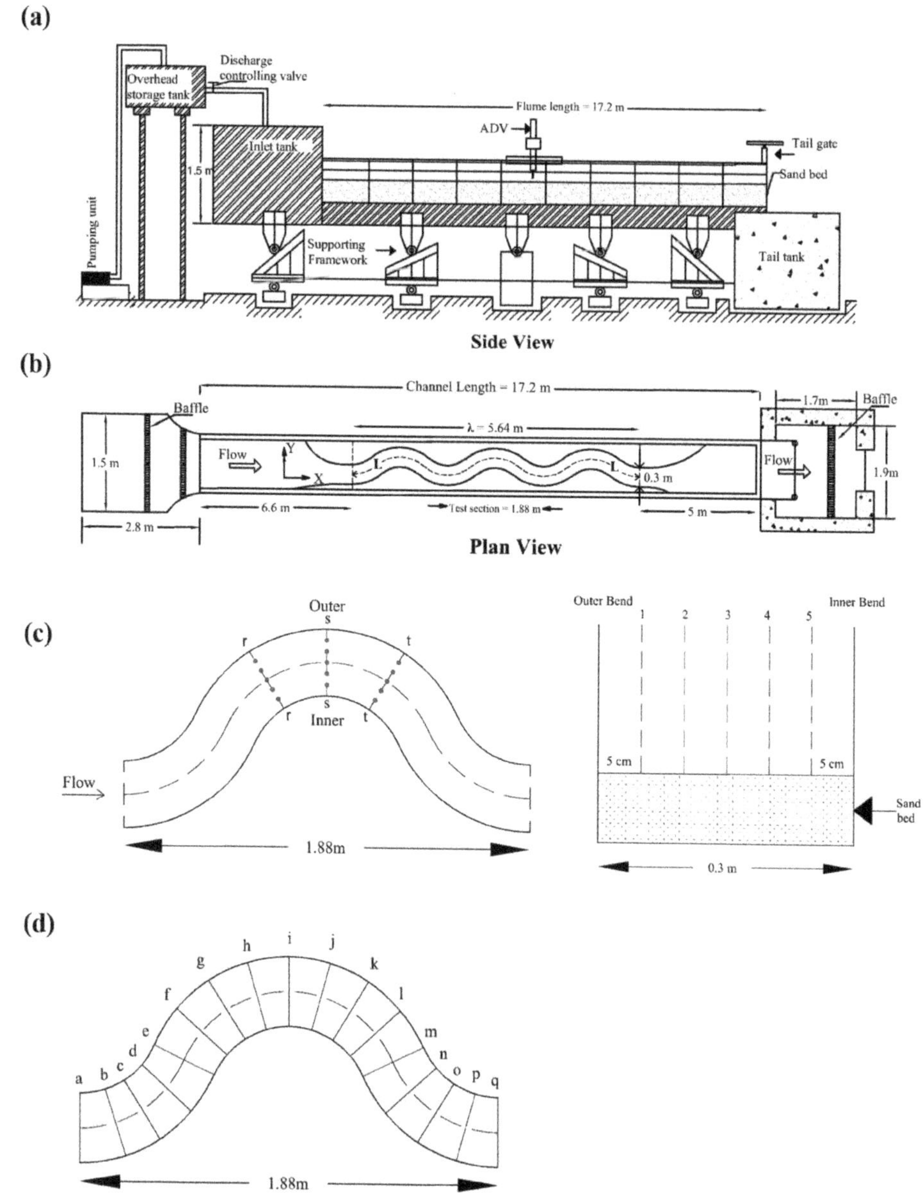

Figure 1. Schematic diagram of the experimental setup showing: (**a**) Side view of the experimental flume; (**b**) plan view of the experimental setup; (**c**) section r − r, s − s, and t − t where velocities were measured at locations 1, 2, 3, 4, and 5; (**d**) sections from "a" to "q" where ultrasonic ranging system (URS) readings were made to track morphological changes.

Table 1. Uncertainty test of acoustic doppler velocimeter (ADV) data.*

	$\bar{u}$ (m/s)	$\bar{v}$ (m/s)	$\bar{w}$ (m/s)	$(\overline{u'u'})^{0.5}$ (m/s)	$(\overline{v'v'})^{0.5}$ (m/s)	$(\overline{w'w'})^{0.5}$ (m/s)
Standard Deviation	5×10^{-3}	6.18×10^{-5}	4.23×10^{-5}	1.84×10^{-5}	3.65×10^{-5}	1.8×10^{-5}
Uncertainty (%)	0.221	0.175	0.192	0.0177	0.029	0.039

*$\left(\overline{u'u'}\right)^{0.5}$, $\left(\overline{v'v'}\right)^{0.5}$, and $\left(\overline{w'w'}\right)^{0.5}$ are the root mean square values of u′, v′, and w′, respectively.

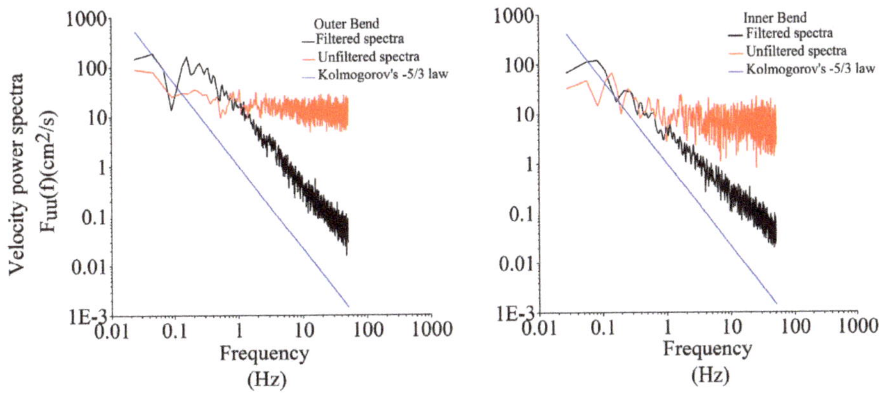

Figure 2. Power spectra $\left[F_{uu}(f)\ cm^2/s\right]$ of unfiltered and filtered velocity time-series at outer and inner bends. $F_{uu}(f)$ is the velocity power spectra of the streamwise velocity u, which is a function of frequency f (in Hz).

To examine the morphological changes along the bend, we used the ultrasonic ranging system (URS). It is a SeaTek®(1.0 cm diameter model, Seatek SPa diesels, Lombardy, Italy,) manufactured instrument consisting of eight transducers. In this instrument, the transducer acts as both transmitter and receiver. The transducer first transmits a pulse of 10-microsecond duration, and then this pulse travels through the water and reflects off a target. The reflected signal travels back to the transducer and is detected back by the electronics. The system has an accuracy of ±0.2 mm. The uncertainty associated with the URS measurements was evaluated by taking 16 sample recordings at the center of the bend (Table 2). The URS measured the bed elevation along the bend in 17 sections (Figure 1d). The transducers were mounted on a trolley and were aligned horizontally in a single line to track the changes. The URS tracks the distance (in centimeters) between the channel bed and the water surface.

Table 2. Uncertainty test for ultrasonic ranging system (URS) measurements.

Standard Deviation	6.19×10^{-3}
Uncertainty (%)	0.051

3. Results and Discussions

The velocity profile in the outer (location 1) and inner (location 5) bend of the section s − s of the sinuous channel is shown in Figure 3. The mean velocities $\overline{u}$, $\overline{v}$, and $\overline{w}$ in streamwise, transverse, and vertical directions are calculated as:

$$\overline{u} = \frac{1}{n}\sum_{i=1}^{n} u_i \tag{2}$$

$$\overline{v} = \frac{1}{n}\sum_{i=1}^{n} v_i \tag{3}$$

$$\overline{w} = \frac{1}{n}\sum_{i=1}^{n} w_i \tag{4}$$

where n is the total sample number. In Figure 3, we observe that the magnitude of inner velocity is more when compared to the outer bend of the channel throughout the flow depth. This result is similar to that found by the authors of the papers [3,12,38] who claimed that the velocity is greater in the inner bend than outer because of velocity redistribution. Rozovskiĭ [1] considered the logarithmic distribution for the streamwise velocity profile expressed as:

$$\frac{u}{\overline{U}} = 1 + \frac{\sqrt{g}}{\kappa C}(1 + \ln \overline{z}) \tag{5}$$

where U is the depth-averaged streamwise velocity, κ is the von Kármán constant, C is the Chézy coefficient, and $\overline{z} = z/h$ (z is the height to the point of measurement above channel bed, and h is the flow depth). The experimental profiles of the streamwise velocity in our study show an acceptable correlation with Equation (5).

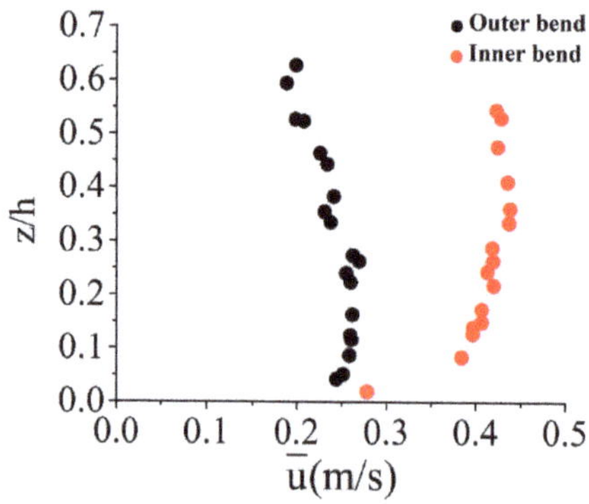

Figure 3. Streamwise velocity profile at the outer (location 1) and inner bends (location 2) of the sinuous channel.

Contour plots show the mean velocity $\overline{u}$ (m/s) distribution across three sections r − r, s − s, and t − t (Figure 4). The mean velocity is maximal towards the inner bend. In section t − t, the velocity is distributed throughout the channel width. This finding reveals that the mean velocity is not always greater towards the outer bend. Due to the inward skewing, the main velocity may take longer time to reach the outer bend, and therefore the maximum velocity is redistributed mostly towards the inner bend. Shams et al. [38] observed higher streamwise velocity towards the inner boundary in their study on a physical and laboratory-scale model.

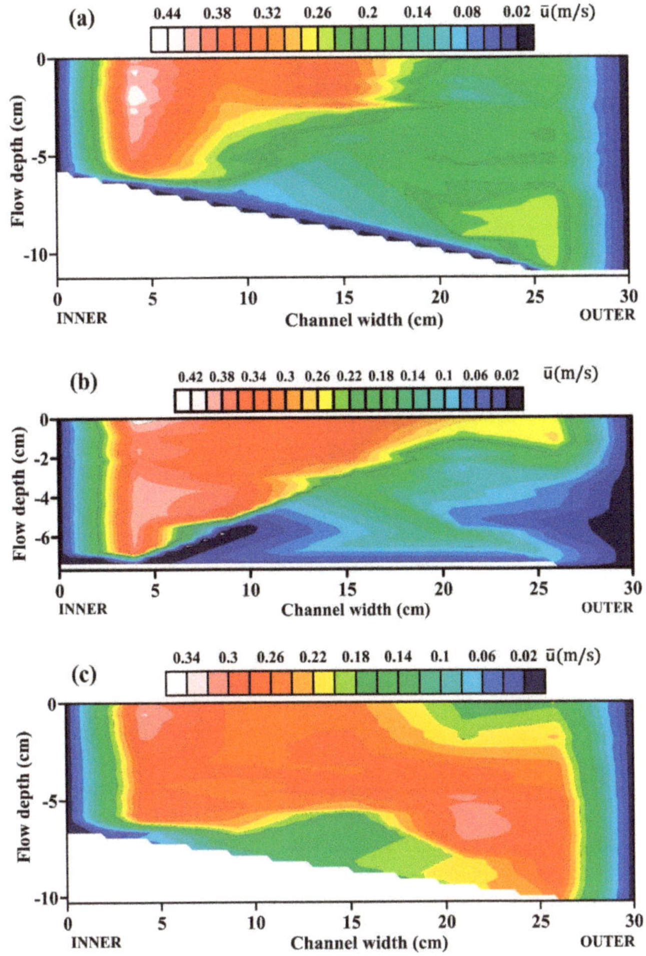

Figure 4. Contour plots of streamwise velocity at the bend cross-sections (**a**) r − r, (**b**) s − s, and (**c**) t − t of the sinuous channel.

The morphological changes are also analyzed along a bend of the sinuous channel. The variations in an alluvial bed are visible in the channel. The contour plots of the morphological changes were represented using the Surfer®(Golden Software, Colorado, US) [39] at different time intervals (Figure 5). After the desired discharge was achieved in the channel, the morphology readings were taken at intervals of 2, 6, and 10 h Readings were taken up to 12 h, and after which no significant changes were noticed in the bend. It can be seen that the outer bend experiences erosion, which is increasing over time [40,41]. With the acquired flow discharge, visible transport of the sediment took place. The flow interaction in bends allows the sediment to move in a transverse or radial direction (perpendicular to the direction of flow) across the bend. This motion of sediment is due to the established secondary currents in bends. Previous studies [8,10,15] have already focused on the average turbulent parameters such as the bed shear stresses, Reynolds stresses, and secondary currents to explain the scour and deposition in bends. Here, we have focused on the high-order turbulence characteristics and how they affect the morphological processes in a sinuous river. Investigation at 2 h shows the initial development of variation in the bed along the outer and inner bend. With time (6 and 10 h), the scour depth at the outer bend is prominent.

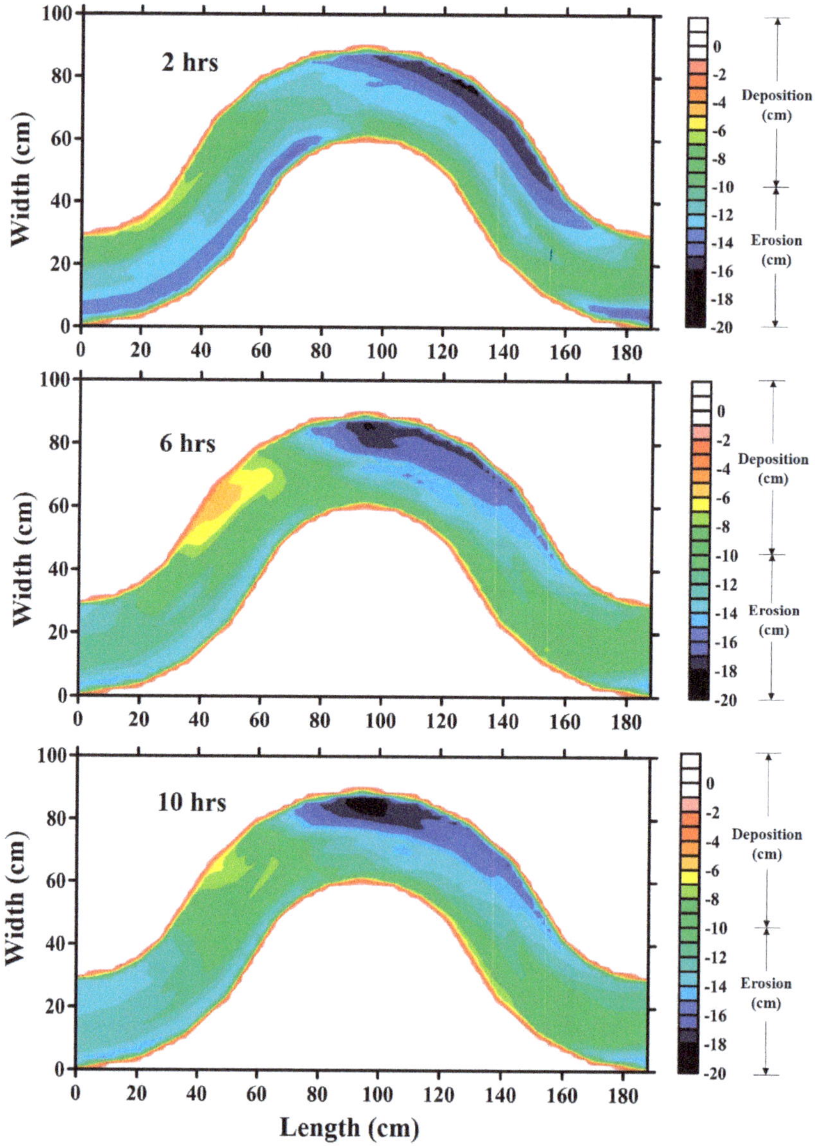

Figure 5. Morphological changes along the second bend of the sinuous channel after 2, 6, and 10 h. This bend was selected because it was unaffected by the entry and exit conditions.

The cross-sectional bed elevation across the bend apex (section i) is shown in Figure 6. Considering the initial level as the datum, the outer bend experiences erosion, which has increased with time. At the outer bend of section i, the depth of scour after 2 h run was found to be 4.35 cm. After 10 h run, the scour depth was estimated to be 7.99 cm. The extreme lower point in the vertical axis is the maximum depth the scour has reached. There is sediment below this point, and thus the scour has not touched the rigid bed of the channel. The flow characteristic and the channel planform play a significant role in the development of morphological changes in a channel. However, the observed morphological changes in the sinuous channel do not conform to the velocity distribution given in Figures 3 and 4. Hence, this conflicting behaviour might be explained by higher-order velocity moments.

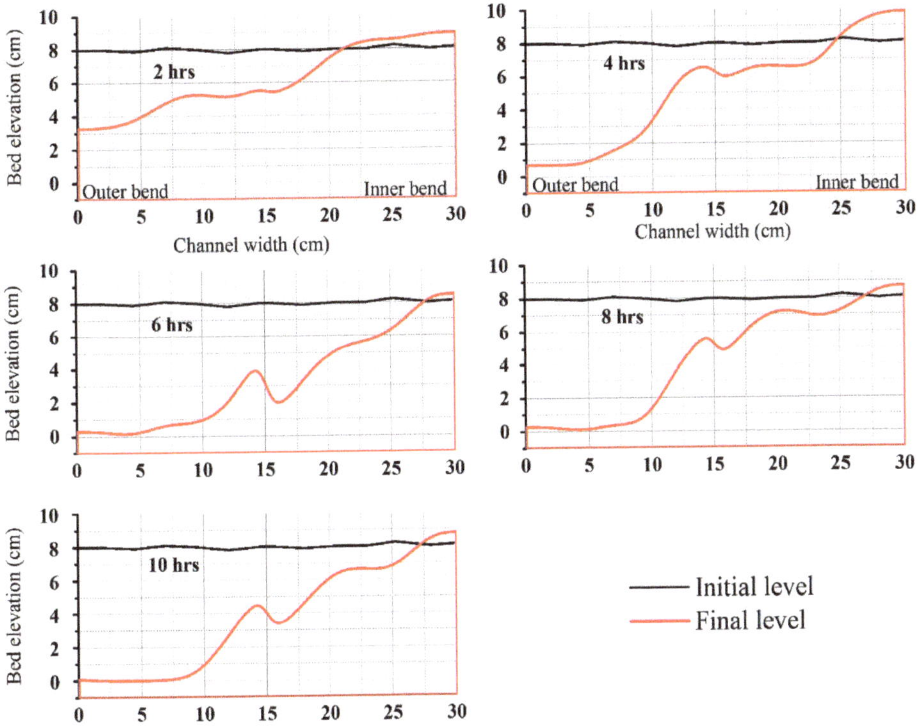

Figure 6. Cross-sectional morphological changes across the bend apex (section i) after 2, 4, 6, 8, and 10 h.

The velocity profile in both the outer and inner banks varies, as the pressure forces on the banks are different. The logarithmic law of mean velocity profile in the inertial region (i.e., inner flow zone) is given as:

$$\frac{\langle u \rangle}{u_*} = \langle u^+ \rangle = \kappa^{-1} \ln\left(\frac{zu_*}{\nu}\right) + B \tag{6}$$

where u_* [$=(\tau_o/\rho)^{0.5}$] is the shear velocity and τ_0 (bed shear stress) is found out by using the TKE (turbulent kinetic energy) method, υ is the kinematic viscosity of water, z is the wall distance (i.e., distance from the bed), κ is the von Kármán constant, and B is the constant. The value of u_* and τ_0 for the channel are 0.034 m/s and 1.2 N/m^2, respectively.

Many previous efforts by various researchers were put forward to understand the erosion and deposition behaviour in the outer and inner bend of a sinuous channel. In this section, this behaviour of a sinuous channel will be looked upon from the perspective of structure function. Here, we establish a relation between higher-order velocity moments with the erosion and deposition across sinuous bend. The erosion is a result of the transport of sediment particles. From the granular perspective, the motion of a sediment particle depends upon the balance of drag force exerted by the fluid flow and submerged weight of the particle. The classical turbulence parameter associated with transport is the average bed shear stress τ_0. When τ_0 exceeds τ_c (critical bed shear stress), erosion is expected. In our experiments, the sand bed was in motion. Moreover, the tractive stress depends upon the fluid velocity above the particle $\tau_o \sim u^2$. Therefore, the behaviour of instantaneous component of velocity fluctuations becomes significant to erosion, as the instantaneous drag force is directly dependent on it. Hence, we have analyzed higher moments of turbulence in the inner and outer bend.

Studies by various researchers [42–44] have found a logarithmic nature of velocity fluctuations, which is given as:

$$\left\langle \left(u'^{+}\right)^{2} \right\rangle = B_1 - A_1 \ln(z/\delta) \tag{7}$$

where $u'^{+} = \frac{(u - \langle u \rangle)}{u_*}$ is the nondimensional fluctuating component of streamwise velocity and δ is the boundary layer thickness. The velocity fluctuating moments raised to the pth root follows logarithmic nature as proposed by Meneveau and Marusic [28]:

$$\left\langle \left(u'^{+}\right)^{2p} \right\rangle^{1/p} = B_p - A_p \ln(z/\delta) = D_p(Re_*) - A_p \ln z^{+} \tag{8}$$

where $z^{+} = \left(\frac{z u_*}{\nu}\right)$ is the distance to the wall and $D_p = B_p + A_p \ln Re_*$ where $Re_* = \left(\frac{u_* \delta}{\nu}\right)$ and A_P can be theoretically expressed as $A_P = A_1\left[(2p-1)!!\right]^{1/p}$, as per Gaussian statistics.

Figure 7 shows the variance in streamwise velocity comparing the inner and outer bends of a sinuous river channel. Both the profiles tend to follow a logarithmic profile after a depth of $z^{+} > 1000$ and tend to approach zero asymptotically after a particular depth ($z^{+} > 4000$). The magnitude of outer bend variance is higher when compared to the inner bend of the channel. This point indicates the erosional behaviour in a sinuous bend. The outer bend of the sinuous channel possesses more magnitude in fluctuating velocity from the mean velocity than the inner bend. However, the mean velocity of the inner bend is more in magnitude than the outer bend, as shown in Figures 3 and 4. This point indicates that outer bend erosion mostly depends on the velocity fluctuation or deviation from mean velocity rather than the mean velocity. The constants for Equation (7) are: $(A_1, B_1) = (3.875, 36.61)$ for outer bend and $(A_1, B_1) = (1.74, 16.024)$ for inner bend of the channel. From this, we can understand that as we move from outer to the inner bend, the constants also decrease. This point indicates that the logarithmic law constants are dependent on the flow impact location of the sinuous channel.

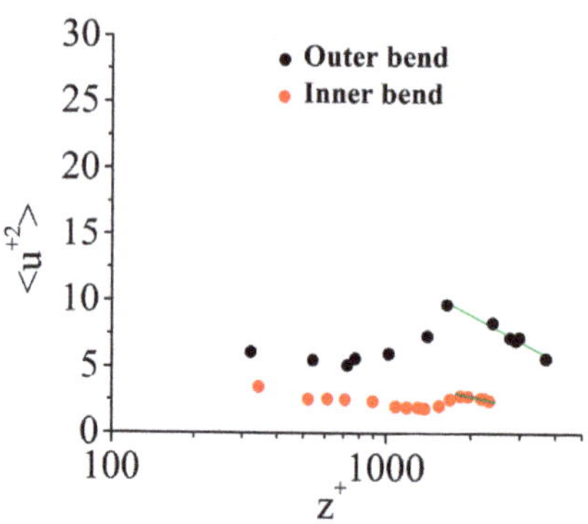

Figure 7. Streamwise velocity variance in turbulent boundary layers for flow at inner and outer bends of the sinuous river channel.

The relationship between structure function and outer bend erosion can be further established by analyzing the higher-order velocity moments (Equation (8)). Before investigating the higher-order moments of velocity fluctuations, we have to check the convergence of higher-order moments. It was achieved by multiplying the marginal probability density function (PDF) by the velocity fluctuation moments in the near-bed region. Marginal PDF is defined as the probability of values of continuous random variable (say P) without referring to the values of the other variable (say Q). In the present study, marginalized PDF of streamwise velocity (u) is considered. Figure 8a,b show premultiplied PDFs for 2p = 2 and 4, respectively at flow depth z/h = 0.08. Here, we notice that the area covered by the moments of order 2p = 2 and 2p = 4 for both outer and inner bend, respectively, are captured by the available data. In other words, we can say that there is convergence at higher-order moments. Furthermore, the premultiplied velocity fluctuation PDFs is greater in the case of outer bend, indicating that the outer bend suffers from erosion.

Higher-order moments for 2p = 4 and 6 were calculated with respect to flow depth. Figure 8c shows higher-order moments for 2p = 4. The result is similar to that of 2p = 2, where both outer and inner bend profiles follow logarithmic law. The outer bend fluctuation is more as compared to inner bend, which indicates more erosion chances in the outer bend. The fitted constants values for 2p = 4 are $(A_2, B_2) = (8.569, 80.962)$ for the outer bend and $(A_2, B_2) = (3.986, 35.091)$ for the inner bend of the sinuous channel. This result was similar to 2p = 2 moments where constants are greater in the outer bend of the channel. The magnitude of higher-order moments was also compared for outer and inner bends of the sinuous channel. Figure 8d,e show higher-order moments for 2p = 2, 4, and 6 at different flow depths for outer and inner bends in a sinuous channel, respectively. We observe that as the order of moments increases, the magnitude also tends to increase. Both the profiles for outer and inner bends at different order moments tend to follow the logarithmic nature and asymptotically tend to zero after some distance. For higher moment, i.e., 2p = 8, 10, etc., it follows the same trend.

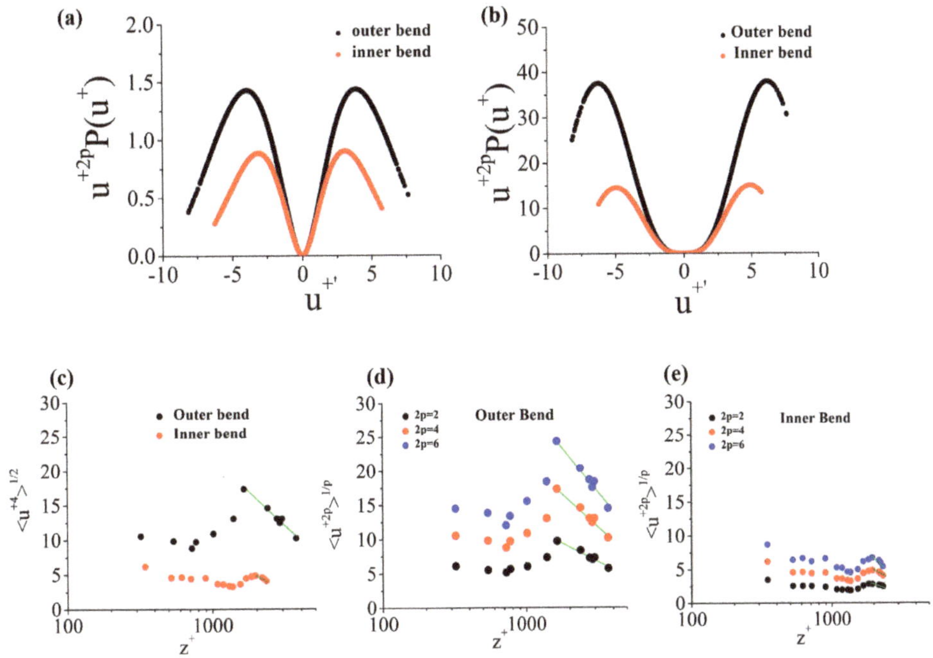

Figure 8. Premultiplied probability density functions (PDF) of normalized velocity fluctuations $u^{+2p}P(u^{+})$ at z/h = 0.08 with moments (**a**) 2p = 2 and (**b**) 2p = 4. (**c**) Moments of order 2p = 4 for streamwise velocity as a function of wall-normal distance. Moments of different orders of streamwise velocity fluctuation as a function of wall normal distance for flow subjected to (**d**) outer bend and (**e**) inner bend.

Further investigation was carried to see the flatness or kurtosis in the inner and outer bend of the channel (Figure 9). The flatness factor of streamwise velocity was calculated as:

$$F_4 = \frac{\langle u^{+4} \rangle}{< u^{+2} >^2} \tag{9}$$

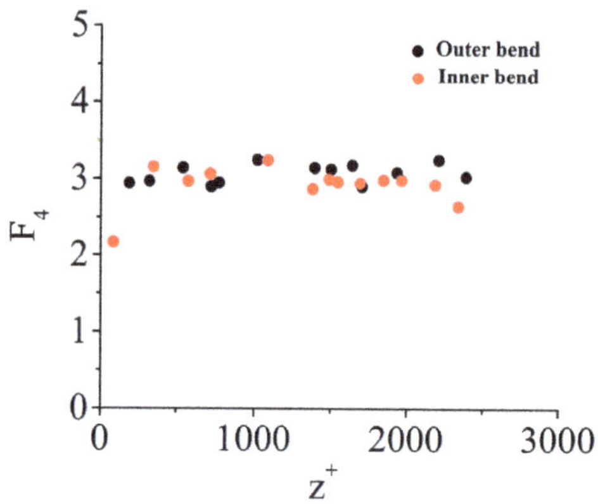

Figure 9. Flatness factor as a function of the wall distance for outer and inner bend of the sinuous channel.

The flatness factor in the outer bend of the channel is greater than that in the inner bend. From Figure 9, we can comment that from $z^+ \approx 1000$, the flatness in the outer bend has increased in comparison to the inner bend. Most points in the outer bend follow $F_4 > 3$ which represents distribution with a peaky signal characteristic. On the other hand, points in the inner bend follow $F_4 < 3$ representing distribution with a flat characteristic. The average kurtosis of the outer bend is 3.06, and that of inner bend is 2.91 throughout the flow depth. This characteristic indicates the erosional behaviour in the outer bend.

4. Conclusions

Higher-order moments of streamwise velocity fluctuations were studied to explain the erosional behaviour of the outer bend of a sinuous river channel. The higher-order moments in association with the morphological changes of a channel contribute a novel approach to understand the turbulent nature in a sinuous channel. The stresses are found higher in the outer bend, which are well reported earlier in the literature. Analyzing the high-order velocity fluctuations has provided clear insights into the scour mechanism near the outer wall of the sinuous bend. The profile of the velocity fluctuations for inner and outer bends at higher moments $2p = 2, 4$, and 6 show logarithmic nature from a particular depth $(z^+ > 1000)$. In all these cases, the magnitude of outer bend velocity fluctuation is more when compared to the inner bend. Premultiplied PDFs for $2p = 2$ and 4 for outer bend are greater than those of the inner bend. Both the points indicate that though the mean velocity at inner bend is higher than the outer bend, its erosional behaviour mainly depends on the fluctuations from the mean velocity. Further, the constants (A_p, B_p) of the logarithmic law for velocity fluctuations depend on the location of the flow impact in the sinuous channel. The flatness factor or kurtosis of streamwise velocity was also found to be slightly higher in the outer bend as compared to inner bend. A similar investigation can be carried out with field data. The analysis of structure function can also be applicable to different sediment particles, flow condition, and numerical modeling.

Author Contributions: J.T. and J.B. did the experimentation, analysis, and wrote the first draft of the manuscript; B.K. supervised the work; J.T., J.B., B.K., and G.O. completed the final draft. All authors have read and agreed to the published version of the manuscript.

Funding: This research received no external funding.

Conflicts of Interest: The authors declare no conflict of interest.

References

1. Rozovskiĭ, I.L. *Flow of Water in Bends of Open Channels*; Academy of Sciences of the Ukrainian SSR: Kiev, Ukraine, 1957.
2. Anwar, H.O. Turbulent structure in a river bend. *J. Hydraul. Eng.* **1986**, *112*, 657–669. [CrossRef]
3. De Vriend, H.J.; Geldof, H.J. Main flow velocity in short river bends. *J. Hydraul. Eng.* **1983**, *109*, 991–1011. [CrossRef]
4. Blanckaert, K.; Graf, W.H. Mean flow and turbulence in open-channel bend. *J. Hydraul. Eng.* **2001**, *127*, 835–847. [CrossRef]
5. Booij, R. Modeling the flow in curved tidal channels and rivers. In Proceedings of the International Conference on Estuaries and Coasts, Hangzhou, China, 9–11 November 2003; pp. 786–794.
6. Sukhodolov, A.; Kaschtschejewa, E. Turbulent flow in a meander bend of a lowland river: Field measurements and preliminary results. In Proceedings of the River Flow 2010, Braunschweig, Germany, 8–10 September 2010; Dittrich, A., Koll, K., Aberle, J., Geisenhainer, P., Eds.; River Flow 2010. Karlsruhe: Bundesanstalt für Wasserbau. S. 309-316.
7. Engel, F.L.; Rhoads, B.L. Velocity profiles and the structure of turbulence at the outer bank of a compound meander bend. *Geomorphology* **2017**, *295*, 191–201. [CrossRef]
8. Graf, W.H.; Blanckaert, K. Flow around bends in rivers. In Proceedings of the 2nd International Conference New Trends in Water and Environmental Engineering for Safety and Life: Eco-Compatible Solutions for Aquatic Environments, Capri, Italy, 24–28 June 2002; pp. 1–9.
9. Blanckaert, K. Topographic steering, flow recirculation, velocity redistribution, and bed topography in sharp meander bends. *Water Resour. Res.* **2010**, *46*. [CrossRef]
10. Blanckaert, K. Analysis of coherent flow structures in a bend based on instantaneous-velocity profiling. In Proceedings of the 3rd International Symposium on Ultrasonic Doppler Methods for Fluid Mechanics and Fluid Engineering, EPFL, Lausanne, Switzerland, 9–11 September 2002.
11. Esfahani, F.S.; Keshavarzi, A. Effect of different meander curvatures on spatial variation of coherent turbulent flow structure inside ingoing multi-bend river meanders. *Stoch. Environ. Res. Risk Assess.* **2011**, *25*, 913–928. [CrossRef]
12. Xu, D.; Bai, Y. Experimental study on the bed topography evolution in alluvial meandering rivers with various sinuousnesses. *J. Hydro-Environ. Res.* **2013**, *7*, 92–102. [CrossRef]
13. Binns, A.D.; da Silva, A.M.F. Meandering bed development time: Formulation and related experimental testing. *Adv. Water Resour.* **2015**, *81*, 152–160. [CrossRef]
14. Huggett, R.J. *Fundamentals of Geomorphology*, 4th ed.; Routledge—Taylor & Francis Group: New York, NY, USA, 2017.
15. Da Silva, A.M.; Ebrahimi, M. *Meandering Morphodynamics: Insights from Laboratory and Numerical Experiments and Beyond*; American Society of Civil Engineers: Reston, VA, USA, 2017. [CrossRef]
16. Whiting, P.J.; Dietrich, W.E. Experimental studies of bed topography and flow patterns in large-amplitude meanders: 2. Mechanisms. *Water Resour. Res.* **1993**, *29*, 3615–3622. [CrossRef]
17. Whiting, P.J.; Dietrich, W.E. Experimental studies of bed topography and flow patterns in large-amplitude meanders: 1. Observations. *Water Resour. Res.* **1993**, *29*, 3605–3614. [CrossRef]
18. Termini, D. Experimental observations of flow and bed processes in large-amplitude meandering flume. *J. Hydraul. Eng.* **2009**, *135*, 575–587. [CrossRef]
19. Luchi, R.; Zolezzi, G.; Tubino, M. Modelling mid-channel bars in meandering channels. *Earth Surf. Process. Landf.* **2010**, *35*, 902–917. [CrossRef]
20. Zhang, C.; Xu, M.; Hassan, M.A.; Chartrand, S.M.; Wang, Z.; Ma, Z. Experiment on morphological and hydraulic adjustments of step-pool unit to flow increase. *Earth Surf. Process. Landf.* **2019**, *45*, 280–294. [CrossRef]
21. Abidin, R.Z.; Sulaiman, M.S.; Yusoff, N. Erosion risk assessment: A case study of the Langat River bank in Malaysia. *Int. Soil Water Conserv. Res.* **2017**, *5*, 26–35. [CrossRef]

22. Choi, C.E.; Cui, Y.; Au, K.Y.K.; Liu, H.; Wang, J.; Liu, D.; Wang, H. Case study: Effects of a partial-debris dam on riverbank erosion in the Parlung Tsangpo River, China. *Water* **2018**, *10*, 250. [CrossRef]
23. Engel, F.L.; Rhoads, B.L. Interaction among mean flow, turbulence, bed morphology, bank failures and channel planform in an evolving compound meander loop. *Geomorphology* **2012**, *163*, 70–83. [CrossRef]
24. Heller, V. Scale effects in physical hydraulic engineering models. *J. Hydraul. Res.* **2011**, *49*, 293–306. [CrossRef]
25. Peakall, J.; Ashworth, P.J.; Best, J.L. Meander-bend evolution, alluvial architecture, and the role of cohesion in sinuous river channels: A flume study. *J. Sediment. Res.* **2007**, *77*, 197–212. [CrossRef]
26. Braudrick, C.A.; Dietrich, W.E.; Leverich, G.T.; Sklar, L.S. Experimental evidence for the conditions necessary to sustain meandering in coarse-bedded rivers. *Proc. Natl. Acad. Sci. USA* **2009**, *106*, 16936–16941. [CrossRef]
27. Coz, J.L.; Michalkova, M.; Hauet, A.; Comaj, M.; Dramais, G.; Holubová, K.; Piégay, H.; Paquier, A. Morphodynamics of the exit of a cutoff meander: Experimental findings from field and laboratory studies. *Earth Surf. Process. Landf. J. Br. Geomorphol. Res. Group* **2010**, *35*, 249–261. [CrossRef]
28. Meneveau, C.; Marusic, I. Generalized logarithmic law for high-order moments in turbulent boundary layers. *J. Fluid Mech.* **2013**, *719*. [CrossRef]
29. De Silva, C.M.; Marusic, I.; Woodcock, J.D.; Meneveau, C. Scaling of second-and higher-order structure functions in turbulent boundary layers. *J. Fluid Mech.* **2015**, *769*, 654–686. [CrossRef]
30. Sharma, A.; Kumar, B. High-Order Velocity Moments of Turbulent Boundary Layers in Seepage Affected Alluvial Channel. *J. Fluids Eng.* **2018**, *140*, 81204. [CrossRef]
31. Leopold, L.B.; Langbein, W.B. River meanders. *Sci. Am.* **1966**, *214*, 60–73. [CrossRef]
32. Yalin, M.S. *River Mechanics*; Pergamon Press: Oxford, UK, 1992.
33. Yalin, M.S.; Da Silva, A.M.F. *Fluvial Processes*; IAHR Monograph: Delft, The Netherlands, 2001.
34. Schwarz, A.C.; Plesniak, M.W.; Murthy, S.N.B. Response of turbulent boundary layers to multiple strain rates. *J. Fluid Mech.* **2002**, *458*, 333–377. [CrossRef]
35. Goring, D.G.; Nikora, V.I. Despiking acoustic Doppler velocimeter data. *J. Hydraul. Eng.* **2002**, *128*, 117–126. [CrossRef]
36. Dey, S.; Das, R.; Gaudio, R.; Bose, S.K. Turbulence in mobile-bed streams. *Acta Geophys.* **2012**, *60*, 1547–1588. [CrossRef]
37. Deshpande, V.; Kumar, B. Turbulent flow structures in alluvial channels with curved cross-sections under conditions of downward seepage. *Earth Surf. Process. Landf.* **2016**, *41*, 1073–1087. [CrossRef]
38. Shams, M.; Ahmadi, G.; Smith, D.H. Computational modeling of flow and sediment transport and deposition in meandering rivers. *Adv. Water Resour.* **2002**, *25*, 689–699. [CrossRef]
39. Golden Software, LLC. 809 14th Street, Golden, Colorado 80401. Available online: https://www.goldensoftware.com/ (accessed on 10 March 2020).
40. Rahman, M.; Nagata, N.; Hosoda, T.; Muramoto, Y. Experimental study on morphological process of meandering channels with bank erosion. *Proc. Hydraul. Eng.* **1996**, *40*, 947–952. [CrossRef]
41. Da Silva, A.M.F.; El-Tahawy, T.; Tape, W.D. Variation of flow pattern with sinuosity in sine-generated meandering streams. *J. Hydraul. Eng.* **2006**, *132*, 1003–1014. [CrossRef]
42. Nezu, I. Turbulent Structure in Open-Channel Flows. English Translation of the Japanese Dissertation of Iehisa Nezu. 1977. Available online: https://repository.tudelft.nl/islandora/object/uuid%3Aa41f39c2-fce6-4647-bd7a-1d412c720ed7 (accessed on 10 March 2020).
43. Marusic, I.; Kunkel, G.J. Streamwise turbulence intensity formulation for flat-plate boundary layers. *Phys. Fluids* **2003**, *15*, 2461–2464. [CrossRef]
44. Hultmark, M.; Vallikivi, M.; Bailey, S.C.C.; Smits, A.J. Turbulent pipe flow at extreme Reynolds numbers. *Phys. Rev. Lett.* **2012**, *108*, 94501. [CrossRef] [PubMed]

water

MDPI

Review

Scour at Bridge Piers Protected by the Riprap Sloping Structure: A Review

Antonija Harasti *, **Gordon Gilja**, **Kristina Potočki** and **Martina Lacko**

Department of Hydroscience and Engineering, Faculty of Civil Engineering, University of Zagreb, 10000 Zagreb, Croatia; gordon.gilja@grad.unizg.hr (G.G.); kristina.potocki@grad.unizg.hr (K.P.); martina.lacko@grad.unizg.hr (M.L.)
* Correspondence: antonija.harasti@grad.unizg.hr

Abstract: Bridge piers on large rivers are often protected from scouring using launchable stone, such as a riprap sloping structure. While such scour countermeasures are effective for pier protection, they significantly alter flow conditions in the bridge opening by overtopping flow and flow contraction, deflecting the formation of the scour hole downstream and exposing the downstream riverbed to additional scour. This paper provides a comprehensive and relevant review of bridge scour estimation methods for piers with a riprap sloping structure installed as a scour countermeasure. Research on empirical methods for bridge scour estimation is reviewed and analyzed with formulae used for comparable structures—complex pier formulae and formulae for river training structures. A summary of relevant formulae applicable to piers with installed scour countermeasures is provided, as well as a discussion on the possible future research directions that could contribute to the field.

Keywords: bridge scour; empirical formulae; riprap sloping structure; flow contraction; overtopping flow

Citation: Harasti, A.; Gilja, G.; Potočki, K.; Lacko, M. Scour at Bridge Piers Protected by the Riprap Sloping Structure: A Review. *Water* **2021**, *13*, 3606. https://doi.org/10.3390/w13243606

Academic Editor: Mouldi Ben Meftah

Received: 17 November 2021
Accepted: 13 December 2021
Published: 15 December 2021

1. Introduction

The majority of bridges have been built to provide an effective connection between the banks over the waterways, impacting society both economically and politically, and at the same time interacting with the waterway flow regime [1]. During the lifespan of the bridge crossing rivers, changes in the flow and sediment regime are anticipated, as well as consequent change in the variable action-imposed loads on the structure [2], or even complete undermining of the foundation soil [3], which can result in bridge failure or even collapse. Economic losses resulting from traffic disruption following the bridge failure exceed its construction value [4], making bridges critical infrastructure assets [5].

The bridge failures data recorded world-wide indicate that hydraulic causes (scour, floods, stream instability, lateral migration and floating debris) are prevailing among the factors causing bridge collapses: Imhof [6] reports that flooding/scour was the most frequent natural hazard (66%) causing bridge failure in Europe and North America; Muñoz Diaz et al. reported scour as causing 35% of bridge failures in Columbia, with an additional 7% failing from overtopping/floods; Schaap and Caner [7] reported 45% of failures in Turkey were hydraulic-caused, and 22% were directly caused by scour; available data for United States suggests that hydraulic-caused failures account for >50% of the collapses, and scour >20% [8–11]. Common materials used for construction of bridges are concrete/reinforced concrete or steel, with a 75-year design life in service [12]. According to the US bridge failure data, the majority of failed bridges are steel bridges (>60%), although their portion in the National Bridge Inventory is significantly smaller than the concrete ones, at 30% compared with 65%, respectively [13]. The average age of the bridges at the time of the failure was 64 years, and <50 years for steel and concrete material types, respectively.

Among the various types of erosion processes that occur in the riverine environment, the ones influencing bridges are generally divided into long-term general scour, contraction scour and local scour [14], acting independently or in combination with other hydraulic causes. Bridge piers obstruct the flow, forcing it to accelerate around the pier and generate large-scale turbulence structures, consequently increasing the turbulence in the flow downward towards the bed (horseshoe vortex) and flow downstream behind the pier (wake vortex) [15], causing local scouring. Therefore, local scour is inherently associated with the hydraulic structures interfering with the natural flow field [16] and its occurrence is often simultaneous with on one (or both) of other two types of scour, causing deepening of the entire riverbed in the vicinity of piers and abutments.

Bridge piers can be protected from scouring by either indirect or direct methods [17]. Indirect methods, such as collars [18], vanes [19], sacrificial piers [20], or self-protected piers with openings within the pier [21], change the flow pattern around the bridge pier and thus reduce the strength of downflow and horseshoe vortex [22]. Direct methods (riprap, geobags, cable-tied blocks or similar launchable material) increase the riverbed resistance in order to withstand the turbulent flow around the piers [23]. They are less expensive and easier to construct in comparison to indirect methods [24], and flexible which allows them to be retrofitted if necessary. The most common bed-armoring protection against scour is riprap [25], which consists of a material larger and heavier than riverbed sediment with the main purpose of preventing removal of the riverbed sediment downstream [26]. There are several types of riprap countermeasures: horizontal riprap layer [17] placed just above the riverbed level, riprap sloping structure [27] formed by stones conically mounded around the pier, gabion mattresses [28], sacks filled with stones [29], etc. Although riprap is considered stable, there is evidence of riverbed particle erosion during floods and consequent riprap slumping and sliding. Breusers et al. [30] pointed out that riprap protection can also induce scour, thus not fully fulfilling its purpose. Nielsen et al. [31] investigated the velocity distribution and critical bed shear stress under the riprap protection. Fredsoe et al. [25] described the horseshoe vortex as the primary mechanism that causes undermining the stones at the junction between the riprap and the riverbed. This way, riprap protection causes formation of the deflected scour hole next to the riprap scour protection, introducing hazard to the pier or adjacent river structures [32].

Scouring locally deepens the riverbed next to bridge piers, exposing a larger area of the pier to direct oncoming flow, thus subjecting them to greater hydrodynamic forces in comparison to nonscoured conditions [33]. Changes in hydrodynamic loading coinciding with other natural hazards, e.g., earthquakes, can have devastating consequences for both infrastructure and society [34]. Morphodynamic analysis of the river reach influencing the bridge is integrated into the bridge management system [35] in order to log changes developed since the historical reports and assess the current condition of the riverbed. Riverbed condition assessment is not unified—each county has adapted inspection procedures to the diversity of their infrastructure, respecting viable methods and instruments for gathering relevant and reliable data [36].

Scour holes reach their maximum depth during floods [37], after which they are backfilled by the bedload of the receding flow, reducing the apparent scour depth [33,38–40]. Most research in recent decades has focused on describing scour depth based on maximum equilibrium scour depth and for steady flow conditions [41,42]. Morphodynamic changes in the riverbed and consequently the scour process around bridge structures, when considered on a larger spatial scale, are not only caused by anthropogenic changes in the watershed, but also by changes in the natural flood regime and extreme climatic events such as floods, that affect the overall stability of bridge structures. Therefore, bathymetric data acquired post-flood are not reliable enough to determine the full extent of scouring, and analyses based on such data can lead to erroneous estimates of bridge safety. Real-time monitoring of scour development during floods is recognized as essential [43] to enable safety monitoring and systematic inspection of the riverbed condition and avoid unnecessary maintenance. Some of the monitoring equipment used for scour detection are magnetic sliding collars [44], sonar devices [45], float-out devices [46] or tilt sensors [47]. The current scour-monitoring systems are being constantly enhanced through development of advanced monitoring techniques, mainly to achieve more robust devices, reduce maintenance cost and enable integration with other sensors. Recent advances in scour monitoring include assessment based on vehicle-induced vibrations measured by acceleration sensors [48], influence of soil characteristics on the change in the predominant natural frequency of a bridge pier [49], scour depth measurement using vibration-based micro-electro-mechanical systems [50], instrumented particles [51], interferometric synthetic aperture radar stacking techniques [52], amongst others.

Regardless of the data acquirement method, in order to complete the risk assessment for preliminary hazard analysis in the bridge management system, hydraulic analysis must be performed. Hydraulic analysis will combine the scour data with other relevant variables: flow environment, pier geometry and riverbed composition, to quantify the scour hazard in comparison to foundation depth. Flow velocity, flow constriction and scour protection measures are the most important factors that should be considered when assessing scour risk to bridges [53]. For this purpose, empirical formulae are mostly used that calculate the maximum scour depth, taking into account all relevant variables as input. These empirical formulae are reliable if their usage adheres to conditions under which they were developed and validated on field-measured data, if possible.

Previous studies have focused on the scour development for a wide range of pier geometries, arrangements and spacing, riverbed compositions, sediment transport conditions, etc. However, there exists an evident gap in knowledge regarding the scour occurring at piers protected with riprap sloping structure and associated scour at the periphery of the riprap mound. The aim of this paper is to provide a comprehensive and relevant review of bridge scour estimation methods and provide discussion of their applicability for piers with riprap sloping structure installed as a scour countermeasure. Research on empirical methods for bridge scour estimation is reviewed and compared to complex pier formulae and formulae for river training structures similar to riprap sloping structure scour protection. Scour formulae in this paper are categorized considering the following: the data used for their development (field vs. experimental); pier geometry (single vs. complex piers); scour type (local vs. contraction), and riprap type (riprap sloping structure vs. layer riprap). A summary of relevant formulae applicable for piers with installed scour countermeasures is provided, as well as a discussion on the possible future research directions.

2. Bridge Scour Assessment

Most often, the approach for developing a scour depth formula is the application of dimensional analysis technique to an experimental dataset. The overall formula is assessed by validation with field-measured data and incorporation of one of the mathematical fitting models. Dataset for establishing formula can be obtained in the field and experimental environment.

2.1. Field and Experimental Data

Compared to the field data, the experimental data obtained in the laboratory provide a controlled flow environment, which is why the maturity of the scour hole is reliable only from the laboratory experiments [54]. However, the experimentally developed formulae have numerous disadvantages: a limited applicability in a range of input data due to scaling issues between geometric and sediment scales; oversimplified conditions that do not correspond to the field conditions (rectangular flume, uniform sediment gradation, clear-water conditions, and steady uniform flow), and time-consuming simulations to acquire the desired dataset. Time to reach equilibrium scour depth condition is reached asymptotically, so duration of the experimental test needs to be much more than at least 48 h to reach at least 90% of the equilibrium [55,56]. Melville and Chiew [43] define a formula for time needed for development of the equilibrium scour hole as s rate of 5% of the pier diameter in a 24 h period, which means that based on first 24 h-long experiment duration, equilibrium scour depth can be calculated. Their results showed that after 10% of the total time, scour depth is in the range of 50% to 80% of the equilibrium scour depth. To shorten time-consuming simulations, some authors accept semiequilibrium conditions, so each experiment lasted 4 to 6 h, and subsequently extend the scour depth values to 100% of the equilibrium scour depths.

Sheppard and Melville [55] noticed why scour depths are overestimated in small-scale experiments—because the forces of the pressure gradients on sediment particles near the smaller bridge piers are much greater than for larger piers. A similar observation was made by Huang et al. [57] in a scale effects study, where a comparison of results from a large-scale numerical model and a small-scale physical model was conducted. They found better prediction of turbulence flow pattern and sediment scour by the numerical model in comparison to the physical model, because the physical model introduces Froude similarity and ignores the effect of turbulent Reynolds number.

The numerical model or computational fluid dynamic (CFD) model is powerful tool to simulate interaction between the sediment transport process and the vortex system in the complex geometry environment. In contrast to time-consuming laboratory experiments and field investigations, numerical simulations are inexpensive and fast way to collect a large and comprehensive amount of the essential scour data. Numerous turbulent models have been already developed to calculate turbulent velocity and sediment scour ($k-\varepsilon$, $k-\omega$, LES, RSM, RNG and others). For example, Aly and Dougherty [58] examined different turbulence models in predicting the bed shear stress around different pier scour countermeasures. They concluded that $k-\varepsilon$ performed better than the $k-\omega$ model in approximating the experimental data, since $k-\omega$ simulates adverse pressure gradients, while the $k-\varepsilon$ realizable model successfully simulates velocity components. Alemi et al. [59] used the LES turbulence model to investigate the flow around a complex bridge pier on the scoured fixed bed. They validated the numerical model based on laboratory experiments by Graf and Istiarto [60] and achieved high correspondence with their results. By combining LES and wall function, a fast simulation process is achieved since the wall function simplifies fully developed flow with zero velocity near the wall region. Developing a scour numerical model can be challenging due to establishing mesh on complex riverbed geometry, which changes gradually as scour occurs. Once the computational grid is established, the flow field will scour the bed form from the first timestep, and the computational domain needs to be re-meshed. Zhu and Liu [61] presented three re-meshing strategies: z-level grid, shaved grid, and σ grid. Jalal and Hassan [62] examined various types and sizes of cells to identify optimum cell (5–10 mm) that balances the accuracy of results and reduction in computation time. Song et al. [63] developed a 3D scour model, ibScourFoam, based on an immersed boundary method, validated against available data from the literature and their own flume experiment. Their model showed advantage over previous scour models in accurate simulation of the scour process around complex structures.

Experimental and numerical data must be validated by field data to ensure the reliability of experimental results [64]. Field measurements of hydraulic parameters during a flood are essential to understand the morphodynamic evolution of the local scour phenomenon. Despite today's sophisticated survey techniques, it is not easy to accurately measure velocity and depth during floods in real-time due to instrument functionality under oscillating flow depths and instrument durability, as it is exposed to impact damage by floating debris. Scour monitoring is challenging from the perspective of instrument placement, where approximate location of the scour hole needs to be known. Surveying implies collecting data at a certain point in time, which means that the maturity of the local scour hole is unknown at the time of measurement. In addition to the difficulty of obtaining accurate scour depth data, the flow field upstream and downstream of the bridge pier must also be collected. Maneuvering a boat during a flood is difficult and hazardous in vicinity of the structures. Moreover, there are some difficulties in estimation of bed material size which used to be highly variable on site, and inaccuracy due to neglecting scour-hole backfilling effect [65].

For these reasons, in the present state-of-the-art data from field surveys are limited and do not encompass a sufficient range of hydraulic properties and characteristics of the river channel to represent a solid basis for high-quality research. Therefore, most of the local scour formulae have been developed based on laboratory experiments [14,22,30,66,67]. Yet, there are several empirical formulae developed from field data only. Governing parameters for scour depth (d_s) formula developed from field data are width of bridge pier (b), water depth (y_0) and Froude number (Fr) at the approach section and correction factor for pier-nose shape (K_1) as presented in Table 1.

Table 1. Local scour data obtained by field survey and corresponding formulae.

Field Data by	Number of Data	Formulae	
Froehlich [68]	83	$d_s = 0.32 \cdot b \cdot K_1 \cdot {Fr_1}^{0.2} \cdot [b_e/b]^{0.62} \cdot [y_0/b]^{0.46} \cdot [b/d_{50}]^{0.08}$	(1)
Gao et al. [69]	137	$d_s = 1.141 \cdot K_1 \cdot b^{0.6} \cdot {y_0}^{0.15} \cdot {d_{50}}^{-0.07} \cdot \left[\frac{v_0 - v_c'}{v_c - v_c'}\right]$	(2)
Ansari and Qadar [70]	100	$d_s = 0.86 \cdot b^{3.0}\ \ b < 2\ \mathrm{m}$ $d_s = 3.6 \cdot b^{0.4}\ \ b > 2.2\ \mathrm{m}$	(3)
Richardson and Davis [41]	-	$d_s = 2 \cdot K_1 \cdot K_2 \cdot K_3 \cdot K_4 \cdot \left[\frac{b}{y_0}\right]^{0.65} \cdot Fr^{0.43}$	(4)

Many authors have expressed concern about the lack of information available in the published studies regarding the source of dataset used to develop the empirical formulae [55,59,71] and insufficient information about the conditions that were present during the measurements. For example, there is no information on the methods that Zhuravlyov [72] used to obtain the data, neither the dimensions of the structures nor their shapes, and a field survey by Froehlich [68] does not contain a sediment gradation parameter.

2.2. Variables Governing the Scour Process

Numerous studies have aimed to evaluate the accuracy of the relevant formulae for scour depth estimation [73–77]. These comparative studies generally summarize that most existing formulae tend to overestimate the scour depth equilibrium. However, evaluating reliability of scour formulae must be done with caution because single variables within each scour depth formula may be defined with several curves, which implies dependency of a third variable [73].

Scour can be greatly affected by sediment size distribution which can be uniform or nonuniform. Initial movement of smaller particle sizes could be achieved at lower velocities, so scour may occur more readily in sandy riverbeds than in coarser sediment material [58]. For nonuniform sediment environments bed armoring occurs where the coarser gravel size protects finer particles from being transported by the flow. Pandey et al. [78] emphasize that the influence of the armored layer regarding local scour is poorly investigated because sediment transport in the armored riverbed is generally challenging to explicitly define. However, the authors analyzed how the formation of the armored layer in the scour hole stops further scouring processes during clear-water conditions. They proposed a new graphical approach for estimating maximum scour depth in an armor riverbed, showing that changes in dimensionless scour depth ($d_s / \left(b^2 \cdot y_0\right)^{1/3}$) with armor ratio ($d_{50a}/d_{50}$) are minor when it exceeds 0.5, where d_{50a} is the median diameter of armor layer d_{50} is median sediment size.

Most of scour formulae differentiate clear-water or live-bed conditions respective to bedload sediment transport: under clear-water conditions no bedload transport is present, while live-bed condition considers initiated sediment bedload transport. Landers and Mueller [73] evaluated the scour formulae by Gao et al. [69], comparing them to field measurements, and concluded that there was underprediction in deeper live-bed scour conditions (for $d_{50} > 3$ mm). On the contrary, the HEC-18 [41] formula overestimates scour depth under live-bed conditions, especially for greater flow intensities, while it shows good agreement with clear-water threshold scour [79]. Scour depth at the clear-water scour is considered to be 10% greater than the scour depth in live-bed conditions [58], but that finding is based on small-scale models. However, recent studies have shown that the reverse phenomena may occur in the field or in large-scale numerical models [79]. Sheppard and Melville [55] assembled field and experimental datasets of scour data from different authors in the literature and eliminated unreasonable ones by quality control screening, where 441 are experimental and 791 are field data. These data were used to test the accuracy of different scour formulae. The best performance showed the new Sheppard–Melville formula which was originated from modification of Sheppard and Miller's [67] and Melville's formulae [80]. Hereafter, Yang et al. [81] proposed a modified Sheppard–Melville formula regarding the influence of flow shallowness ratio (y_0/D), where D is pier diameter. They prove the considerable influence of flow shallowness ratio scour on scour depth, which increases with an increasing (y_0/D). Although modified the Sheppard–Melville formula still retains a small level of overestimation, Yang et al. demonstrate better accuracy of the formula than the HEC-18 formula. Overestimation is greater with smaller shallowness (y_0/D) ratio. Yang et al. collected live-bed field data during short-peak flood waves and concluded that live-bed scour peak may exceed the threshold peak.

Changes in the natural flow regime of a river can be defined in terms of ecological [82] and morphological processes [83], by magnitude, frequency, timing, duration, and unsteadiness. In contrast to engineering design practice, where typical flood events with a return period of 100 or 200 years are chosen to account for the scour susceptibility of bridges [84], Tubaldi et al. [85] propose a more rigorous assessment of scour susceptibility, that examines not only a single flood event but also subsequent events, as floods of short duration or low intensity can lead to partial erosion of the riverbed, making it easier for a subsequent flood to reach a maximum scour depth. Therefore, although the flow magnitude is the main factor influencing scour depth development, hydrograph duration and shape also have an important influence on the scour process and should be taken into account. The maximum scour depth is significantly lower when actual hydrograph shape is considered than the equilibrium scour depth for the constant flow rate. According to Plumb et al. [83], the effects of hydrograph shape, addressing the number of cycled hydrographs and duration of each hydrograph, as well as changes in the flow regime, should be examined in the scour development analysis. In addition, the inclusion of more complex sediment grain size distributions in the estimation of scour next to the bridges, along with the hydrograph shape parameters, results in a potentially higher threshold for bed motion state, complicating the prediction of instantaneous flux using formulae for steady-flow conditions [86,87]. Some empirical formulae have been developed for evaluation of time-dependent scour depth, i.e., considering hydrograph shape characteristics [43,88]. However, these formulae were mostly developed for steady-state flow conditions or adapted to unsteady conditions using various approaches, such as superposition of the hydrograph as a sequence of steady-state discharge steps [89–91], introducing a mathematical function for form of the hydrograph [92], or defining a dimensionless effective flow [93,94]. Apart from the fact that real-time field measurements are rarely available at high turbidity during a flood, the main challenge in studying the effects of flow characteristics, i.e., unsteady flow characteristics and the influence of multiple flood waves [95] on bedload transport and the scour process, is to separate the effects of flow conditions from the effects of bed material properties and sediment supply conditions [96]. It can be concluded that in future research it is necessary to investigate the effects on the development of maximum scour depth not only during a single flood event and the associated flood wave characteristics, but also previous flood events should be considered. Particular attention should be paid to the development of formulae and field measurement techniques that would allow the determination and measurement of the temporal evolution of scour depth.

2.3. Local Scour Formulae for Complex Piers

Many studies of local scour formulae refer to the single pier formula [30,70,97] which is a function of pier width (b), flow depth (y_0), Froude number (Fr), critical flow velocity (v_c), and sediment bed size (d_{50}). Sometimes the formulae are expanded to include the shape pier factor (K_1), the angle of flow attack (α) (if the pier is skewed), or the sediment characteristic (ρ_s) [98,99] and generally the formula can be written as follows:

$$d_s = f\ [\text{flow}(y,\ Fr, v_c);\ \text{pier characteristics}(b, l, \alpha);\ \text{bed material}(d_{50},\ \rho_s)]. \tag{5}$$

Physical, economic, and geotechnical considerations usually indicate the need for bridge piers to be constructed with a column founded on a pile group with a pile cap or on a caisson, so-called complex piers [100]. The process of scour development at complex piers differs from that on uniform piers because lee-wake vortices dominate sediment transport, whereas on uniform piers horseshoe vortex has governing influence on scouring [101]. In recent decades, a number of studies on scouring around complex bridge piers have been conducted. Melville and Raudkivi [102] proposed an approach to calculate the scour around complex piers composed of a cylindrical pier (with diameter D) and of a foundation (with diameter $D*$ and with the top elevation H). In their study, an alternative approach using the effective diameter (D_e) is employed. Effective diameter represents a diameter of a circular pile that induces the same scour depth as the scour depth at the complex pier. Melville and Raudkivi [102] examined the influence of effective diameter concept and proposed following formula:

$$D_e = \frac{D\cdot(y_0 + H) + D * \cdot(d_s - H)}{d_s + y_0} \tag{6}$$

Finally, Melville and Raudkivi's scour depth formula for complex piers introduces three scour zones dependent of the top of the foundation elevation, as presented in Table 2.

Table 2. Scour depth formula around complex piers based on influence of the top of the foundation.

Zone	Condition	Formula	
Foundation below the bottom of the scour hole	$H > 2.4{\cdot}D$	$d_s = 2.4{\cdot}D$	(7)
Foundation is above the bottom of the scour hole and below the bed level	$2.4{\cdot}D \le H \le 0$ [1] $d_s < 2.4{\cdot}D$ [1]	$\frac{d_s}{D} = 2.4{\cdot}\left(\frac{D}{D*}\right)^{0.4+0.5(D/D*)}$	(8)
Foundation is above the bed level	$H < 0$ $2.4{\cdot}D < d_s < 2.4{\cdot}D*$	$\frac{d_s}{D} = 2.4{\cdot}\left(\frac{D}{D*}\right)^{\left(\frac{D}{D*}\right)^3+0.1-0.47\sqrt{0.75-\frac{H}{D}}}$	(9)

[1] Except $H < 0.7{\cdot}D$ and $D/D^* < 0.6$

Even though using the equivalent single pier simplification shortens the computational time considering the complex flow field and turbulence around the pier, Melville and Raudkivi [102] showed that it could lead to conservative estimates of scour depth.

One of the most commonly used [65,75,103,104] scour formulae is the HEC-18 formula, developed by the Federal Highway Administration (FHWA) [41]. To account for the effects of foundation geometry on scour depth, the superposition of scour components is proposed. In this method, each pier component is calculated separately and superimposed on the total scour depth, as follows:

$$d_s = d_{s,pier} + d_{s,\ pile\ cap} + d_{s,pile\ group} \tag{10}$$

Afterwards, Coleman [105] expanded scour estimation of complex piers (comprising column, pile cap and pile group) into the five following cases dependent of the top of the foundation elevation: (I) buried pile cap; (II) pile cap at the bed level of the scour hole with no exposure of the pile group; (III) pile cap and pile group extending over the base of the scour hole while water level is above the pile cap; (IV) pile cap and pile group extending over the base of the scour hole while water level is at the top of the pile cap; and (V) pile cap and pile group extending over the base of the scour hole while water level is below the pile cap. Coleman also provides formulae for equivalent pier diameter (b_e) based on variation of the pile cap elevation (H) relative to the bed level (Table 3.):

Table 3. Equivalent pier diameter formulae based on influence of the pile cap elevation.

Formula	Condition	
$b_e = b_{col}$	for $H \geq b_{col}$	(11)
$b_e = b_{col} \cdot \left(\frac{b_{col}}{b_{pc}}\right)^{\{(b_{col}/b_{pc})^3 + 0.1 - [0.47(0.75 - H/b_{col})^{0.5}]\}}$	for $H_T \leq H < 0$	(12)
$b_e = \left(\frac{0.52 \cdot T \cdot b_{pc} + (y_0 - 0.52 \cdot T) b_{pd}}{y_0}\right)$	for $H = (-y_0)$	(13)
$b_e = b_{pd}$	for $H \leq (-y_0 - T)$	(14)

A similar study by Sheppard and Glasser [66] resulted in a detailed iterative procedure for determining effective diameter for the geometry case where the top of the pile cap is above the bed level with no exposure of the pile group—equivalent to the piers founded on the exposed caisson. The outlined procedure was developed in clear-water and live-bed conditions, whereas effective diameter is a function of the shapes, locations, and orientations of each pier component, and as a function of the sediment and flow properties (Table 4).

Table 4. Procedure for calculating effective width of complex piers for estimating scour depth.

Element	Formulae	
column	$y_{max} = \begin{cases} 5 \cdot b_{col} & \text{for } y_0 \geq 5 \cdot b_{col} \\ y_0 & \text{for } y_0 < 5 \cdot b_{col} \end{cases}$	
	$K_1 = \begin{cases} 1 & \text{for circular columns} \\ 0,86 + 0,97 \left|\frac{\pi}{\alpha}\right|^4 & \text{for square columns} \end{cases}$	
	$K_2 = \frac{b_{col} \cos(\alpha) + L_{col} \sin(\alpha)}{b_{col}}$	
	if $H_{col} \geq y_{max}$ $D_{col}{}^* = 0$, else $D_{col}{}^* =$ $K_1 K_2 K_5 b_{col} \left[0.16\left(\frac{H_{col}}{y_{max}}\right)^2 - 0.39\left(\frac{H_{col}}{y_{max}}\right) + 0.25\right]$	(15)
caisson [1]	$H'_{pc} = -d_{s(col)} T' = T' + H'_{pc}$	
	$y_{max} = \begin{cases} 2 \cdot b_{pc} \left(\frac{T'}{y_0 + \lvert H'_{pc} \rvert}\right)^{0.4} & y_0 \geq 2 \cdot b_{pc} \left(\frac{T'}{y_0 + \lvert H'_{pc} \rvert}\right)^{0.4} \\ y_0 + \lvert H'_{pc} \rvert & y_0 < 2 \cdot b_{pc} \left(\frac{T'}{y_0 + \lvert H'_{pc} \rvert}\right)^{0.4} \end{cases}$	
	$D_{pc}{}^* = K_1 K_2 b_{pc} \exp\left[-1 - 1.8 \exp\left(\frac{H'_{pc}}{y_{max}}\right) + 1.7\left(\frac{T'}{y_{max}}\right)^{1/2}\right]$	(16)
total	$D^* = D_{col}{}^* + D_{pc}{}^*$	(17)

[1] Caisson is interpreted as buried pile cap.

Furthermore, Jannaty et al. [77] conducted an investigation to determine the cause of the Adinan complex bridge failure. They examined the performance of several formulae and demonstrated substantial distinction between calculated and measured scour depth values. Several stages of scouring can be noticed in their study: (I) if the bottom of scour hole is above the up edge of the pile cap, the pile cap has a protective effect against scour; (II) if the bottom of scour hole is between the up and the down edge of the pile cap, the column is the prominent component to produce scour; (III) if the scour hole induced by the column reaches the down edge of the pile cap, the pile cap will play the most important role in scouring. Analysis of results show that the cause of Adinan Bridge failure is the large width of the foundation in comparison to the column width. Additionally, scouring was intensified due to lot of accumulated debris. Based on study by Jannaty et al., it can be concluded that scour depth increases if obstruction area (either foundation or debris) increases. This leads us to the conclusion that in estimating scour depth, the component of flow area reduction should be considered in addition to local scour. Similarly, Yang et al. [106] found that for complex bridge piers in close proximity, the upstream pier significantly affects scouring at the downstream pier in comparison to individual piers, reducing scour rate for clear-water conditions and increasing the bedform celerity for live-bed conditions.

2.4. Contraction Scour Formulae

One of the first contraction scour formulae was presented by Laursen and Toch [107]. They evolve Straub's contraction scour for the long contractions to the local scour at the pier. The formula is valid for bridges where the spacing between the piers is much larger than the piers' width (for contraction of cross-section approximately 10%). The effect of contraction scour will become present when scour depth achieves a value that corresponds to scour depth which would occur in a long contraction, or when scour holes from adjacent piers overlap. The formula is as follows:

$$\frac{d_{s,c}}{y_0} = \frac{1}{(1-\beta)^{9/14}} - 1, \tag{18}$$

where β is width ratio of the contracted to the uncontracted profile. Briaud et al. [108] verified contraction scour data using previous experimental results, employed a new parameter v_1, that presents averaged velocity at the contracted section, and provided the following formula:

$$\frac{d_{s,c}}{y_0} = 2.21 \cdot \left(1.31 \cdot \frac{v_1}{\sqrt{g y_0}} - \frac{v_c}{\sqrt{g y_0}}\right). \tag{19}$$

Furthermore, one of the most commonly used contraction scour formulae is HEC-18 [109], that assumes contraction scour as a long contraction, where the length of the contracted section is longer than the section width. The HEC-18 contraction scour formula is developed as a part of total scour formula, which presents superposition of both separately calculated processes. The formula contains flow, bottom channel width, and Manning's roughness coefficient ratios as provided below:

$$\frac{d_{s,c}}{y_0} = \left(\frac{Q_1}{Q_0}\right)^{\frac{6}{7}} \left(\frac{L_0}{L_1}\right)^{k_1} \left(\frac{n_1}{n_0}\right)^{k_2} - 1. \tag{20}$$

Like many other contraction scour formulae in the literature, HEC-18 also overestimates the scour depth [75,110]. One of the possible reasons for that inaccuracy could be neglecting the possible dependence of local and contraction scour [65].

Recently, several researchers have observed uncertainty in the scour formulae due to the interaction of contraction and local pier scour estimation. Hong [110] conducted laboratory experiments comparing the bridge cross-section with piers and abutments with the cross-section without piers. The results demonstrate that the model without piers has 25% greater scour depth. A decrease in contraction scour depth implies possible interactions between local scour and contraction scour processes. Contraction scour development requires a longer time to reach equilibrium than local scour, as well as a possible consequence of local and contraction scour dependance. Additionally, experimental results of contraction scour depth were compared to the values obtained by Laursen's formula and results shows overestimation of Laursen's formula results of about 30–60%. Recently, Saha et al. [104] carried out laboratory experiments in the presence of flow contraction and compared results with reference local scour depth data obtained by new theoretical method developed from a combination of HEC-18 and M-S formulae. Additional contraction scour estimation showed that the discharge contraction ratio q_2/q_1 has a great influence on the contraction effect on the total scour depth, where the cross-section at the upstream face of bridge is labeled with number 2 and approach section with number 1. However, their experiments were only conducted under clear-water conditions, and the nonuniform sediment size is neglected. Mueller and Wagner [64] emphasize that long-term in situ measurements are essential to separate local scour from contraction to identify reference bed elevation. They suggested that long-term scour in the uncontracted section can be determined by adding the general slope of the line to eliminate general aggradation or degradation. Afterwards, short-term scour determination should be performed by comparing the uncontracted and contracted cross-sections in the preflood and flood conditions.

3. Riprap Sloping Structure

When riprap is selected as a scour countermeasure, it is usually placed flushed with the riverbed level if flow conditions are favorable for detailed installation, or as a riprap sloping structure when flow conditions do not allow fine maneuvering of the machinery.

Some authors consider layer riprap as a more suitable protection than a riprap sloping structure [26,30,107] because of excessive exposure of the sloping structure that induces unwanted contraction. Recommended placement of the layer riprap depends on the hydraulic environment and sediment transport, i.e., the occurrence of bedforms. Riprap should be placed deep enough so it does not protrude above the bed level and disturb the flow. Installation in rivers with significant amounts of bedload should be buried below the estimated through of the dunes occurring during floods [10]. The number of studies in the literature describing layer riprap and its failure mechanisms is higher than that of riprap sloping structures, which have been rarely investigated. Generally, hydraulic processes that occur downstream of the riprap sloping structure and its consequences are ambiguous—flow obstruction, debris accumulation and scouring at the periphery of the riprap mound. Since there is no such research about estimating scouring around the riprap sloping structure, an analogy with complex piers can be drawn. Contours of the piers protected with riprap sloping structure (pier + riprap) are similar to the contours of complex piers (pier + caisson), as can be seen in Figure 1.

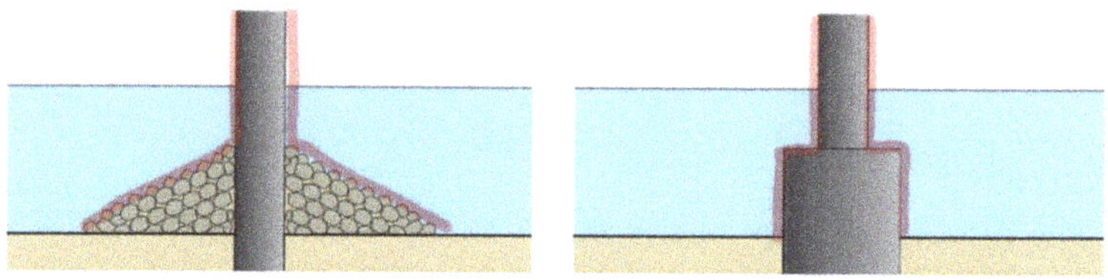

Figure 1. Similarity of outer contours between the riprap sloping structure (**left**) and complex piers (**right**), outlined in red.

If analogy between the riprap sloping structure shape and complex piers geometry is considered valid, then the scour formulae applicable for the riprap sloping structure also have to be case-specific depending on the top elevation of the riprap, similar to the scour formulae for complex piers that are distinguished depending on the caisson submergence [66,77,102,105]. According to Melville [111], there are three cases for scour around complex piers: case I—where the top of the caisson foundation is below general bed level; case II—where the top of the caisson foundation is above general bed level; case III—where the top of the caisson foundation is above the water surface, as depicted by Figure 2.

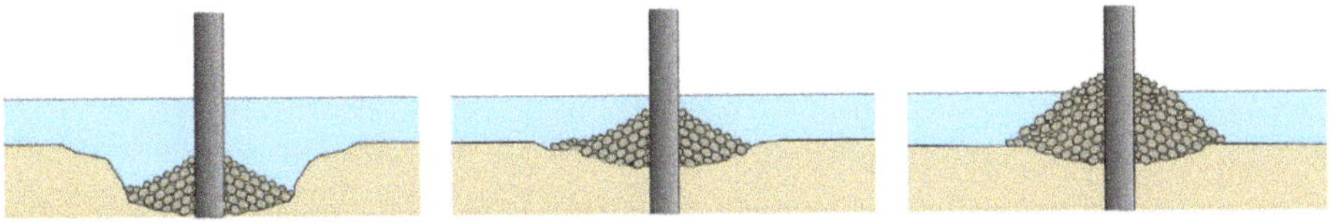

Figure 2. Pier scour countermeasure retrofitting cases distinctive by the position of the top of the riprap sloping structure relative to the riverbed level: top of the riprap below the general bed level (**left**); top of the riprap above the general bed level (**middle**); and top of the riprap above the mean water level (**right**).

Melville [111] and Ghorbani [112] draw further conclusions: if the top of the caisson (or riprap sloping structure) is exposed above general bed level, the scour depth is increased; if the overflow above the top of the pile cap (crest of the riprap) is less shallow, the scour depth is increased. The riprap sloping structure is a wide structure, in terms of reducing the flow area, so conclusions about wide piers might be comparable. Yang et al. [81] stated that the downflow in front of the wider piers ($y_0/D \leq 1.4$) is weakened, as well as there being horseshoe vortex. Thus, wake vortices consequently remain the main turbulence structures that produce scour hole, the maximum of which will occur at side of the pier.

The design of riprap considers calculating the size of riprap stone that can withstand flow attack. One of the first formulae for riprap sizing was developed by Izbash [113]:

$$d_r = y_0 \cdot 0.347 \cdot \frac{Fr^3}{(S_r - 1)}. \quad (21)$$

As can be seen from previous formula, water depth (y_0), Froude number (Fr), and specific gravity of riprap (S_r) were governing parameters in riprap sizing formulae [30,114,115], until Parola [116] introduced new characteristic factors. Factors are proportional to the size of the riprap stones (K_r) and related to pier geometry—pier shape (K_s) and pier width (K_b). Subsequently, several formulae with pier characteristic factors were developed [46,117]. Furthermore, Froehlich [118] compared the range of riprap sizes obtained by formulae from different authors, and came out with conclusion of quite a wide range of results, and therefore proposed a new formula for calculating minimum diameter of loose rock riprap, introducing new parameters that introduce crossflow shear (K_w), transverse pier spacing factor (K_p) and approach flow influence (K_α) as follows:

$$d_r = y \cdot K_r \cdot K_b \cdot K_\omega \cdot K_p \cdot K_s \cdot K_\alpha \cdot Fr^3. \quad (22)$$

Besides riprap sizing, the design of the riprap sloping structure stability includes calculating an appropriate slope to prevent the riprap stones from sliding into the scour hole, considering protection at the toe due to undermining, and determining the riprap stone diameter to resist the hydrodynamic force of the flow. For riprap side slope, Park et al. [27] used the U.S. Army Corps of Engineering (USACE) method for calculating riprap size as follows:

$$d_{r,30\%} = y_0 \cdot \frac{1.1 \cdot C_s \cdot C_v \cdot C_t}{{K_{sl}}^{1.25} \cdot (S_r - 1)^{1.25}} \cdot Fr^{2.5}, \quad (23)$$

which was developed primarily for bank and channel protection. Parameter $d_{r,30\%}$ represents particle diameter, corresponding to a 30% finer grain-size of the granulometric curve, C_s is stability coefficient for incipient failure, C_v is vertical velocity distribution coefficient, C_t is blanket thickness coefficient and K_{sl} is side-slope correction factor. According to Breusers et al. [30], the horizontal dimension of riprap protection depends on the pier diameter, and it needs to be at least two times larger.

3.1. Riprap Failure Mechanisms

Riprap failure mechanisms in clear-water conditions can be divided into: (I) shear failure—dislodging of individual riprap stones due to hydrodynamic forces; (II) edge failure—undermining of the riverbed at the toe of the riprap; and (III) winnowing failure—movement of finer material through voids between riprap stones principally initiated by turbulence [117]. In addition, two more failure mechanisms have been identified under live-bed conditions: (IV) bedform-induced failure—fluctuations in flow of the bed level prompted by bed features, such as ripples and dunes, causes lack of stone stability; (V) bed-degradation-induced failure—due to general bed degradation, riprap stones protrude above the general bed level causing a reduction in bed shear stress and consequently disintegration of the riprap structure [119]. Fredsoe et al. [25] described horseshoe vortices in front of the individual stones as the primary reason for undermining at the junction between the riprap countermeasure and the bed in a steady parallel current. The following conclusion is determined—scour depth increases if the Shields parameter increases, a slope of the revetment increases, and if the stone shape is angular. Vasquez et al. [24] conducted physical and numerical investigation of scour around the riprap layer protection in Golden Ears Bridge. In their study, the riprap layer is called riprap apron. The bridge is located in a wide section of a sand-bed river just upstream from a bifurcation, subjected to migrating dunes under live-bed conditions, and with piers oriented at an angle to the flow attack, except the pile group that is exposed above the natural riverbed level. The study proves that a 16 m-wide and 2 m-thick riprap layer successfully eliminates local scour around the pier, even in the presence of an unfavorable combination of passing dunes and general scour. The riprap, however, will partially destabilize and reallocate individual riprap stones, especially at the edges of the layer to the form of a semiconical mound around the pier, but it will still prevent scour hole from forming near piers.

However, at the contacts of the riprap and sediment bed in terms of live-bed conditions, the edges of the riprap will inevitably start to destabilize. The upper stones at the edges roll down the riprap slope or towards the dunes through into the scour hole. Vasquez et al. [24] noticed that after the dune passes by, the riprap side slope formed a conical mound with an angle of repose of about 30°. This process continues downstream until the riprap becomes scattered and forms a riprap mound. Chiew [120] states that such a riprap mound still serves its function of protecting the pier from erosion, but nevertheless the riprap mound shifts the scour area away from the riprap toe and scour continues downstream. This phenomenon has been noticed in Croatian rivers [121]. Gilja et al. establish a sediment transport model to investigate morphological characteristics of a section of the Sava River where a bridge with three piers remediated with riprap sloping structure are placed. It has been noticed that erosional processes affect downstream sections of the river due to increased flow velocities at the constricted bridge opening. The location of the final scour hole is unknown, because it is formed based on the interaction between the flow and the structure under site-specific conditions [122]. Petersen et al. [123] noticed scour of the riprap in the farther downstream area. They conducted a detailed experimental and field investigation on the three-dimensional flow field via Particle Image Velocimetry (PIV) measurements to improve the understanding of scour in the interaction of the riprap protection and the sand. The study was based on monopiles in the marine environment in terms of steady current, waves and the combination of current and waves, where analogy with sea current and river flow can be correlated. They concluded that a pair of symmetrical, counter-rotating vortices induced by a steady current cause significant scour hole downstream of the scour protection. Whitehouse et al. [71] support the statement that the riprap placed around the pier and above the level of the natural bed initiates the development of the secondary scour response. As the scour continues downstream at the edge of the riprap, a deeper scour hole is created than around the unprotected bridge pier. In addition, it has been shown that with this type of riprap protection, scour wakes extend to distance of 100 times the pier diameter. The riprap scour countermeasure is considered to be inadequate, because after secondary scour, more vortices appear near the piers and accelerate the scour process [124].

3.2. Similar Structures Analogy

Groynes, spur dikes, and sloping abutments are geometrically similar to the riprap sloping structures, and thus comparable complex interactions between the flow and riverbed material occur—shallow and on occasion supercritical flow overtopping the structure and hydraulic jump forming on the downstream slope, undermining the toe.

Affecting parameters in local scour formulae at the groyne are same as those for local scour at bridge piers (hydraulic and sediment parameters), except for geometrical parameters that encompass scour depth's dependency on the distance from the first groyne or from the entrance of the bend, as can be seen in Przedwojski [125] formula:

$$\frac{d_s}{y_0} = \left[\frac{y_{un}}{y_0} + \frac{Q_s}{Q_0} cos\left(\frac{2\pi}{L_m}s\right) + sin(\alpha - 90)\right]^{\mu-2}. \quad (24)$$

Another formula developed by Rashak and Khassaf [126] in clear-water conditions for T-shape submerged groynes is as follows:

$$\frac{d_s}{y_0} = 1.489 \cdot e^{(4.117 \cdot Fr)} + 4.117 \cdot e^{\left(-3.292 \cdot \frac{v_0}{v_c}\right)} - 3.292 \cdot e^{\left(0.002 \cdot \frac{y_{gr}}{y_0}\right)} + 0.002 \cdot e^{0.747 \cdot \frac{s_g}{y_0}} + 0.747 \cdot e^{(-1.092 \cdot n_g)}. \quad (25)$$

As can be seen in abovementioned formulae, local scour around groynes depend on parameter of spacing between groynes (s_g). Since one groyne case is comparable to the riprap sloping structure, Rashak and Khassaf's formula can be considered for the riprap sloping structure in the case where spacing between groynes is zero ($s_g = 0$). In a sequence of groynes, maximum scour depth occurs at the nose of the upstream groyne.

According to Han and Lin [127], groynes are most vulnerable when the upcoming flood reaches an elevation just above the groyne crest and a shallow submerged condition is reached. Such overtopping flow has a vertical stream direction down the lee side of the groyne, and thus causes small sediment particles to be carried away at relatively low threshold velocities. Rashak and Khassaf [128] noticed that the magnitude of the groyne submergence affects scour depth inversely, which means that an increase in submerged ratio (y_1/y_0) will result in smaller scour depth ratios (d_s/y_0). During relatively deep submergence, the overtopping flow near the free surface is parallel to the flow direction, while recirculatory motions are present at the bottom around the groyne. Meaning that in the low submergence level case, the recirculation region affects stones at the groyne surface.

McCoy et al. [129] investigated the horseshoe vortex system and shear distribution around the groyne to clarify evolution of the scour hole around the groyne region. They showed that deflection of the horizontal, detached shear layer will be magnified if the groyne crest elevation is higher. Furthermore, they support statement that mass exchange is larger in the submerged conditions than for emerged groynes due to a large increase in the inflow velocities downstream of the groyne. Melville and Coleman [14] mentioned that the vortex system will not change in higher water levels, which means that flow depth has a limited influence on scour depth.

Pandey et al. [130] have tested the accuracy of formulae for scour depth around groynes developed by different authors, concluding that although the position of scour depth varies remarkably, a maximum value typically occurs near the nose of the groyne on the upstream side and spreads up to a width of three times the spur dike length. The extent of the scour region is greater on the downstream side compared to the upstream, but the volume of the upstream scour hole is about 65% of the total volume. They indicate that Froude number is a significant parameter in scour depth around the groyne, and other parameters play secondary role, which was later also confirmed by Rashak and Khassaf [126]. Giglou et al. [131] conducted an analysis of flow pattern around spur dikes by a 3D numerical model, and showed that vortex length behind the spur dike is four times longer than the length of spur dike, and with approximately 1.2 times the spur dike length. Pandey et al. [132] developed two novel methods to estimate maximum scour depth around the spur dike in a uniform sediment condition that consists of three standalone machine-learning approaches. Experimentally collected data were obtained in a clear-water condition with uniform sediment material. The result of their model is presented as the ratio of scour depth and spur dike (d_s/L_s), while input parameters were ratio of average to critical velocity (v_0/v_c), ratio of water depth to spur dike length (y_0/L_s), ratio of spur dike length to mean sediment diameter (L_s/d_{50}) and densimetric Froude number (Fd_{50}). Based on sensitivity analysis of the input parameters, L_s/d_{50} showed the most significant influence on model performance. Eventually, statistical metrics showed the good performance of the developed model in assessing maximum scour depth near the spur dike.

Scour hole geometry at the abutment is similar to those at groynes by means of strong scour activity as a consequence of contraction scouring, which causes a scour hole to be located slightly downstream along the centerline of the watercourse, complementary with turbulent intensity pattern and the time-averaged velocity pattern [109]. However, at long abutments, the scour hole is slightly more elongated to the downstream in comparison to the short abutments [14]. It can be concluded that increasing the obstruction of the flow area will cause larger principal vortex that deflects the scour hole away until it reaches a more elongated shape.

4. Discussion

The purpose of the bridge pier scour design formulae are to determine the required depth of the foundation before the bridge is built, or to estimate bridge safety margins during its design life, with the goal of identifying the need for maintenance or scour protection measures' installment. Numerous pier scour formulae are available for scour depth estimation—however, they do not provide reliable results as most of them are developed from limited datasets, experimentally acquired or measured directly in case-specific conditions. Formulae developed experimentally may result in under- or over-estimation of scour depth if applicated at bridge piers in prototype scale [77,133]. Thus, formula must be cautiously selected based on conditions that are similar to those for which the formula is developed [17]. In order to find out the most applicable empirical formulae, sensitivity analysis is often required [134].

Any kind of obstruction in the river, such as a bridge pier, inevitably leads to morphodynamic changes, especially in sand-bed rivers with median particle size that can be easily moved during floods [135]. Since the occurrence of scour hole is only a matter of time, it is inevitable that some sort of scour protection will have to be installed during the design life of the bridge. Riprap scour protection using launchable stone almost exclusively requires the riprap to be placed above the riverbed level in a form of riprap sloping structure, which represents additional rigid obstacles in contact with the erodible sediment bed. On the other hand, scour countermeasures could have adverse effect if not installed properly. Therefore, the effect of the proposed scour countermeasures on flow environments should be investigated in detail to achieve design effectiveness during use [136].

The riprap sloping structure contracts the flow through the bridge opening, increasing the velocity and shear stress, resulting in additional lowering of the riverbed elevation next to piers. The installment of the riprap sloping structure has an effect on the bridge hydraulics similar to the flow constrictions resulting from other bridge-related structures—abutments placed in the main channel or long embankments traversing the overbanks. Sudden contraction of the oncoming flow concentrates the flow through the bridge opening, increasing the flow velocity, shear stress and turbulence in the proximity of bridge piers. Depending on the contraction ratio and associated afflux, increased flow velocity and shear stress can significantly surpass the threshold value. Thus, the riverbed deepening will be continuous until the equilibrium is reached, undermining the bridge piers. The effect of the flow contraction is local, i.e., the extent of the morphodynamic changes is rarely evident on the wider river reach, contributing to late identification of the potential hazard to the bridge. While contraction scour is not always present in the bridge locations, depending on the flow approach section layout, it will always accompany the riprap sloping structure, as the volume of the mounded stone presents significant flow constriction. Therefore, the effect of the constriction scour must not be neglected as an influencing factor in studies focusing on scour at riprap sloping structures [137]. For many bridges protected by riprap sloping structures, scour has been found to occur further downstream, often causing a deeper scour hole [121]. Yet, the final depth and relative position of the scour hole next to the riprap sloping structure has not been investigated. Scarce data about scour monitoring next to piers with installed riprap sloping structures do not allow us to draw conclusions regarding the potentially adverse effects of such pier protection. The actual position of the scour hole could be deflected from the pier as a result of toe undermining and subsequent gradual collapsing of the riprap stones into the erosion zone, propagating the scouring process downstream of the unprotected riverbed [61].

Research focusing on the riprap primarily addresses the riprap failure mechanisms [25,64,123,138], while formulae which would estimate the scour depth downstream of the riprap protection are lacking. Similarly, the formulae used for pier scour are not applicable for piers protected by riprap sloping structures. Considering that a number of existing bridges have been retrofitted with riprap sloping structures following scour hazard occurrence, future research should be oriented towards determining the effectiveness of such scour countermeasures as well as their adverse effects on the riverbed and adjacent structures. Similarities between riprap sloping structures and river training structures exist, highlighting the need for in-detail research, as flow overtopping and additional contraction scour can significantly reduce the bridge safety under more frequent flood events driven by climate change. Although many researchers have investigated the effects of flow events on bedload transport and scour processes in laboratory studies [83,85,139–142] (for a systemic review, see also [96]), studies with field measurements are still rare [39,40], especially those with continuously measured flow waves and associated scour depths [95,143].

Based on the literature review, this study hypothesizes the following: if a large database with a sufficient range of independent hydraulic parameters, riprap geometries, and resulting scour hole depth is established, an empirical formula that provides a reliable estimation of scour depth next to bridge piers protected by riprap sloping structure can be derived. This hypothesis can be tested using field data, experimental data, numerical simulation data, as well as hybrid modeling approaches. Field data are the most reliable, but occurrence of flood events often surpasses the timeframe available for the research. Therefore, field data can be used for calibration and verification of experimental and numerical simulations on more frequent flood events. Experimental and numerical simulations provide means to simulate flood events of longer return periods, such as 1000-year floods. However, preparation of a physical model may be both time-consuming and challenging from a scaling perspective. Therefore, a physical model can be used to investigate local turbulent flow field in the vicinity of the pier. Interaction between the complex geometry and the vortex system can be explored utilizing 3D numerical simulations. The 3D CFD model overperforms the physical model in predicting turbulent velocities and sediment scour, especially if the physical model relies on a single similarity method for scaling [57]. Compared to time-consuming laboratory experiments and field investigations, numerical simulations are a relatively inexpensive and fast way to collect a large and comprehensive dataset for analysis of the scour process. Numerical models can be established for a larger flow area influencing the bridge, providing representative flow conditions at the bridge opening for the boundary conditions of the physical model. Data should be complementary in the sense that each dataset addresses the shortcomings of the others, obtaining the relevant variables across the entire range.

The state-of-the-art review and discussion on the potential research direction are the basis of the project R3PEAT (Remote Real-time Riprap Protection Erosion AssessmenT on large rivers), aiming to contribute to the field by developing empirical formulae for scour estimation next to the bridges with installed scour countermeasures in the form of the riprap sloping structure. Currently, this field is not adequately researched, while similarities with other river training structures exist. Therefore, it can be hypothesized that reliable scour equations can be developed that would combine the influencing variables used for pier scour, contraction scour and toe scour.

5. Conclusions

The bridge failures data recorded worldwide indicate that hydraulic causes are the most common causes of the bridge collapses. Therefore, scour monitoring in real-time is crucial for efficient bridge management. Maximum scour depth for the design flood must be estimated, which can be compared to scour development under specific flood events and associated risk calculated. Once the scour risk is determined in the life cycle of bridges over large rivers, piers are usually retrofitted with riprap sloping structures as scour protection measures. In the literature there are numerous empirical equations developed for different pier shapes and sizes, but the ones taking into account complex flow conditions in the bridge opening with installed riprap sloping structure as scour countermeasure are lacking. This paper provides a comprehensive and relevant review of bridge scour estimation methods for piers with riprap sloping structure installed as scour countermeasure. From the state-of-the-art review, hydrodynamic conditions characteristic for such structures are singled out, separating them from single pier equations. These are shallow and on occasion supercritical flow overtopping the structure and hydraulic jump forming on the downstream slope, undermining the toe. The contributions of this work to the research field are the following:

(1) Systematic formulation of the scouring problem occurring next to the riprap sloping structure—flow contraction as the driving mechanism, including the flow overtopping and associated relevant hydraulic variables;
(2) Overview of the currently available empirical formulae for pier scour estimation and analysis of their applicability for the piers protected with riprap sloping structure—similarity to the complex piers regarding the pier shape and river training structures regarding the flow pattern;
(3) Proposal of the future research direction—utilizing the advantages of available numerical models to investigate complex flow patterns in the bridge opening.

The presented research framework is the basis of the project R3PEAT (Remote Real-time Riprap Protection Erosion AssessmenT on large rivers), aiming to contribute to the field by developing empirical formulae for scour estimation next to the bridges with installed scour countermeasures in the form of the riprap sloping structure.

Author Contributions: Conceptualization, A.H. and G.G.; resources, A.H., G.G., K.P. and M.L.; writing—original draft preparation, A.H., G.G., K.P. and M.L.; writing—review and editing, A.H., G.G., K.P. and M.L.; visualization, A.H.; supervision, G.G.; project administration, G.G.; funding acquisition, G.G. All authors have read and agreed to the published version of the manuscript.

Funding: This work has been funded in part by the Croatian Science Foundation under the project R3PEAT (UIP-2019-04-4046).

Institutional Review Board Statement: Not applicable.

Informed Consent Statement: Not applicable.

Data Availability Statement: Not applicable.

Acknowledgments: This work has been supported in part by Croatian Science Foundation under "Young Researchers' Career Development Project—Training New Doctoral Students" (DOK-01-2020).

Conflicts of Interest: The authors declare no conflict of interest.

Glossary

d_s	scour depth
$d_{s,c}$	contraction scour depth
y_0	water depth at the approach section
y_1	water depth at the contracted section
y_{un}	unperturbed water depth
y_{gr}	water depth above the submerged groyne crest
y_{max}	maximum water depth
b	width of bridge pier
b_e	effective width of bridge pier
b_{col}	column width
L_{col}	column length
H_{col}	column height from the bed level
b_{pc}	pile-cap width
b_{pd}	pile-diameter
d_{50}	mean particle size of bed material
$d_{50,a}$	mean particle size of armor layer
n	Manning's roughness coefficient
T	pile-cap thickness
H_T	exposure of pile cap at which cap is scoured.
v_0	averaged approach velocity
v_1	averaged velocity at the contracted section
v_c	critical velocity
$v_c{}'$	approach velocity that induces incipient motion
τ_c	critical shear stress
H	the top elevation of the foundation (or pile cap)
D	pier diameter
$D*$	foundation diameter
D_e	effective pier diameter
d_r	riprap rock diameter
S_r	specific gravity of riprap
C_s	stability coefficient for incipient failure
C_v	vertical velocity distribution coefficient
C_t	blanket thickness coefficient
K_{sl}	side-slope correction factor
α	angle between the groyne or pier and flow attack
β	ratio of the contracted to the uncontracted width
k	turbulent kinetic energy
ε	turbulent dissipation rate
ω	specific dissipation rate
μ	the exponent in the sediment transport formula.
Fr	Froude number
L_0	bottom channel width at the approached section
L_1	bottom channel width at the contracted section
L_m	meander length
s	distance from the entrance of the bend
s_g	spacing between groynes
n_g	number of groynes
Q_0	discharge at the approached section
Q_1	discharge at the contracted section
Q_s	part of discharge stopped by groyne
k_1, k_2	medial fall velocities of the sediment d_{50}
K_1	the correction factor for pier-nose shape
K_2	the correction factor for angle of flow attack
K_3	the correction factor for bed condition
K_4	the correction factor for armoring of coarse bed material

K_5	the correction factor for pile extension
K_r	rock properties' slope effects factor for riprap sizing
K_b	pier width factor for riprap sizing
K_ω	crossflow shear factor for riprap sizing
K_p	transverse pier spacing factor for riprap sizing
K_s	pier shape factor for riprap sizing
K_α	approach flow alignment factor for riprap sizing

References

1. Sholtes, J.S.; Ubing, C.; Randle, T.J.; Fripp, J.; Cenderelli, D.; Baird, D.C. *Managing Infrastructure in the Stream Environment*; U.S. Department of the Interior & U.S. Department of Agriculture: Denver, CO, USA, 2017; p. 65.
2. Lee, M.; Yoo, M.; Jung, H.-S.; Kim, K.H.; Lee, I.-W. Study on dynamic behavior of bridge pier by impact load test considering scour. *Appl. Sci.* **2020**, *10*, 6741. [CrossRef]
3. Gilja, G.; Kuspilić, N.; Bekić, D. Impact of morphodynamical changes on the bridge stability: Case study of Jakuševac bridge in Zagreb. In *Current Events in Hydraulic Engineering*; Sawicki, J.M., Zima, P., Eds.; Gdansk University of Technology: Gdansk, Poland, 2011; pp. 112–122.
4. Zhang, Y. Economic Impact of Bridge Damage in a Flood Event. Master's Thesis, RMIT University, Melbourne, Australia, 2016.
5. Honfi, D.; Lange, D.; Pursiainen, C.; Rød, B. On the contribution of technological concepts to the resilience of bridges as critical infrastructure assets. In Proceedings of the IABSE Congress: Challenges in Design and Construction of an Innovative and Sustainable Built Environment, Stockholm, Sweden, 21–23 September 2016; pp. 975–982.
6. Imhof, D. Risk Assessment of Existing Bridge Structures. Ph.D. Thesis, University of Cambridge, Cambridge, UK, 2004.
7. Schaap, H.S.; Caner, A. Bridge collapses in Turkey: Causes and remedies. *Struct. Infrastruct. Eng.* **2021**, 1–16. [CrossRef]
8. Cook, W. Bridge Failure Rates, Consequences, and Predictive Trends. Ph.D. Thesis, Utah State University, Logan, UT, USA, 2014.
9. Wardhana, K.; Hadipriono, F.C. Analysis of recent bridge failures in the United States. *J. Perform. Constr. Facil.* **2003**, *17*, 144–150. [CrossRef]
10. Lagasse, P.F.; Clopper, P.E.; Zevenbergen, L.W.; Girard, L.G. *Countermeasures to Protect Bridge Piers from Scour*; National Cooperative Highway Research Program—Transportation Research Board: Fort Collins, CO, USA, 2007; p. 284.
11. Yao, C.; Briaud, J.-L.; Gardoni, P. Risk Analysis on Bridge Scour Failure. In Proceedings of the International Foundations Congress and Equipment Expo, San Antonio, TX, USA, 17–21 March 2015; pp. 1936–1945.
12. Barker, R.M.; Puckett, J.A. *Design of Highway Bridges: An LRFD Approach*; Wiley: Hoboken, NJ, USA, 2021.
13. Lee, G.C.; Mohan, S.B.; Huang, C.; Fard, B.N. *A Study of U.S. Bridge Failures (1980–2012)*; University at Buffalo, State University of New York: Buffalo, NY, USA, 2013; p. 128.
14. Melville, B.W.; Coleman, S.E. *Bridge Scour*; Water Resources Publications: Colorado, CO, USA, 2000.
15. Unger, J.; Hager, W.H. Down-Flow and horseshoe vortex characteristics of sediment embedded bridge piers. *Exp. Fluids* **2007**, *42*, 1–19. [CrossRef]
16. Breusers, H.N.C.; Raudkivi, A.J. *Scouring: Hydraulic Structures Design Manual Series*; CRC Press: London, UK, 1991; Volume 2.
17. Dey, S. Hydrodynamic Principles. In *Fluvial Hydrodynamics: Hydrodynamic and Sediment Transport Phenomena*; Springer: Berlin/Heidelberg, Germany, 2014; Volume 52, pp. 870–871.
18. Chen, S.-C.; Tfwala, S.; Wu, T.-Y.; Chan, H.-C.; Chou, H.-T. A hooked-collar for bridge piers protection: Flow fields and scour. *Water* **2018**, *10*, 1251. [CrossRef]
19. Fathi, A.; Zomorodian, S.M.A. Effect of submerged vanes on scour around a bridge abutment. *KSCE J. Civ. Eng.* **2018**, *22*, 2281–2289. [CrossRef]
20. Elnikhely, E.A. Minimizing scour around bridge pile using holes. *Ain Shams Eng. J.* **2017**, *8*, 499–506. [CrossRef]
21. Abudallah Habib, I.; Wan Mohtar, W.H.M.; Muftah Shahot, K.; El-Shafie, A.; Abd Manan, T.S. Bridge failure prevention: An overview of self-protected pier as flow altering countermeasures for scour protection. *Civ. Eng. Infrastruct. J.* **2021**, *54*, 1–22. [CrossRef]
22. Ettema, R.; Nakato, T.; Muste, M. *An Illustrated Guide for Monitoring and Protecting Bridge Waterways Against Scour*; Iowa Highway Research Board: Ames, IA, USA, 2006.
23. Agrawal, A.K.; Khan, M.A.; Yi, Z. *Handbook of Scour Countermeasures Designs*; University Transportation Research Center: New York, NY, USA, 2005; p. 201.
24. Vasquez, J.; McLean, D.; Walsh, B. Modeling Scour and Riprap Protection in Golden Ears Bridge. In Proceedings of the 18th Canadian Hydrotechnical Conference, Winnipeg, MB, Canada, 22–24 August 2007.
25. Fredsoe, J.; Sumer, B.; Bundgaard, K. Scour at a riprap revetment in currents. In Proceedings of the 2nd IAHR Symposium on River, Coastal and Estuarine Morphodynamics, Obihiro, Japan, 10–14 September 2001; pp. 245–254.
26. Lauchlan, C.S.; Melville, B.W. Riprap protection at bridge piers. *J. Hydraul. Eng.* **2001**, *127*, 412–418. [CrossRef]
27. Park, S.K.; Julien, P.Y.; Ji, U.; Ruff, J.F. Case Study: Retrofitting Large Bridge Piers on the Nakdong River, South Korea. *J. Hydraul. Eng.* **2008**, *134*, 1639–1650. [CrossRef]
28. Craswell, T.; Akib, S. Reducing bridge pier scour using gabion mattresses filled with recycled and alternative materials. *Eng* **2020**, *1*, 188–210. [CrossRef]

29. Yoon, T.; Kim, D.-H. Bridge Pier Scour Protection by Sack Gabions. In Proceedings of the World Water and Environmental Resources Congress 2001, Orlando, FL, USA, 20–24 May 2001; pp. 1–8.
30. Breusers, H.N.C.; Nicollet, G.; Shen, H.W. Local scour around cylindrical piers. *J. Hydraul. Res.* **1977**, *15*, 211–252. [CrossRef]
31. Nielsen, A.W.; Liu, X.; Sumer, B.M.; Fredsøe, J. Flow and bed shear stresses in scour protections around a pile in a current. *Coast. Eng.* **2013**, *72*, 20–38. [CrossRef]
32. Cikojević, A.; Gilja, G. Research approach for spatial and temporal development of deflected scour hole next to riprap scour protection. In Proceedings of the 5. Simpozij Doktorskog Studija Građevinarstva, Zagreb, Croatia, 9–10 September 2019; pp. 59–68.
33. Hung, C.-C.; Yau, W.-G. Behavior of scoured bridge piers subjected to flood-induced loads. *Eng. Struct.* **2014**, *80*, 241–250. [CrossRef]
34. Argyroudis, S.A.; Mitoulis, S.A. Vulnerability of bridges to individual and multiple hazards- floods and earthquakes. *Reliab. Eng. Syst. Saf.* **2021**, *210*, 107564. [CrossRef]
35. FHWA. *National Bridge Inspection Standards*; 23 CFR Part 650 [FHWA Docket No. FHWA–2001–8954]; FHWA: Washington, DC, USA, 2004; pp. 74419–74439.
36. Aberle, J.; Rennie, C.D.; Admiraal, D.M.; Muste, M. *Experimental Hydraulics: Methods, Instrumentation, Data Processing and Management*; Taylor & Francis Group: London, UK, 2017; Volume II, p. 448.
37. Borghei, S.M.; Kabiri-Samani, A.; Banihashem, S.A. Influence of unsteady flow hydrograph shape on local scouring around bridge pier. *Proc. Inst. Civ. Eng. Water Manag.* **2012**, *165*, 473–480. [CrossRef]
38. Manfreda, S.; Link, O.; Pizarro, A. A theoretically derived probability distribution of scour. *Water* **2018**, *10*, 1520. [CrossRef]
39. Lu, J.-Y.; Hong, J.-H.; Su, C.-C.; Wang, C.-Y.; Lai, J.-S. Field measurements and simulation of bridge scour depth variations during floods. *J. Hydraul. Eng.* **2008**, *134*, 810–821. [CrossRef]
40. Su, C.-C.; Lu, J.-Y. Comparison of sediment load and riverbed scour during floods for gravel-bed and sand-bed reaches of intermittent rivers: Case study. *J. Hydraul. Eng.* **2016**, *142*, 05016001. [CrossRef]
41. Richardson, E.V.; Davis, S.R. *Evaluating Scour at Bridges*, 4th ed.; FHWA NHI 01-001; Federal Highway Administration Hydraulic Engineering Circular No. 18: May 2001; FHWA: Washington, DC, USA, 2001; p. 378.
42. Ettema, R.; Constantinescu, G.; Melville, B. *Evaluation of Bridge Scour Research: Pier Scour Processes and Predictions*; Transportation Research Board of the National Academies: Washington, DC, USA, 2011.
43. Melville, B.W.; Chiew, Y.-M. Time scale for local scour at bridge piers. *J. Hydraul. Eng.* **1999**, *125*, 59–65. [CrossRef]
44. Lagasse, P.F.; Richardson, E.V.; Schall, J.D.; Price, G.R. *Instrumentation for Measuring Scour at Bridge Piers and Abutments*; NCHRP Report No 396; Transportation Research Board: Washington, DC, USA, 1997.
45. De Falco, F.; Mele, R. The monitoring of bridges for scour by sonar and sedimetri. *NDT E Int.* **2002**, *35*, 117–123. [CrossRef]
46. Lagasse, P.F.; Clopper, P.E.; Pagán-Ortiz, J.E.; Zevenbergen, L.W.; Arneson, L.A.; Schall, J.D.; Girard, L.G. *Bridge Scours and Stream Instability Countermeasures: Experience, Selection and Design Guidance*; Hydraulic Engineering Circular No. 23; Federal Highway Administration: Washington, DC, USA, 2009.
47. Orsak, J. *A New Method for Detecting the Onset of Scour and Managing Scour Critical Bridges*; SENSR Monitoring Technologies: Washington, DC, USA, 2019; p. 93.
48. Fitzgerald, P.C.; Malekjafarian, A.; Cantero, D.; Obrien, E.J.; Prendergast, L.J. Drive-By scour monitoring of railway bridges using a wavelet-based approach. *Eng. Struct.* **2019**, *191*, 1–11. [CrossRef]
49. Bao, T.; Liu, Z.L.; Bird, K. Influence of soil characteristics on natural frequency-based bridge scour detection. *J. Sound Vib.* **2019**, *446*, 195–210. [CrossRef]
50. Lin, Y.-B.; Lee, F.-Z.; Chang, K.-C.; Lai, J.-S.; Lo, S.-W.; Wu, J.-H.; Lin, T.-K. The artificial intelligence of things sensing system of real-time bridge scour monitoring for early warning during floods. *Sensors* **2021**, *21*, 4942. [CrossRef]
51. Al-Obaidi, K.; Xu, Y.; Valyrakis, M. The design and calibration of instrumented particles for assessing water infrastructure hazards. *J. Sens. Actuator Netw.* **2020**, *9*, 36. [CrossRef]
52. Selvakumaran, S.; Plank, S.; Geiß, C.; Rossi, C.; Middleton, C. Remote monitoring to predict bridge scour failure using Interferometric Synthetic Aperture Radar (InSAR) stacking techniques. *Int. J. Appl. Earth Obs. Geoinf.* **2018**, *73*, 463–470. [CrossRef]
53. Lamb, R.; Aspinall, W.; Odbert, H.; Wagener, T. *Vulnerability of Bridges to Scour: Insights from an International Expert Elicitation Workshop*; University of Bristol: Bristol, UK, 2016; pp. 1393–1409.
54. Pizarro, A.; Manfreda, S.; Tubaldi, E. The science behind scour at bridge foundations: A review. *Water* **2020**, *12*, 374. [CrossRef]
55. Sheppard, D.M.; Melville, B.W. *Scour at Wide Piers and Long Skewed Piers*; The National Academies Press: Washington, DC, USA, 2011; p. 55.
56. Hamidifar, H.; Zanganeh-Inaloo, F.; Carnacina, I. Hybrid scour depth prediction equations for reliable design of bridge piers. *Water* **2021**, *13*, 2019. [CrossRef]
57. Huang, W.; Yang, Q.; Xiao, H. CFD modeling of scale effects on turbulence flow and scour around bridge piers. *Comput. Fluids* **2009**, *38*, 1050–1058. [CrossRef]
58. Aly, A.M.; Dougherty, E. Bridge pier geometry effects on local scour potential: A comparative study. *Ocean Eng.* **2021**, *234*, 109326. [CrossRef]

59. Alemi, M.; Pêgo, J.P.; Maia, R. Numerical simulation of the turbulent flow around a complex bridge pier on the scoured bed. *Eur. J. Mech. B/Fluids* **2019**, *76*, 316–331. [CrossRef]
60. Graf, W.H.; Istiarto, I. Flow pattern in the scour hole around a cylinder. *J. Hydraul. Res.* **2002**, *40*, 13–20. [CrossRef]
61. Zhu, Z.-W.; Liu, Z.G. CFD prediction of local scour hole around bridge piers. *J. Cent. S. Univ.* **2012**, *19*, 273–281. [CrossRef]
62. Jalal, H.K.; Hassan, W.H. Three-dimensional numerical simulation of local scour around circular bridge pier using Flow-3D software. *IOP Conf. Ser. Mater. Sci. Eng.* **2020**, *745*, 012150. [CrossRef]
63. Song, Y.; Xu, Y.; Ismail, H.; Liu, X. Scour modeling based on immersed boundary method: A pathway to practical use of three-dimensional scour models. *Coast. Eng.* **2021**, *171*, 104037. [CrossRef]
64. Mueller, D.S.; Wagner, C. *Field Observations and Evaluations of Streambed Scour at Bridges*; Federal Highway Administration: MsLean, VA, USA, 2005.
65. Johnson, P.A.; Clopper, P.E.; Zevenbergen, L.W.; Lagasse, P.F. Quantifying Uncertainty and Reliability in Bridge Scour Estimations. *J. Hydraul. Eng.* **2015**, *141*, 04015013. [CrossRef]
66. Sheppard, M.D.; Glasser, T.L. Sediment Scour at Piers with Complex Geometries. In Proceedings of the International Conference on Scour and Erosion (ICSE-2), Singapore, 14–17 November 2004.
67. Sheppard, D.M.; Miller, W. Live-Bed local pier scour experiments. *J. Hydraul. Eng.* **2006**, *132*, 635–642. [CrossRef]
68. Froehlich, D.C. Analysis of onsite measurements of scour at piers. In Proceedings of the Hydraulic Engineering: Proceedings of the 1988 National Conference on Hydraulic Engineering, Colorado Springs, CO, USA, 8–12 August 1988; pp. 534–539.
69. Gao, D.G.; Pasada, L.G.; Nordin, C.F. *Pier Scour Equations Used in the People's Republic of China: Review and Summary*; FHWA-SA-93-076; U.S. Federal Highway Administration, Office of Technology Applications: Washington, DC, USA, 1993; p. 66.
70. Ansari, S.A.; Qadar, A. Ultimate Depth of Scour Around Bridge Piers. In Proceedings of the ASCE National Hydraulic Conference, Buffalo, NY, USA, 1–5 August 1994; pp. 51–55.
71. Whitehouse, R.J.; Harris, J.; Sutherland, J.; Rees, J. The nature of scour development and scour protection at offshore windfarm foundations. *Mar. Pollut. Bull.* **2011**, *62*, 73–88. [CrossRef]
72. Zhuravlyov, M.M. *New Method for Estimation of Local Scour Due to Bridge Piers and Its Substantiation*; State All Union Scientific Research Institute on Roads: Moscow, Russia, 1978.
73. Landers, M.N.; Mueller, D.S. Evaluation of selected pier-scour equations using field data. *Transp. Res. Rec.* **1996**, *1523*, 186–195. [CrossRef]
74. Gaudio, R.; Grimaldi, C.; Tafarojnoruz, A.; Calomino, F. Comparison of formulae for the prediction of scour depth at piers. In Proceedings of the First European IAHR Congress, Edinburgh, UK, 4–6 May 2010; pp. 6–12.
75. Zevenbergen, L.W. Comparison of the HEC-18, Melville, and Sheppard Pier Scour Equations. In Proceedings of the International Conference on Scour and Erosion (ICSE-5) 2010, San Francisco, CA, USA, 7–10 November 2010.
76. Sheppard, D.; Melville, B.; Demir, H. Evaluation of existing equations for local scour at bridge piers. *J. Hydraul. Eng.* **2014**, *140*, 14–23. [CrossRef]
77. Jannaty, M.H.; Eghbalzadeh, A.; Hosseini, S.A. Using field data to evaluate the complex bridge piers scour methods. *Can. J. Civ. Eng.* **2016**, *43*, 218–225. [CrossRef]
78. Pandey, M.; Chen, S.-C.; Sharma, P.K.; Ojha, C.S.P.; Kumar, V. Local scour of armor layer processes around the circular pier in non-uniform gravel bed. *Water* **2019**, *11*, 1421. [CrossRef]
79. Sheppard, D.M.; Renna, R. *Bridge Scour Manual*; Florida Department of Transportation: Tallahassee, FL, USA, 2010.
80. Melville, B.W. Pier and abutment scour: Integrated approach. *J. Hydraul. Eng.* **1997**, *123*, 125–136. [CrossRef]
81. Yang, Y.; Melville, B.W.; Sheppard, D.M.; Shamseldin, A.Y. Live-Bed scour at wide and long-skewed bridge piers in comparatively shallow water. *J. Hydraul. Eng.* **2019**, *145*, 06019005. [CrossRef]
82. Poff, N.L.; Allan, J.D.; Bain, M.B.; Karr, J.R.; Prestegaard, K.L.; Richter, B.D.; Sparks, R.E.; Stromberg, J.C. The natural flow regime. *BioScience* **1997**, *47*, 769–784. [CrossRef]
83. Plumb, B.D.; Juez, C.; Annable, W.K.; McKie, C.W.; Franca, M.J. The impact of hydrograph variability and frequency on sediment transport dynamics in a gravel-bed flume. *Earth Surf. Process. Landf.* **2020**, *45*, 816–830. [CrossRef]
84. Kirby, A.; Roca, M.; Kitchen, A.; Escarameia, M.; Chesterton, J. *Manual on Scour at Bridges and Other Hydraulic Structures*; CIRIA: London, UK, 2015.
85. Tubaldi, E.; Macorini, L.; Izzuddin, B.A.; Manes, C.; Laio, F. A framework for probabilistic assessment of clear-water scour around bridge piers. *Struct. Saf.* **2017**, *69*, 11–22. [CrossRef]
86. Mao, L. The effect of hydrographs on bed load transport and bed sediment spatial arrangement. *J. Geophys. Res. Space Phys.* **2012**, *117*, F03024. [CrossRef]
87. Guney, M.S.; Bombar, G.; Aksoy, A.O. Experimental study of the coarse surface development effect on the bimodal bed-load transport under unsteady flow conditions. *J. Hydraul. Eng.* **2013**, *139*, 12–21. [CrossRef]
88. Oliveto, G.; Hager, W.H. Temporal evolution of clear-water pier and abutment scour. *J. Hydraul. Eng.* **2002**, *128*, 811–820. [CrossRef]
89. Kothyari, U.C.; Garde, R.C.J.; Raju, K.G.R. Temporal variation of scour around circular bridge piers. *J. Hydraul. Eng.* **1992**, *118*, 1091–1106. [CrossRef]
90. López, G.; Teixeira, L.; Ortega-Sánchez, M.; Simarro, G. Estimating final scour depth under clear-water flood waves. *J. Hydraul. Eng.* **2014**, *140*, 328–332. [CrossRef]

91. Waters, K.A.; Curran, J.C. Linking bed morphology changes of two sediment mixtures to sediment transport predictions in unsteady flows. *Water Resour. Res.* **2015**, *51*, 2724–2741. [CrossRef]
92. Hager, W.H.; Unger, J. Bridge pier scour under flood waves. *J. Hydraul. Eng.* **2010**, *136*, 842–847. [CrossRef]
93. Link, O.; Castillo, C.; Pizarro, A.; Rojas, A.; Ettmer, B.; Escauriaza, C.; Manfreda, S. A model of bridge pier scour during flood waves. *J. Hydraul. Res.* **2016**, *55*, 310–323. [CrossRef]
94. Pizarro, A.; Ettmer, B.; Manfreda, S.; Rojas, A.; Link, O. Dimensionless effective flow work for estimation of pier scour caused by flood waves. *J. Hydraul. Eng.* **2017**, *143*, 06017006. [CrossRef]
95. Link, O.; García, M.; Pizarro, A.; Alcayaga, H.; Palma, S. Local scour and sediment deposition at bridge piers during floods. *J. Hydraul. Eng.* **2020**, *146*, 04020003. [CrossRef]
96. Mrokowska, M.M.; Rowiński, P.M. Impact of unsteady flow events on bedload transport: A review of laboratory experiments. *Water* **2019**, *11*, 907. [CrossRef]
97. Laursen, E.M. Scour at bridge crossings. *Trans. Am. Soc. Civ. Eng.* **1962**, *127*, 166–179. [CrossRef]
98. Raudkivi, A.; Ettema, R. Clear-Water scour at cylindrical piers. *J. Hydraul. Eng.* **1983**, *109*, 338–350. [CrossRef]
99. Melville, B.W.; Sutherland, A.J. Design method for local scour at bridge piers. *J. Hydraul. Eng.* **1988**, *114*, 1210–1226. [CrossRef]
100. Moreno, M.; Maia, R.; Couto, L.; Cardoso, A. Contribution of complex pier components on local scour depth. In Proceedings of the 3rd IAHR Europe Congress, Porto, Portugal, 14–16 April 2014.
101. Umeda, S.; Yamazaki, T.; Yuhi, M. An Experimental Study of Scour Process and Sediment Transport around a Bridge Pier with Foundation. In Proceedings of the International Conference on Scour and Erosion (ICSE-5), San Francisco, CA, USA, 7–10 November 2010; pp. 66–75.
102. Melville, B.W.; Raudkivi, A.J. Effects of foundation geometry on bridge pier scour. *J. Hydraul. Eng.* **1996**, *122*, 203–209. [CrossRef]
103. Namaee, M.R.; Li, Y.; Sui, J.; Whitcombe, T. Comparison of three commonly used equations for calculating local scour depth around bridge pier under ice covered flow condition. *World J. Eng. Technol.* **2018**, *06*, 50–62. [CrossRef]
104. Saha, R.; Lee, S.O.; Hong, S.H. A Comprehensive method of calculating maximum bridge scour depth. *Water* **2018**, *10*, 1572. [CrossRef]
105. Coleman, S.E. Clearwater local scour at complex piers. *J. Hydraul. Eng.* **2005**, *131*, 330–334. [CrossRef]
106. Yang, Y.; Melville, B.W.; Macky, G.H.; Shamseldin, A.Y. Local scour at complex bridge piers in close proximity under clear-water and live-bed flow regime. *Water* **2019**, *11*, 1530. [CrossRef]
107. Laursen, E.M.; Toch, A. *Scour around Bridge Piers and Abutments*; Iowa Institute of Hydraulic Research: Iowa City, IA, USA, 1956.
108. Briaud, J.L.; Chen, H.C.; Chang, K.A.; Oh, S.J.; Chen, X. *Abutment Scour in Cohesive Materials*; National Cooperative Highway Research Program, Transportation Research Board: Washington, DC, USA, 2009; p. 435.
109. Arneson, L.A.; Zevenbergen, L.W.; Lagasse, P.F.; Clopper, P.E. *Evaluating Scour at Bridges*, 5th ed.; FHWA-HIF-12-003 HEC-18; U.S. Department of Transportation Federal Highway Administration: Washington, DC, USA, 2012; pp. 71–73.
110. Hong, S. Interaction of Bridge Contraction Scour and Pier Scour in a Laboratory River Model. Master's Thesis, Georgia Institute of Technology, Atlanta, GA, USA, 2005.
111. Melville, B.W. The Physics of Local Scour At Bridge Piers. In Proceedings of the Fourth International Conference on Scour and Erosion, Chuo University, Tokyo, Japan, 5–7 November 2008.
112. Ghorbani, B. A Field Study of Scour at Bridge Piers in Flood Plain Rivers. *Turk. J. Eng. Environ. Sci.* **2008**, *32*, 189–199.
113. Izbash, S.V. *Construction of Dams by Dumping Stones into Flowing Water*; US Engineering Office, War Department: Eastport, ME, USA, 1935.
114. Faraday, R.V.; Charlton, F.G. *Hydraulic Factors in Bridge Design*; Hydraulics Research: Wallingford, UK, 1983.
115. Parola, A.C.; Jones, J.S. Sizing riprap to protect bridge piers from scour. *Transp. Res. Rec.* **1991**, *1290*, 276–280.
116. Parola, A.C. Stability of riprap at bridge piers. *J. Hydraul. Eng.* **1993**, *119*, 1080–1093. [CrossRef]
117. Chiew, Y.-M. Mechanics of riprap failure at bridge piers. *J. Hydraul. Eng.* **1995**, *121*, 635–643. [CrossRef]
118. Froehlich, D.C. Protecting bridge piers with loose rock riprap. *J. Appl. Water Eng. Res.* **2013**, *1*, 39–57. [CrossRef]
119. Chiew, Y.M. Failure Mechanisms of Riprap Layer Around Bridge Piers. In Proceedings of the First International Conference on Scour of Foundations, Texas A&M University, College Station, TX, USA, 17–20 November 2002; pp. 70–91.
120. Chiew, Y.M. Riprap Protection Around Bridge Piers In A Degrading Channel. In Proceedings of the First International Conference on Scour of Foundations, Texas A&M University, College Station, TX, USA, 17–20 November 2002; pp. 707–718.
121. Gilja, G.; Kuspilić, N.; Tečić, D. Morphodynamic impact of scour countermeasures on riverbed topography. In Proceedings of the 15th International Symposium Water Management and Hydraulics Engineering, Primošten, Croatia, 7–8 September 2017; pp. 176–183.
122. Gilja, G.; Cikojević, A.; Potočki, K.; Varga, M.; Adžaga, N. Remote Real-time Riprap Protection Erosion Assessment on large rivers. In Proceedings of the EGU General Assembly 2020, Vienna, Austria, 3–8 May 2020.
123. Petersen, T.U.; Sumer, B.M.; Fredsøe, J.; Raaijmakers, T.C.; Schouten, J.-J. Edge scour at scour protections around piles in the marine environment—Laboratory and field investigation. *Coast. Eng.* **2015**, *106*, 42–72. [CrossRef]
124. Veerappadevaru, G.; Gangadharaiah, T.; Jagadeesh, T.R. Vortex scouring process around bridge pier with a caisson. *J. Hydraul. Res.* **2011**, *49*, 378–383. [CrossRef]
125. Przedwojski, B. Bed topography and local scour in rivers with banks protected by groynes. *J. Hydraul. Res.* **1995**, *33*, 257–273. [CrossRef]

126. Rashak, B.M.; Khassaf, S.I. Local scour around t-shape submerged groynes in clearwater conditions. *Int. J. Geomate* **2020**, *20*, 163–172. [CrossRef]
127. Han, X.; Lin, P. 3D numerical study of the flow properties in a double-spur dikes field during a flood process. *Water* **2018**, *10*, 1574. [CrossRef]
128. Rashak, B.M.; Khassaf, S.I. Study the local scour around different shapes of single submerged groyne. *J. Water Land Dev.* **2020**, *47*, 1–9. [CrossRef]
129. McCoy, A.; Constantinescu, G.; Weber, L. A numerical investigation of coherent structures and mass exchange processes in channel flow with two lateral submerged groynes. *Water Resour. Res.* **2007**, *430*, 1–26. [CrossRef]
130. Pandey, M.; Ahmad, Z.; Sharma, P. Estimation of maximum scour depth near a spur dike. *Can. J. Civ. Eng.* **2016**, *43*, 270–278. [CrossRef]
131. Giglou, A.N.; Mccorquodale, J.A.; Solari, L. Numerical study on the effect of the spur dikes on sedimentation pattern. *Ain Shams Eng. J.* **2018**, *9*, 2057–2066. [CrossRef]
132. Pandey, M.; Jamei, M.; Karbasi, M.; Ahmadianfar, I.; Chu, X. Prediction of maximum scour depth near spur dikes in uniform bed sediment using stacked generalization ensemble tree-based frameworks. *J. Irrig. Drain. Eng.* **2021**, *147*, 04021050. [CrossRef]
133. Tabarestani, M.K.; Salamatian, S.A. Physical modelling of local scour around bridge pier. *Malays. J. Civ. Eng.* **2018**, *28*, 349–364. [CrossRef]
134. Cikojević, A.; Gilja, G.; Kuspilić, N. Sensitivity analysis of empirical equations applicable on bridge piers in sand-bed rivers. In Proceedings of the 16th International Symposium on Water Management and Hydraulic Engineering WMHE 2019, Skopje, Macedonia, 5–7 September 2019; pp. 100–109.
135. Lee, S.O.; Sturm, T.W. Effect of sediment size scaling on physical modeling of bridge pier scour. *J. Hydraul. Eng.* **2009**, *135*, 793–802. [CrossRef]
136. Valela, C.; Nistor, I.; Rennie Colin, D.; Lara Javier, L.; Maza, M. Hybrid modeling for design of a novel bridge pier collar for reducing scour. *J. Hydraul. Eng.* **2021**, *147*, 04021012. [CrossRef]
137. Fischenich, C.; Landers, M. *Computing Scour*; U.S. Army Engineer Research and Development Center: Vicksburg, MS, USA, 1999; p. 5.
138. Ji, U.; Yeo, W.; Kang, J. Subsidence of riprap protection without filters for different installation types of riprap around a pier in sands. *J. Hydro-Environ. Res.* **2013**, *7*, 41–49. [CrossRef]
139. Mao, L. The effects of flood history on sediment transport in gravel-bed rivers. *Geomorphology* **2018**, *322*, 196–205. [CrossRef]
140. Redolfi, M.; Bertoldi, W.; Tubino, M.; Welber, M. Bed load variability and morphology of gravel bed rivers subject to unsteady flow: A laboratory investigation. *Water Resour. Res.* **2018**, *54*, 842–862. [CrossRef]
141. Wang, L.; Cuthbertson, A.J.S.; Pender, G.; Cao, Z. Experimental investigations of graded sediment transport under unsteady flow hydrographs. *Int. J. Sediment Res.* **2015**, *30*, 306–320. [CrossRef]
142. Phillips, C.B.; Hill, K.M.; Paola, C.; Singer, M.B.; Jerolmack, D.J. Effect of flood hydrograph duration, magnitude, and shape on bed load transport dynamics. *Geophys. Res. Lett.* **2018**, *45*, 8264–8271. [CrossRef]
143. Sturm, T.; Sotiropoulos, F.; Landers, M.; Gotvald, T.; Lee, S.O.; Ge, L.; Navarro, R.; Escauriaza, C. *Laboratory and 3D Numerical Modeling with Field Monitoring of Regional Bridge Scour in Georgia*; Georgia Department of Transportation: Atlanta, GA, USA, 2004.

water

MDPI

Review

Flow Resistance in Open Channel Due to Vegetation at Reach Scale: A Review

Antonino D'Ippolito *, **Francesco Calomino, Giancarlo Alfonsi and Agostino Lauria**

Dipartimento di Ingegneria Civile, Università della Calabria, 87036 Arcavacata, Italy; francesco.calomino@unical.it (F.C.); giancarlo.alfonsi@unical.it (G.A.); agostino.lauria@unical.it (A.L.)
* Correspondence: antonino.dippolito@unical.it; Tel.: +39-0984-496550

Abstract: Vegetation on the banks and flooding areas of watercourses significantly affects energy losses. To take the latter into account, computational models make use of resistance coefficients based on the evaluation of bed and walls roughness besides the resistance to flow offered by vegetation. This paper, after summarizing the classical approaches based on descriptions and pictures, considers the recent advancements related to the analytical methods relative both to rigid and flexible vegetation. In particular, emergent rigid vegetation is first analyzed by focusing on the methods for determining the drag coefficient, then submerged rigid vegetation is analyzed, highlighting briefly the principles on which the different models are based and recalling the comparisons made in the literature. Then, the models used in the case of both emergent and submerged rigid vegetation are highlighted. As to flexible vegetation, the paper reminds first the flow conditions that cause the vegetation to lay on the channel bed, and then the classical resistance laws that were developed for the design of irrigation canals. The most recent developments in the case of submerged and emergent flexible vegetation are then presented. Since turbulence studies should be considered as the basis of flow resistance, even though the path toward practical use is still long, the new developments in the field of 3D numerical methods are briefly reviewed, presently used to assess the characteristics of turbulence and the transport of sediments and pollutants. The use of remote sensing to map riparian vegetation and estimating biomechanical parameters is briefly analyzed. Finally, some applications are presented, aimed at highlighting, in real cases, the influence exerted by vegetation on water depth and maintenance interventions.

Keywords: river hydraulics; vegetation; flow resistance; turbulence; numerical methods

Citation: D'Ippolito, A.; Calomino, F.; Alfonsi, G.; Lauria, A. Flow Resistance in Open Channel Due to Vegetation at Reach Scale: A Review. *Water* **2021**, *13*, 116. https://doi.org/10.3390/w13020116

Received: 19 November 2020
Accepted: 29 December 2020
Published: 6 January 2021

1. Introduction

As it is well known, vegetation is an important issue on the viewpoint of catchment hydrology [1–3], since rain drops interception, evapotranspiration, and infiltration are elements to consider in surface and sub-surface water balance.

Moreover, riparian vegetation plays a key role both on the ecologic and habitat viewpoints, as well as a source of biodiversity. Indeed, vegetation creates micro-environments that can host birds, small mammals, and insects helpful to agricultural purposes, prevents fertilizers and pollutants from getting to the watercourses [4] and, because of the effect on landscape, has a significant recreational function.

On a more strictly technical viewpoint, riparian vegetation interacts with water flow, with effects both on the bank stability and on the river hydraulics. Vegetation acts on bank stability since it mechanically strengthens the soil because of the presence of roots [5–8]; moreover, it reduces the soil water content because of evapotranspiration, with the consequence of reducing interstitial pressures [9].

As for river hydraulics, vegetation clutters up part of the river cross-section [7,10,11], increases the roughness, and reduces the velocity; all this results in increased water levels and reduced water conveyance. Moreover, while the smaller average velocity on one hand reduces the erosion of the riverbed and banks, on the other one, it increases the sediment deposition, what makes the water cross-sections smaller and raises flooding risk. On the scale of the hydrographic network, the general velocity reduction influences the travel time of water particles, making the peak flow control easier [12,13].

Therefore, one cannot know the general effect of vegetation in advance, but every case should be considered singularly, using proper procedures. Indeed, this effect depends on both the hydraulic and mechanical properties of the water cross-section, as of the present vegetation, that may be different according to species, phenological stage, age and, possibly, maintenance.

Numerous studies are presented in the literature on the experimental, theoretical, and computational points of view [4,14–19]. The major difficulty lies in the impossibility of studying the problem under common conditions or to draw conclusions of a general value from the experiments or from the case studies. The recent progresses in the numerical solution of the flow equations make it possible to treat single problems, but it remains difficult to adapt the codes to different conditions.

Usually, in the literature, the vegetation is considered as rigid or flexible, and according to the water level, as emergent or submerged. Flexible vegetation refers to grass, reeds, and shrubs, or, when speaking about trees, to the branch and leaf system. Combinations of the above categories, really found in natural streams and channels, are still difficult to treat.

One should note that, commonly, the river cross-sections present variable roughness along the wetted perimeter, and a typical example is given by vegetated cross-sections; moreover, one river station can be formed by more sub cross-sections, differently shaped. When using one-dimensional models of water flow, one needs a roughness coefficient representative enough that it can be computed as a weighted average of local roughness coefficients. To this purpose, several equations are present in [20] on the basis of the assumptions made on how a particular variable (discharge, velocity, contour shear stress) in the subsections is related to that in the total section.

Although the subject of this review is the flow resistance due to vegetation, however, it should be noted that research on the interaction between vegetation and flow is currently focusing on a more correct assessment of the action exerted by the shear stress on the bed and the banks [21,22], on velocity distribution [23,24], on sediment transport [25–31], on finite-sized vegetation patches [32–39], on the interaction between jets and vegetation [40,41], on processes of transport and dispersion [42–44], on evolution of patches of vegetation [45], and on one-line emergent vegetation [46].

Correctly evaluating the resistance to flow is then a major aspect of not only the river safety, regulation, and maintenance, but also of flood model calibration and validation. In the following, we will present the methods found in the literature, allowing estimation of flow resistance coefficients to input into models for flood simulation, based on different types of vegetation in the river banks and floodplains. We will then analyze briefly the three-dimensional modeling of free surface flow in the presence of vegetation and the use of remote sensing for mapping riparian vegetation and estimating biomechanical parameters. Finally, the issue of vegetation management will be dealt with, presenting a number of points of view when cutting or pruning of plants is required for safety or maintenance reasons.

2. Flow Resistance Equations

According to Chow [47], the resistance to flow in artificial channels and watercourses is influenced by several factors, i.e., size and shape of the grains of the material forming the wetted perimeter, vegetation, channel irregularity, channel alignment, silting and scouring, obstruction, size and shape of channel, stage and discharge, seasonal change, suspended material, and bed load. More classifications exist, among which the best known are those by Rouse [48] and Yen [20].

As is well known, the resistance to flow can be expressed by the Darcy-Weisbach f friction factor, the Chézy's C, or the Gauckler-Strickler k velocity coefficients and the Manning n roughness coefficient; the relation among these coefficients is the following:

$$\sqrt{\frac{8}{f}} = \frac{R^{1/6}}{n\sqrt{g}} = \frac{kR^{1/6}}{\sqrt{g}} = \frac{C}{\sqrt{g}} = \frac{V}{\sqrt{gRJ}} \tag{1}$$

where V is the mean flow velocity, R the hydraulic radius, J the energy line slope, and g the gravity acceleration.

Cowan [49], considering the case of fixed bed and ignoring the effects of the suspended solid flux on the turbulence, believes that the resistance to flow in natural water-courses depends on one hand on the shape, the dimension, and arrangement of the elements that form the roughness; and on the other hand, it depends on further dissipative effects, caused by macro-vortices produced by the flow separation in the abrupt changes of direction, cross-section shape, and vegetation; and that the overall Manning coefficient can be expressed as the sum of the relevant values.

The contribution of vegetation to the roughness coefficient can be evaluated by means of descriptive approaches, photographic comparison approaches, or by analytical methods. Even though the descriptive and photographic approaches should be considered empirical in nature, nevertheless they add to the knowledge of the river conditions and in many cases are a good way to assess a value of a friction factor or roughness/velocity coefficient.

3. Descriptive and Photographic Comparison Approaches

In these methods, one roughness coefficient is chosen on the basis of the class to which the river reach belongs. Among the descriptive methods, the best known is Chow's [47]. The author gives, for every class of channels, the minimum, average, and maximum values of Manning n coefficient, warning that when the channel is artificial, the average values should be used in case of good maintenance only. In Tables 5 and 6 of Chow's book ([47], p. 110–113), one can observe that the Manning coefficient is 0.018 $sm^{-1/3}$ in case of the excavated channel, straight, clean, uniform cross-section with no vegetation, and 0.035 $sm^{-1/3}$ in case of dense weeds. In natural streams, its values are 0.030 $sm^{-1/3}$ when the cross-section is clean, and 0.045 $sm^{-1/3}$ in case of weeds. We should underline that for banks with many trees, the normal values of the Manning coefficient vary in a rather large field, between 0.040 $sm^{-1/3}$ and 0.150 $sm^{-1/3}$.

The photographic comparison approach consists of evaluating the Manning coefficient of a given river reach on the basis of similarity with the pictures of other similar cases, for which the coefficient was estimated in ordinary or flood conditions. Chow [47] shows some pictures that can be considered as a first example, representing 24 cases concerning typical channels and one natural river, with a short description of the channel conditions and the corresponding value of Manning coefficients.

Some more manuals show color pictures carrying data and description of rivers in the USA and Australia [50–53], with features very different from each other and Manning coefficients ranging in a wide field. In addition to a qualitative description of the channel, there is some information about cross-sections and hydraulic characteristics. Almost all the watercourses exhibit riparian vegetation over the banks, but estimates of the Manning coefficients very often refer to the main channel only; indeed, the values pertaining to flooding cases are limited.

Arcement and Schneider [54] show the pictures of 15 areas subject to flooding and densely vegetated, for which roughness coefficients are evaluated. This is one of the few works trying to define the contribution of vegetation to the total flow resistance. Starting from the Cowan [49] approach and taking into account the method of vegetation density proposed by Petryk and Bosmajian [55], the authors make use of density measurements and of an effective drag coefficient, to compute the total roughness and the different contribution to it. The contribution due to vegetation ranges between percentages of the total roughness from 64% to 81%, that is very significant values.

Some researchers think that the above methods are probably more reliable than the analytical, since, in field observations, the heterogeneity is accounted for. Nevertheless, pictures of vegetated channels are more or less limited, and, consequently, choosing a reference channel is often difficult, not to say impossible.

4. Analytical Methods

In the analytical methods, plants are generally described by a characteristic diameter, D, and height, h_v, or by a characteristic area of plants, A_c. These methods are mainly suitable in laboratory experiments, where reference is made to both natural and artificial vegetation, the latter represented by cylinders of different materials or strips of plastic. Much more limited, in this approach, are the field experiments.

In laboratory experiments, the elements representing vegetation are often distributed with parallel (called linear by some authors) or staggered patterns; there are cases where the distribution is random, as vegetation usually is in nature (Figure 1).

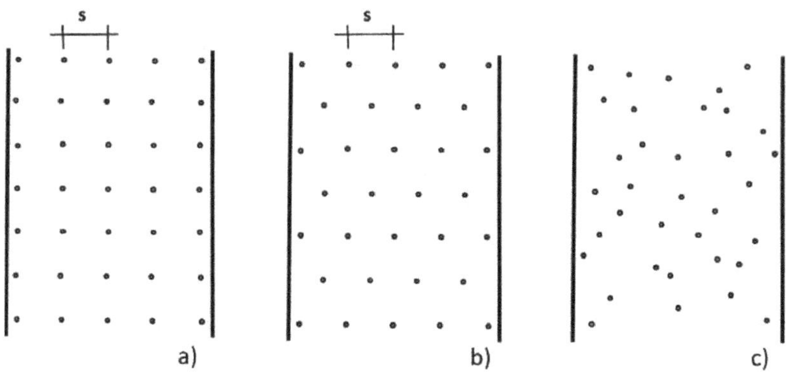

Figure 1. Plan view of (**a**) parallel, (**b**) staggered, and (**c**) random patterns.

In the case of rigid vegetation represented by cylinders, the density of vegetation λ is often expressed as

$$\lambda = \frac{m\pi D^2}{4} \tag{2}$$

where m is the number of cylinders per unit bed area; it is also used as a density measure of the projected plant area per unit volume, a, with

$$a = mD \tag{3}$$

as well as the ratio of stem diameter to spacing between the stems D/s.

4.1. *Rigid Vegetation*

When the vegetation is made up by trees, in an analogy of the resistance to flow due to immersed bodies, the roughness coefficient is expressed as a function of the drag force exerted by the flow on the body, depending then on the number of trees, their arrangement, the diameters of their trunks and, where appropriate, the branch system. Usually, in the laboratory tests, the rigid vegetation is simulated by cylinders, like in Figure 2, which is practically correct when the flow does not touch the leafage.

Figure 2. Laboratory flume with arrays of cylinders representing rigid vegetation, University of Calabria.

4.1.1. Emergent Rigid Vegetation

In case of one isolated vertical cylinder whose axis is orthogonal to the flow direction, the resistance to flow is expressed by the drag force F_D, computed as

$$F_D = \frac{\rho C_D h D V^2}{2} \tag{4}$$

where ρ is the water density, h is the depth of the immersed part of the cylinder, V is the approach velocity, and C_D is a drag coefficient. C_D is a function of a stem Reynolds number computed by means of the approach velocity V and the cylinder diameter D, $Re_D = \frac{VD}{\nu}$, ν being the water kinematic viscosity.

When the cylinder is a part of a group of elements (see Figure 3), one cannot ignore that the longitudinal and transversal interference make considerably more difficult the study of the resistance to flow [56–58]. In Figure 3, x is the streamwise coordinate, z is the vertical coordinate above the river bed, u_z is the local time-averaged velocity, u is the mean velocity along the vertical, h is the water depth, and h_v the vegetation height.

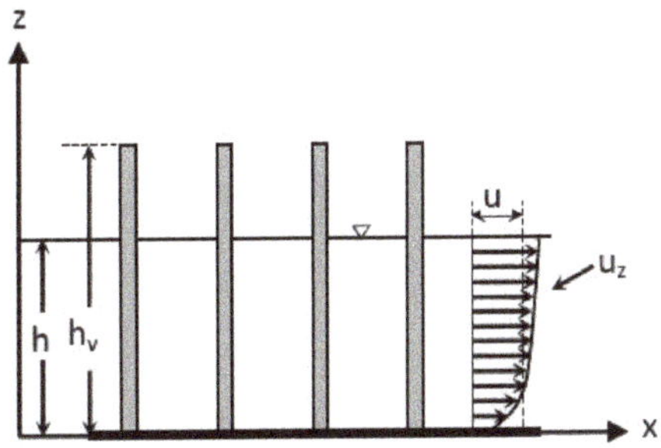

Figure 3. Side view of emergent vegetation and velocity profile.

Liu et al. [59], based on an experimental survey, described in detail the flow characteristics through rigid vegetation both in emergent and submerged conditions; they considered a staggered and a linear arrangement and for the latter referred to different densities; they also considered bottoms with different roughness. Liu et al. [59] measured the horizontal component of the velocity on several verticals between two dowels and also in the free field.

With reference to the emergent vegetation in the verticals immediately downstream of the dowels, the lower velocities are obtained; the velocities are progressively increasing, proceeding downstream and approaching the next dowel. In the free field there are higher velocities. Along each vertical, the velocities are practically constant for most of the height and gradually increase as they approach the free surface. In the case of verticals placed between two dowels, the velocity distribution near the bed has a spike which is particularly pronounced in the vertical immediately downstream of the dowel and which attenuates in the flow direction.

The velocity spike is probably caused by the horseshoe vortex that forms at the base of the dowel, attracting the liquid of the surrounding region, which is faster, towards the base of the dowel. The fluid masses near the bottom, characterized by high speed values, mix with the higher, characterized by low speeds, creating vortices that rotate counterclockwise. These vortices, moving downstream, bring the fluid masses upwards, and this determines, as the abscissa increases, the reduction of the velocity spike. The staggered distributions determine a greater resistance than the linear [56,59].

The tests carried out with a rough bed have shown a significant change in the velocity distribution only as regards the profile immediately downstream of the dowel where there have been reductions in velocity from 30 to 130%, and also an increase of about 20% for the velocity spike near the channel bed. The measurements of the vertical component of the velocity have shown, in the vertical immediately downstream to the dowel and in the vicinity of the bed, a relatively large value which is attenuated by proceeding in the flow direction. In the vertical downstream to the dowel, Liu et al. [59] found weak downward velocity components.

As to the turbulence intensity relative to the streamwise component of the velocity, Liu et al. [59] found almost constant values along each vertical, however, its magnitude varied significantly with the location, presenting the highest values in the verticals immediately downstream to the dowels and the smaller ones in the free stream region. The turbulence generated by the dowels has a length scale much smaller than the shear generated turbulence and is quickly dissipated. In the free stream region, the turbulence intensity increases near the bed due to the shear.

To find the values of C_D in case of sparse emergent arrays, both experimental tests and numerical simulations have been carried out, with the vegetation arrangements defined before as linear, or staggered, or random. Among the first studies, we will cite only Petryk [60], Li and Shen [56], and Petryk and Bosmajian [55].

Petryk and Bosmajian [55], to determine the Manning coefficient in a vegetated channel, implement the momentum equation for a reach, imposing that the vector sum of the weight of the control volume be equal to zero, projected on the bed direction, plus the contour resistance, plus that opposed by the tree trunk; they conclude with defining the overall Manning coefficient n as a function of the value relative to the soil, n_b, and the one relative to vegetation drag coefficient C_D, by writing

$$n = n_b\sqrt{1 + \frac{C_D \sum A_i}{2gAL}\left(\frac{1}{n_b}\right)^2 R^{4/3}} \tag{5}$$

where L is the reach length, A the area of the water cross-section, ΣA_i the area opposed by the vegetation to the flow. The authors consider the drag coefficient C_D equal to 1.

Stone and Shen [61] developed a method for predicting the apparent channel velocity and the velocities in the stem layer in both emergent and submerged conditions. Four staggered arrangements of stems with varying diameter and density were used. The drag coefficient values compare well with that for a single cylinder, and the authors suggest an average value of 1.05 applicable with the average velocity in the constricted cross-section.

Baptist et al. [62], based on the balance of horizontal momentum in uniform steady flow condition, calculated the Chézy coefficient in the presence of emergent vegetation, C_k, by writing

$$C_k = \sqrt{\frac{1}{(1/C_b^2) + (C_D mDh/2g)}} \tag{6}$$

where C_b is the Chézy coefficient of the bed. Baptist et al. [62] consider $C_D = 1$.

In Table 1 are reported different expressions for the drag coefficient. In the case of Ishikawa et al. [63], Kothiary et al. [64], and D'Ippolito et al. [65] they are based on direct measurements of the drag force. In Kothyari et al. [64], the cylinders are distributed on a grid with a staggered pattern forming angles of 30° with the flow direction, and the action is on one cylinder, while in the case of Ishikawa et al. [63], the angle with the flow direction was 45° and the action was calculated as mean on seven or thirteen cylinders, while instead in D'Ippolito et al. [65], the cylinders are in a linear arrangement and the action was calculated as mean on two to twenty five cylinders.

Table 1. Equations for estimating the drag coefficient C_D in case of arrays of emergent cylinders.

Authors	Relationship
Ishikawa et al. [63]	$C_D = 1.71\lambda^{0.11}$ when $i = 1/50$; $C_D = 2.45\lambda^{0.20}$ when $i = 1/20$; $C_D = 3.89\lambda^{0.31}$ when $i = 1/10$;
Tanino and Nepf [66]	$C_D = 2\left(\frac{\alpha_{0E}}{Re_{D*}} + \alpha_{1E}\right)$
Kothyari et al. [64]	$C_D = 1.53[1 + 0.45\,ln(1+100\lambda)]Re_{D*}^{-3/50}$
Cheng and Nguyen [67]	$C_D = \frac{50}{Re_v^{0.43}} + 0.7\left[1 - exp\left(-\frac{Re_v}{15{,}000}\right)\right]$ for $Re_v = 52 - 5.6 \times 10^5$ with $Re_v = \frac{V_v r_v}{\nu}$ and $r_v = \frac{\pi}{4}\frac{(1-\lambda)}{\lambda}D$
Cheng and Nguyen [67]	$C_D = \frac{130}{r_{v*}^{0.85}} + 0.8\left[1 - exp\left(-\frac{r_{v*}}{400}\right)\right]$ for $r = 24 - 5000$ with $r_{v*} = \left(\frac{gJ}{\nu^2}\right)^{1/3} r_v$
Wang et al. [68]	$C_D = \frac{50}{Re_v^{0.5}} + 4.5\frac{D}{h} - 0.303ln\lambda - 0.9$
Sonnenwald et al. [69]	$C_D = 2\left(\frac{6475D+32}{Re_{D*}} + 17D + 3.2\lambda + 0.5\right)$
D'Ippolito et al. [65]	$C_D = 0.211\,ln(100\lambda) + 0.784$ $0.003 < \lambda < 0.05$, $0.48\% < i < 2.02\%$, $1000 < Re_D < 10.000$

(i is the bed slope, α_{0E} and α_{1E} are the Ergun coefficients, J is the energy slope, V_v is the average pore velocity $V_v = V/(1-\lambda)$, r_v is the vegetation-related hydraulic radius $r_v = \frac{\pi}{4}\frac{(1-\lambda)}{\lambda}D$, r_{v*} is the dimensionless vegetation-related hydraulic radius $r_{v*} = \left(\frac{gJ}{\nu^2}\right)^{1/3} r_v$, Re_v is the vegetation Reynolds number $Re_v = \frac{V_v r_v}{\nu}$, Re_{D*} is the stem Reynolds number calculated with the average pore velocity $Re_{D*} = V_v D/\nu$).

In Ishikawa et al.'s tests [63], the drag coefficients based on the experimental results differ significantly for the same stem Reynolds number, although the dependence is unclear. The authors give three equations for the drag coefficient as a function of the vegetation density, depending on the flume slope.

Tanino and Nepf [66], on the basis of Ergun's [70] studies, proposed the equation reported in Table 1, where α_{0E} is relative to the contribution of viscous forces on the cylinder surface, and α_{1E} to the contribution of inertial forces deriving by the pressure drop downstream to the cylinders. The authors find that α_{1E} is linearly varying with the density, whereas α_{0E} is independent of the cylinder array characteristics for $0.15 \leq a \leq 0.35$. These results are confirmed by Tinoco and Cowen [71] for $Re_D > 1000$.

In the tests performed by Kothyari et al. [64], the density λ is defined as Equation (2) and is ranging between 0.0022 and 0.0885. The mean flow velocity was estimated as the flowrate (Q) divided by the flume cross section and $(1-\lambda)$, obtaining the so-called pore velocity V_v:

$$V_v = \frac{Q}{A(1-\lambda)}, \tag{7}$$

In the case of subcritical flows, the authors obtained the equation reported in Table 1 in which C_D remarkably increases with λ and varies weakly with the stem Reynolds number calculated with the average pore velocity. The logarithmic increase of C_D with λ ensures that, when λ is small, C_D increases rapidly with λ, whereas, as λ becomes large, C_D tends to a constant value.

Cheng and Nguyen [67], when the wall and bottom effects are negligible, introduced the vegetation-related hydraulic radius $r_v = \frac{\pi}{4}\frac{(1-\lambda)}{\lambda}D$, which is a function of the vegetation density and diameter only. This vegetation-related hydraulic radius is used together with the pore velocity to define a new Reynolds number, the vegetation Reynolds number $Re_v = \frac{V_v r_v}{\nu}$. Using experimental data from several authors (random, staggered, only two cases linear), suitably unified, they showed that the drag coefficient decreases monotonically with the increase in the vegetation Reynolds number and propose the two equations reported in Table 1, respectively, function of Re_v and r_{v^*} with r_{v^*} dimensionless vegetation-related hydraulic radius.

Wang et al. [68], in a study for incipient bed shear stress partition in mobile bed channels, investigated the vegetation drag coefficient. By ignoring the bed surface shear stress, an empirical formula was developed by data fitting in which the drag coefficient is a function of the Reynolds number, calculated with the vegetation-related hydraulic radius of Cheng and Nguyen [67], the ratio between vegetation diameter and flow depth (D/h), and the vegetation density λ. The proposed formula is reported in Table 1.

Sonnenwald et al. [69] based on the data of Ben Meftah and Mossa [23], Stoesser et al. [72], Tanino and Nepf [66], and Tinoco and Cowen [71] proposed the equation reported in Table 1, where the coefficient of the D terms must have units m^{-1} to retain non-dimensionality.

D'Ippolito et al. [65], on the basis of 70 tests with emergent stems in a linear arrangement, proposed an equation in which C_D is a function of the density λ only, valid in the field λ from 0.003 to 0.05, bed slope i from 0.48% to 2.02%, and Re_D from 1000 to 10.000.

Figure 4 shows how the drag coefficient varies with density, and Reynolds numbers, for some of the formulas reported in Table 1. The values obtained by D'Ippolito et al. [65] are smaller with respect to those of other authors with the same λ, because of the different rod arrangements (square mesh against triangular mesh).

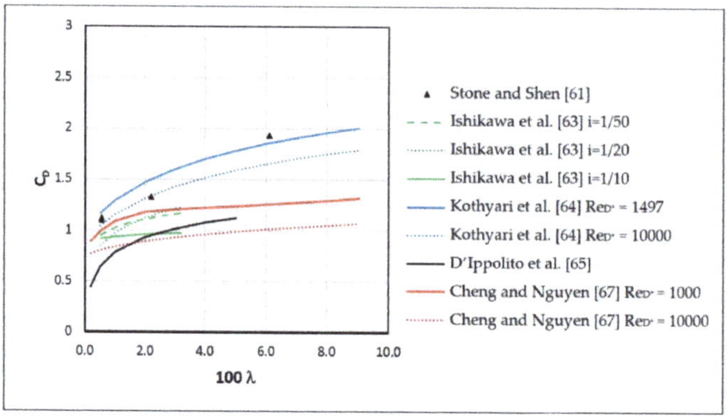

Figure 4. Emergent rigid vegetation—C_D as a function of λ and Re_D.

As it is easily seen, the C_D values most frequently range between 0.5 and 2.0. One must notice that a comparison among the above equations is difficult, because in the equations of Kothyari et al. [64], C_D has the form $C_D = C_D(\lambda,\ Re_{D*})$, while the equations of Cheng and Nguyen [67] have the forms $C_D = C_D(Re_v)$ and $C_D = C_D(r_{v*})$; the equation of Wang et al. [68] has the form $C_D = C_D(\lambda,\ D/h,\ Re_v)$; in the equation of Sonnenwald et al. [69], it is $C_D = C_D(\lambda,\ Re_{D*},\ D)$; and in the equation of D'Ippolito et al. [65], $C_D = C_D(\lambda)$.

4.1.2. Submerged Rigid Vegetation

In the case of submerged vegetation, Baptist et al. [62] have identified in the velocity profile along a vertical four distinct areas, even though very often the profile is schematized as only two interacting zones (two-layer approach): The vegetation layer, containing the cylindrical elements representing the vegetation, and the surface layer, above them, up to the flow surface (Figure 5). In Figure 5, u_s is the mean velocity along the vertical in the surface layer, and u_v is the mean velocity along the vertical in the vegetated layer.

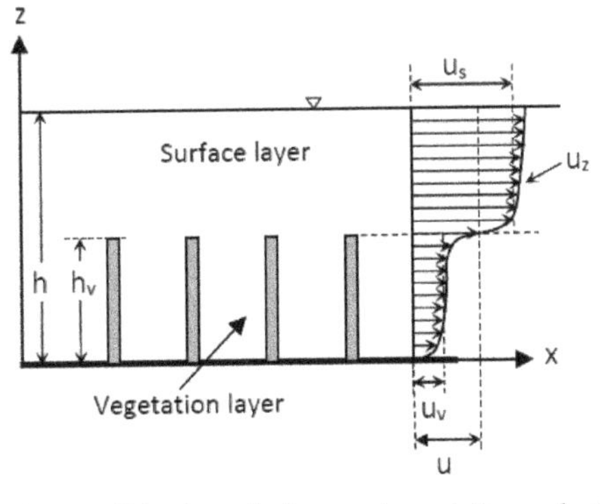

Figure 5. Side view of submerged vegetation and velocity profile.

The flow characteristics within a set of submerged cylinders, with the same arrangement and height of the cylinders, are similar to the case of emergent vegetation [59]. Above the cylinders, the liquid flows with a higher velocity and this, at the top of the cylinders, gives rise to an inflection point. The two coflowing streams, the upper one and the one between the cylinders, give rise to a Kelvin-Helmholtz instability, which causes the liquid to rotate clockwise, causing vortices that become larger in the downstream direction, forcing the inflection point deeper into the array. In the case of sparse vegetation, the vortex affects the entire vegetated layer, whereas in the case of dense vegetation, it affects only a layer limited to the top of the dowels [7]. The longitudinal turbulence intensities reach a maximum near the top of the dowel array. It has the largest values immediately behind the dowels and decreases in the flow direction. In the free stream region, the longitudinal turbulence intensity is lowest.

Rigid submerged vegetation has been the subject of a large number of investigations [7,61,62,73–77] and comparisons [78–83]. Some researchers provided the average velocity values in the two layers, while others derived the velocity distribution and the average values [7,73,74,76]. In the vegetation layer, the streamwise velocity is usually considered constant with the flow depth [62,76], while in the surface layer various expressions were adopted for the velocity distribution [81]: The logarithmic theory [61,73,76], the Kolmogorov theory of turbulence [75], the genetic programming [62], and the representative roughness height [77,84]. Usually, to determine the constants involved in the velocity formulas, the two distributions are required to assume the same values on the separation surface between the vegetated and the surface layer. The average velocity over the entire water depth is obtained as a combination of the velocity of the vegetated layer and that of the surface layer.

Klopstra et al. [73] derive the velocity profile in the vegetated layer starting from the turbulent shear stress, described by means of a Boussinesq-type equation, and adopt the logarithmic law for the surface layer. The different constants within the model are found on the basis of three conditions at the interface (continuity of shear stress, velocity and its vertical gradient) and a condition at the bed (negligible shear stress). The value of the Chézy coefficient, derived by Klopstra et al. [73], is a rather lengthy expression, and for it we suggest referring to the original paper ([73], Equation (9)). The model has only one unknown parameter, the characteristic turbulence length scale, α_{KL}, for which the following expression is proposed:

$$\alpha_{KL} = 0.0793\, h_v \ln \frac{h}{h_v} - 0.0090 \text{ for } \alpha_{KL} \geq 0.001 \tag{8}$$

For this parameter, van Velzen et al. [85] instead proposed the following relationship:

$$\alpha_{KL} = 0.0227\, {h_v}^{0.7} \tag{9}$$

The equations for the determination of the Chézy coefficient in the case of submerged rigid vegetation, C_r, proposed by Baptist et al. [62] and Cheng [84] are shown in Table 2. This table also shows the Chézy coefficient obtained from the value of the average velocity proposed by Huthoff et al. [75] and the Manning coefficient proposed by Yang and Choi [76].

Table 2. Equations for estimating C_r and n in case of submerged rigid vegetation.

Authors	Relationship
Baptist et al. [62]	$C_r = \sqrt{\frac{1}{(1/C_b^2)+(C_D m D h_v/2g)}} + \frac{\sqrt{g}}{\kappa} ln\left(\frac{h}{h_v}\right)$
Huthoff et al. [75]	$C_r = \sqrt{\frac{2g}{C_D m D h}}\left(\sqrt{\frac{h_v}{h}} + \frac{h-h_v}{h}\left(\frac{h-h_v}{1/\sqrt{m}}\right)^{\frac{2}{3}\left(1-\left(\frac{h}{h_v}\right)^{-5}\right)}\right)$
Yang and Choi [76]	$n = \left[\sqrt{\frac{2gh}{C_D a h_v}} + \frac{C_u\sqrt{g(h-h_v)}}{\kappa}\left(ln\frac{h}{h_v} - \frac{h-h_v}{h}\right)\right]^{-1} h^{2/3}$
Cheng [84]	$C_r = \sqrt{\frac{\pi g}{2C_D}\frac{(1-\lambda)^3}{\lambda}\frac{D}{h_v}\left(\frac{h_v}{h}\right)^{3/2}} + 4.54\sqrt{g}\left(\frac{h-h_v}{D}\frac{1-\lambda}{\lambda}\right)^{1/16}\left(\frac{h-h_v}{h}\right)^{3/2}$

For submerged rigid vegetation, Baptist et al. [62] derived an equation by dimensionally aware genetic programming, and this equation has been improved manually.

Yang and Choi [76] proposed a two-layer model. The momentum balance was applied to each layer and an expression for the mean velocities was proposed. The velocity was assumed uniform in the vegetation layer and logarithmic in the upper layer. In the equation reported in Table 2, $C_u = 1$ for $a \leq 5\ m^{-1}$ and $C_u = 2$ for $a > 5\ m^{-1}$.

A representative roughness height characterized by its proportionality to both the stem diameter and vegetation concentration was proposed by Cheng [84] to estimate, with the same approach of Yang and Choi [76], the average flow velocity, and thus the resistance coefficients in vegetated channels. Cheng's [84] approach was developed for submerged rigid vegetation, but also gave good results in the case of flexible vegetation.

López and García [86] employed two numerical algorithms, the $k-\varepsilon$ and $k-\omega$ type, based on two closure equations, to model the mean flow and the turbulence structure in open-channel flows with submerged vegetation. The Manning coefficient, computed on the basis of experimental observations, shows an almost constant value close to the one corresponding to non-vegetated channels up to some threshold plant density and, once this limit is exceeded, a linear increase.

Defina and Bixio [74] extended the model by Klopstra et al. [73] and that by López and García [86] to consider both the plant geometry and drag coefficient variable with depth. A comparison between experimental data of real and artificial vegetation and the results of numerical simulations demonstrates that both models are able to reproduce the vertical profiles of velocity and shear stress within and above vegetation, whereas the turbulence characteristics are poorly predicted.

Li et al. [77] proposed the concept of the auxiliary bed and produced a dynamic two-layer model consisting of a basal layer and a suspension layer (Figure 6). The roughness of the suspension layer is defined as the product of the concentration of vegetation, λ, and the depth of penetration of the suspension layer in the vegetation, δ_e. The authors obtained the following expression for the Manning coefficient:

$$n = \frac{h^{5/3}}{\sqrt{g}}\left[\frac{1.96(h-h_v+\delta_e)^{5/3}}{(\lambda\delta_e)^{1/6}} + (1-\lambda)(h_v-\delta_e)h_*^{1/2}\right]^{-1} \quad (10)$$

where $h_* = 2(1-\lambda)/C_D a$, with C_D calculated with the expression of Cheng and Nguyen [67]. For δ_e, refer to Equation (7) of Li et al. [77].

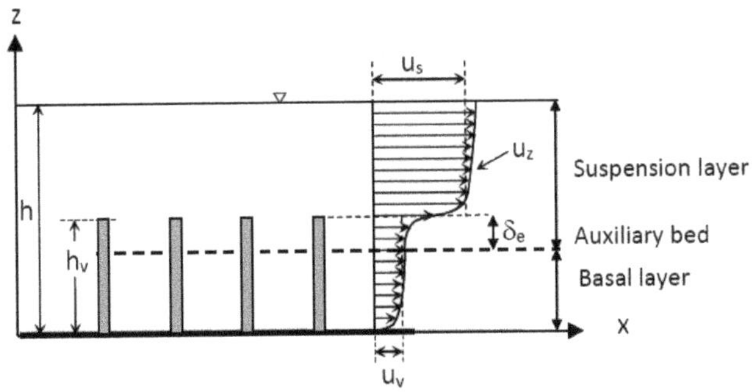

Figure 6. Side view of submerged vegetation with auxiliary bed and velocity profile.

Cheng [84], with reference to the flow rates, in the case of rigid submerged vegetation, compared his model with that of Stone and Shen [61], Baptist et al. [62], Huthoff et al. [75], and Yang and Choi [76]. It emerged that his model gave the best overall results while the others exhibited different behavior with varying flow rates. In particular, the Stone and Shen [61] model underestimated the flow rates especially for the higher values, while the model by Baptist et al. [62] works well for high flow rates, but yields overprediction for low flow rates, whereas Huthoff et al.'s [75] model gives the best prediction for high flow rates, but underestimates low flow rates.

Based on the Vargas-Luna et al. [79] analysis, with reference to the comparison between the measured Chézy coefficient and the one estimated with different models [61,62,73,75,76,84,85], it appears that the Klopstra et al. [73], van Velzen et al. [85], and Cheng [84] models show the best results. Although many models are based on a representation of vegetation with rigid cylinders, according to the authors, they also provide good results in the case of flexible vegetation when using the inflected vegetation height.

Pasquino and Gualtieri [81] compared the predicted and measured velocity for the Stone and Shen [61], Huthoff et al. [75], Yang and Choi [76], Cheng [84], Baptist et al. [62], and Li et al. [77] models as the density (sparse vegetation, transitional density, dense vegetation) and the degree of submergence (low submergence and high submergence) varied. They analyzed the performances of the different models on the basis of six different statistical parameters. In the case of transitional density with low submergence, all models give good results; in the case of dense vegetation with low submergence, the Huthoff et al. [75], Cheng [84], Baptist et al. [62], and Li et al. [77] models give very good results. In other cases, the number of data for estimating the statistical parameters, even if this analysis has been carried out, is less than 10.

Morri et al. [80], with the comparison between the measured and simulated mean velocities by different analytical models, highlight how the model by Huthoff et al. [75] provides the best results. It can be used for velocity lower than 0.8 m/s and underestimates the velocity values in the case of sparse vegetation, which could be explained because the bed roughness effect is neglected.

To conclude the analysis of the models proposed for submerged rigid vegetation, it must be said that they refer essentially to laboratory data and give overall good results, even if the models by Cheng [84] and Huthoff et al. [75] seem to provide the best results. One must note that comparison among the equations presented by various authors is more difficult than in the case of rigid emergent vegetation, because of the many parameters or variables involved.

4.1.3. Submerged and Emergent Rigid Vegetation

Katul et al. [87] and Luhar and Nepf [11] proposed expressions to calculate the Manning coefficient for both submerged and emergent vegetation, given in Table 3.

Table 3. Equations for estimating n in case of submerged and emergent rigid vegetation.

Authors	Relationship
Katul et al. [87,88]	$n = \frac{h^{1/6}}{\sqrt{g} C_u f_c(h/h_v, \alpha_{KA})}$ with $f_c(h/h_v, \alpha_{KA}) = 1 + \alpha_{KA} \frac{h_v}{h} ln\left(\frac{\cosh\left(\frac{1}{\alpha_{KA}} - \frac{1}{\alpha_{KA}} \frac{h}{h_v}\right)}{\cosh\left(\frac{1}{\alpha_{KA}}\right)}\right)$
Luhar and Nepf [11]	$n = \frac{k_{LN} h^{1/6}}{g^{1/2}} \left(\frac{C_f}{2}\right)^{1/2} (1 - B_x)^{-3/2}$ for $B_x < 0.8$
Luhar and Nepf [11]	$n = \frac{k_{LN} h^{1/6}}{g^{1/2}} \left(\frac{C_D a h}{2}\right)^{1/2}$ for $B_x = 1.0$
Luhar and Nepf [11]	$n = \frac{k_{LN} h^{1/6}}{g^{1/2}} \frac{1}{\left(\frac{2}{C_f}\right)^{1/2} \left(1 - \frac{h_v}{h}\right)^{3/2} + \left(\frac{2}{C_D a h_v}\right)^{1/2} \left(\frac{h_v}{h}\right)}$ when the vegetation, of height $h_v < h$, fills the entire width.

Katul et al. [87,88] starting from the characteristics of the velocity in turbulent flow within and above rigid vegetation canopies proposed to calculate the Manning coefficient according to the relationship reported in Table 3, where C_u is the similarity constant (empirically estimated as 4.5) and α_{KA} is the characteristic eddy size coefficient (estimated as 1 for gravel bed streams and 0.5 for canopy). The previous relationship can also be used in the case of shallow streams; more generally, it is valid for $0.2 < h/D < 7$.

Luhar and Nepf [11], in agreement with Green [10], consider that the flow resistance due to aquatic vegetation depends on the blockage factor, B_x, which is the fraction of the channel cross-section blocked by vegetation. Additionally, for the same blockage factor B_x, the distribution of vegetation, in terms of distinct patches, affects the hydraulic resistance, since the interfacial area between the patches and the unobstructed flow increases when the patches are divided into smaller units. However, on the basis of observations made in natural rivers, the authors estimate velocities for the case where the blockage factor is known, but the exact distribution pattern is unknown, and it introduces up to 20% uncertainty. Luhar and Nepf [11] propose the relationships reported in Table 3 for $B_x < 0.8$, for $B_x = 1.0$ and for channels where the vegetation, of height $h_v < h$, fills the entire width. In the reported equations, $k_{LN} = 1\ m^{1/3}/s$ is a constant necessary to make the equation dimensionally correct, a is the frontal area per unit volume, C_f is a non-dimensional coefficient proposed by Luhar-Nepf [11] given by $C_f = 2g/C^2$ where C is the Chezy coefficient.

4.2. Flexible Vegetation

4.2.1. Potentially Changing Vegetation Condition

Sometimes during a flood event, as the discharge increases, the vegetation can lay over or be removed [89–92], which leads to a reduction in roughness and to an increase in the flow capacity through the section; therefore, the peak flow, which could occur later, takes lower water-surface elevations than it would have had in the case of upright vegetation. To determine under what conditions the vegetation flattens, Phillips et al. [89] referred to the stream power, defined as $SP = gRSV$, and to the resistance of the vegetation characterized by an index defined as the susceptibility index of the vegetation. This index is given by the product of the vegetation flexibility factor, the vegetation blocking coefficient, the vegetation distribution coefficient and, finally, the flow depth coefficient. The vegetation flexibility factor is characteristic of the type of vegetation and is equal to the bending moment determined by computing the product of the moment arm (distance from the bed to the point of application) and the force required to bend the vegetation to 45 degrees from the vertical. Phillips et al. [89], with reference to four different types of vegetation, by means of regression techniques, proposed the equations relating bending moment to vegetation height. The vegetation blocking coefficient takes into account the combined resistance associated with the vegetation and was determined by assigning a weighted value to the estimated percentage of the cross-sectional area of flow that was blocked by vegetation for pre-flow conditions. The vegetation distribution coefficient takes into account the fact that vegetation aligned to the flow direction determines a velocities redistribution and the vegetation is affected by a smaller action, whereas in the case of uniform distribution every element is affected by the same velocity. Finally, the action of the flow to bend the vegetation depends on the water depth in relation to vegetation height. Five different categories have been identified, and the relative coefficients have been assigned according to the ratio between hydraulic radius (approximated with mean flow depth) and vegetation height. Phillips et al. [89], based on the experimental values, identified a threshold curve from which, for an assigned value of the vegetation susceptibility index, it is possible to determine the minimum value of the stream power beyond which the vegetation is layover.

Francalanci et al. [92], with reference to a river in Italy carried out the hydraulic modelling of one flood event in order to investigate the hydraulic forcing on trees, in terms of flow velocity and water depth. A conceptual model for the stability of riparian vegetation, based on the flow drag, the plant weight, and the resistance force of the root apparatus at the soil-root interface, has been used in a predictive form to investigate the range of the input variables which can promote the plant uprooting or preserve its stability.

For rigid vegetation, the action of the flow varies with the squared velocity. In case of flexible vegetation, this is not true, since the vegetation reconfigures by reducing the area projected on a plane orthogonal to the flow direction and aligning the leaves with it. The relationship between flow velocity and drag force was expressed as $F_D \propto V^{2+b}$, where the Vogel exponent b [58] is a measure of the plant reconfiguration. When is $b = -1$, the drag force varies linearly with the velocity. A linear increase of drag force with the flow velocity was observed for flexible plants by direct measurement in prototype scale by Armanini et al. [93].

The friction factor due to the vegetation, f_V, in a reach of length L is

$$f_v = \frac{8F_D h}{\rho V^2 A L} \tag{11}$$

We will first analyze the studies in the case of submerged flexible vegetation and then those related to the non-submerged vegetation.

4.2.2. Submerged Flexible Vegetation

The first studies on submerged flexible vegetation (Figure 7) are related to the design of irrigation canals [94]. In Figure 7, h_{vf} is the bent vegetation height. Since the resistance depends on the curvature of the vegetation, the link between the Manning coefficient, n, and the product of velocity, V, and hydraulic radius, R, was identified on an experimental basis. This relationship depends on the type of vegetation and is practically independent on the slope of the canal and its shape. Five experimental curves have been obtained relating the Manning coefficients, also called delay coefficients, with the product VR classified as very high, high, moderate, low, or very low, and identified with the letters A, B, C, D, and E. This link was later reviewed and generalized [95–98].

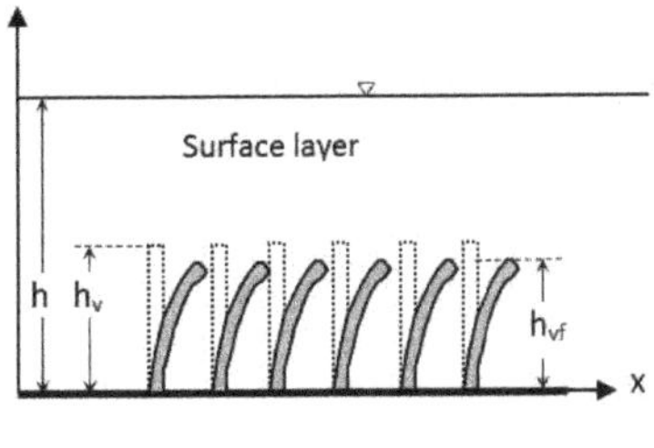

Figure 7. Side view of submerged flexible vegetation.

The aforementioned curves, for $n < 0.2$, were interpolated with the following equation:

$$n = \frac{1}{(2.08 + 2.3x + 6\ln(10.8\ VR))} \tag{12}$$

with x equal to $-0.5, 2, 5, 7, 11$ respectively for curves A, B, C, D, and E [97].

According to Kouwen et al. [99], the friction factor in the case of submerged flexible vegetation can be represented through a semi-logarithmic function of the relative roughness, defined as the ratio of the deflected plant height, h_{vf}, to the flow depth

$$\frac{V}{u_*} = C_0 + C_1 \log \frac{h}{h_{vf}} \tag{13}$$

where the coefficients C_0 and C_1 depend on u_*/u_{*c} with u_{*c}

$$u_{*c} = \min\left(0.23\ MEI^{0.106},\ 0.028 + 6.33MEI^2\right) \tag{14}$$

where M is the stem count per unit area, E is the modulus of elasticity, and I the second moment of the cross-sectional area of the stems. The product MEI is defined as flexural rigidity. The deflected height depends on the drag exerted by the flow and the flexural rigidity of the vegetation. Kouwen and Unny [90] deduced the following relationship:

$$\frac{h_{vf}}{h_v} = 0.14\left[\frac{\left(\frac{MEI}{\gamma hS}\right)^{1/4}}{h_v}\right]^{1.59} \tag{15}$$

where γ is the water specific weight.

Kouwen's [100] model implicitly considers the fact that the flow can cause the vegetation to lay over on the bottom of the channel without using the stream power illustrated in point 4.2.1.

Carollo et al. [91], based on the dimensional analysis, wrote the equation

$$\frac{V}{u_*} = \sqrt{\frac{8}{f_v}} = f_c(m)\left(\frac{h}{h_{vf}}\right)^{\beta_1}\left(\frac{u_* h_{vf}}{v}\right)^{\beta_2}\left(\frac{h_v}{h_{vf}}\right)^{\beta_3} \tag{16}$$

where f_c is a function of the concentration m (number of stems per unit area), h_v the vegetation height in the absence of flow, h_{vf} the height of bent vegetation, and u_* the shear velocity. The authors carried out experimental test to determine the exponents β_1, β_2, β_3 and the function f_c, which allow computation of the velocity V, since the bent vegetation height can be evaluated on the basis of an empirical equation.

Stephan et al. [101] analyzed, through laboratory experiments, the influence of roughness caused by aquatic vegetation on the overall flow field. The analysis of the velocity measurement shows that equivalent sand roughness and zero plane displacement of the logarithmic law are of the same order of magnitude as the mean deflected plant height.

4.2.3. Non-Submerged Flexible Vegetation

Studies on non-submerged flexible vegetation have been carried out, among others, by Kouwen and Fathi Moghadam [102], Freeman et al. [103], Västilä et al. [104], and Jalonen and Järvelä [105]. In particular, Kouwen and Fathi Moghadam [102], Västilä et al. [104], and Jalonen and Järvelä [105] provide an expression for the friction factor, while Freeman et al. [103] provide an expression for the Manning coefficient, given in Table 4.

Table 4. Equations for estimating f_v and n in case of non-submerged flexible vegetation.

Authors	Relationship
Kouwen and Fathi Moghadam [102]	$f_v = 4.06\left(\frac{V}{\sqrt{\xi E/\rho}}\right)^{-0.46}\left(\frac{h}{h_v}\right)$
Västilä et al. [104]	$f_v = 4C_{D\chi}\frac{A_c}{A_b}\left(\frac{V}{V_\chi}\right)^{\chi}$
Jalonen and Järvelä [105]	$f_v = \frac{4}{A_b}\left[A_L C_{D\chi,F}\left(\frac{V}{V_{\chi,F}}\right)^{\chi_F} + A_S C_{D\chi,S}\left(\frac{V}{V_{\chi,S}}\right)^{\chi_S}\right]$
Freeman et al. [103]	$n = 3.487 \times 10^{-5}\left(\frac{EA_s}{\rho A_i^* \ u_*^2}\right)^{0.150}(mA_i^*)^{0.166}\left(\frac{u_* R}{v}\right)^{0.622}\left(\frac{R^{2/3}J}{u_*}\right)$

Kouwen and Fathi Moghadam [102] performed laboratory experiments on four different types of conifers both in water and in air; the authors, on the basis of dimensional analysis and a series of simplifying hypotheses, proposed to calculate the friction factor as reported in Table 4, where ξ takes into account the deformation of the plant, and E is the modulus of elasticity. The term ξE is called vegetation index, it is unique for each species and is obtained from the resonant frequency, mass, and length of a tree [102,106]. The authors then propose to suitably modify the resistance to flow when the projection of a plant area overlaps with that of adjacent plants.

Recently, Västilä et al. [104] proposed to calculate the friction factor, f_v, as shown in Table 4, where A_c represents a characteristic area of plants, A_b is the bed area related to a plant, $C_{D\chi}$ is a species-specific drag coefficient, χ depends on the area A_c, and V_χ is the lowest velocity used in determining χ. As to the area A_c, Västilä et al. [104] compared three areas: The first one, A_L, is obtained by means of the leaf area index (LAI), (defined as the ratio between the one-sided leaf area and the ground area), the second one as the area projected onto a plan orthogonal to the flow direction when the plant is under the flow action (A_p), and the third one as the area projected onto a plan orthogonal to the flow direction when the plant is on free air (A_0). The results obtained by Västilä et al. [104] show that the three above characteristic areas give good results, even though the easiest area to quantify is the one relative to the leaf area index, that can be measured by laser-scanner techniques. Västilä et al. [104] determined the parameter values for Black Poplars ($C_{D\chi} = 0.33$, $V_\chi = 0.1$, $\chi = -1.03$, $0.4 < A_L/A_b < 3.21$) and other foliated deciduous species of Poplars ($0.43 \leq C_{D\chi} \leq 0.53$, $0.10 \leq V_\chi \leq 0.11$, $-0.57 \leq \chi \leq -0.9$, $0.4 \leq A_L/A_b \leq 3.2$).

Jalonen and Järvelä [105], taking into account the reconfiguration properties of branches and leaves, split their contribution in the friction factor proposing the equation reported in Table 4, where the subscript F and S refer to leaves and stem, respectively. The authors give the values of the parameters ($C_{D\chi,F}$, χ_F, $C_{D\chi,S}$, χ_S) for four different species, in the velocity field from 0.1 m/s to 0.6 m/s (low) and for velocities higher than 0.6 m/s (high).

Freeman et al. [103] on the basis of dimensional analysis and extensive experimental investigation, wrote the Manning coefficient as reported in Table 4, where A_i^* is the net submerged frontal area, A_s is the total cross-sectional area of all of the stems of an individual plant, measured at a quarter of its non-deflected height, m is the density. The plant characteristic A_i^* and A_s are density-weighted average values that were measured without any bending or distortion attributable to flow. The modulus of elasticity is calculated starting from the force necessary to bend the plants at an angle of 45° or, more simply, by an empirical relation expressed as a function of the ratio between the height of the plant and its diameter measured at a quarter of the height.

Once again, for submerged or non-submerged flexible vegetation, comparison among the equations proposed by different authors is difficult if not impossible.

5. Numerical Methods

Apart the empirical approach, turbulence study is the basis for flow resistance assessment, at least from Prandtl and von Kàrmàn theory dating back to the beginning of the last century. Moreover, a detailed analysis of the flow fields and turbulence characteristics may be of importance as related to solid and pollutant transport.

Experimental studies of velocity profiles, like those shown in Figures 3–5, are of much interest; in some cases also the shear stress profiles are studied by means of velocity fluctuant components and Reynolds stress theory [19,107–111].

For a detailed analysis of the flow fields and turbulence characteristics, one can refer to measurements with a particle image velocimetry system (PIV) or acoustic Doppler velocimeter (ADV) probe or, furthermore, on numerical simulations, more or less detailed depending on the flow cases at hand [65,112–115]. An excellent review about the numerical models utilized for the analysis of the interaction between flow and vegetation is due to Stoesser et al. [116], and one can refer to this work for a systematic view of the different aspects of this matter. In what follows, attention is given to the more recent works on this subject.

The interaction between flow and vegetation has been studied by means of RANS (Reynolds Averaged Navier-Stokes equations [112]) and LES (Large Eddy Simulation [117]) techniques. The DNS (Direct Numerical Simulation of turbulence [118]) approach was used to analyze the interaction between fluid and cylinders, in which all the fluctuating components are computed and no closure models are used, then the swirling-strength criterion for flow-structure extraction is applied to the velocity field [119]; this method is scarcely used, mainly due to the remarkable computing resources required.

RANS models are operated on coarse grids, and the drag due to vegetation is accounted for through additional source terms in the governing equations [86,120–123]. These models require an a-priori knowledge of the drag coefficient, contain a number of empirical constants, and are fairly accurate as related to the evaluation of the mean flow field, but do not simulate the flow past single trunks or branches. Kim and Stoesser [124] demonstrated the importance of the a-priori knowledge of the drag coefficient for the correct evaluation of the resistance to flow. Their RANS simulations, carried out by adopting actual C_D values, furnished values of the tangential shear stress on the bottom in a perfect agreement with the experimental results, while the adoption of a constant unitary value produced an underestimation of the resistance to flow as the density was greater. The coherent structures that form downstream to the cylinders or over the vegetated layer in the case of submerged vegetation can be obtained by means of the LES approach [72,117,124]. In the latter simulations, one can still refer to the resistance to flow of each single plant through the drag coefficient, or one can explicitly represent the plants by inserting in the mesh some blocked cells representing the vegetation. Stoesser et al. [72] executed a LES with reference to a companion experiment of Liu et al. [59] related to rigid emergent vegetation, obtaining interesting results. They calculated the flow characteristics with reference to reach 0.127 m long and 0.0635 m wide, with 11,604,640 computational cells. The time-averaged velocity profiles along six vertical lines resulted in a satisfactory agreement with the experimental values, both as to the streamwise and the normal components. The velocity field showed a high-velocity zone between the cylinders, the flow separation at 95°, and a recirculation zone downstream to the cylinders with two counter-rotating vortices. Additionally, the turbulence intensities in the streamwise and vertical directions showed a good agreement with the experimental results. Then the authors executed additional simulations with reference to densities and Reynolds numbers different from the experimental case. Starting from mean velocities and pressure fields, Stoesser et al. [72] calculated the drag on the cylinders, as decomposed in pressure and shear, and the shear on the bed. The drag coefficient values resulted in a good agreement with the experimental values of Tanino and Nepf [66]. Turbulence was investigated through the instantaneous pressure fluctuations. Kim and Stoesser [124] proposed a low-resolution LES calculation in which the computing time is considerably reduced compared to a fully resolved LES. In a high-resolution LES, each vegetation element is represented by means of a curvilinear grid, while in the case of low-resolution LES, the vegetation is handled by means of a simplified immersed boundary condition on a Cartesian grid (see Figure 2 of Kim and Stoesser [124]). A square-cell grid is preliminarily built-up. In the area interested by a generic circular obstacle, one can distinguish three types of cells, *(i)* those belonging to the obstacle for which the velocity is zero, *(ii)* those belonging to the fluid, for which the velocity is computed without any treatment, and *(iii)* those belonging to both of them (cut cells), for which the computed velocity is multiplied by the "volume fraction" (the volume of fluid divided by the total cell volume). Kim and Stoesser [124] compared the results of the low-resolution calculations with those of high-resolution calculations (different densities and Reynolds numbers of 500 and 1340 have been considered as related to a single simulation, to 534,681 and 22,994,560 grid points, respectively) and found that, in particular for low densities, the model was able to correctly represent the velocity gradients, the wakes behind the stems, and the secondary flows. In the low-resolution LES, the knowledge of the drag coefficient was not necessary, and the model could be used also to analyze complex rod configurations, namely casual distribution of the vegetation with elements very near to each other.

It should be noted that numerical simulations have been used not only for comparisons with companion experiments, but also in real situations. As an example [116,125], a RANS simulation was carried out to mirror a flood event of the Rhine river with a return period of 100 years. A reach of 3.46 km was considered, as discretized with a mesh of 258 × 64 × 12 computational cells along the streamwise, spanwise, and vertical directions, respectively, for a total of 198,144 cells. The dimension of each cell was about 13 m × 3 m × 0.5 m. The resistance to flow due to the plants was taken into account only in the momentum equation, the characteristic of the vegetation was taken into account starting from a survey related to an area of 77 m × 45 m, the Manning coefficient was calibrated on the basis of a flood event that not involved the floodplain, while the representative diameter of the vegetation was calibrated on the basis of a flood event that actually flooded the whole floodplain. The flow velocity was measured using dilution gauging techniques. The numerical results were in good agreement with the observed levels, and the values of the velocities in the zone with vegetation corresponded well to those measured by the dilution method. Additionally, the velocity distribution in the whole cross section resulted in a good agreement with the expected values. In conclusion, the turbulence study based on numerical approach is promising, even though still very far from practical use.

6. Hydraulic Roughness Assessment

Riparian vegetation, in particular that located on flood areas, has very heterogeneous characteristics, both from a spatial and temporal point of view. These characteristics must be adequately included in the hydraulic–hydrological models. Conventional ground-based monitoring is often unfeasible, as these techniques are time-demanding and expensive [126], especially for large areas and when they are inaccessible. New opportunities are offered by remote sensing, which has developed considerably in recent decades and has been increasingly used in the environmental field. Some reviews have addressed the use of remote sensing in fluvial studies [127–129] and, in particular, for mapping riparian vegetation and estimating biomechanical parameters [130].

Remote sensing is based on satellite images (digital or radar) or aerial platforms (LiDAR (Light Detection and Ranging) and orto-photography). Digital satellite images, in the last decade, have reached a definition similar to those of orthophotos (the pixel sizes in the sensors Quickbird and Ikonos are equal, respectively, to 0.6 m and 1 m) making possible their application to riparian areas that, very often, are of limited size. Image classification is the process of assigning individual pixel or groups of pixels to thematic classes; it is either supervised and unsupervised.

The classification of the images is very often supervised, that is, it is based on a priori knowledge of the type of coverage, but in the case of very high resolutions, it is based on segmentation techniques [131]: Edge-finding, region-growing, knowledge-based segmentation, and a mixture of the last two.

Forzieri et al. [132] proposed a method to estimate the vegetation height and flexural rigidity for the herbaceous patterns and plant density, tree height, stem diameter, crown base height, and crown diameter of high-forest and coppice consociations for arboreal and shrub patterns from satellite multispectral data (SPOT 5). The method is developed through four sequential steps: (1) Classification of pixel surface reflectance into five land cover classes: Mixed arboreal, shrub, herbaceous, bare soil, and water; (2) data transformation based on Principal Component Analysis of the original multispectral bands and use of only the first principal component since it explains a lot of variances; (3) identification of significant correlation structures between the main components and biomechanical properties; (4) identification/estimation/validation of the relationship (simple tri-parametric power laws) between the biomechanical properties and the normalized principal component. The vegetation hydrodynamic maps are also able to well describe the equivalent Manning's roughness coefficient as proved by comparison with simulated water stages [132].

One of the biggest limitations of optical sensors is the inability to penetrate the cloud system. Radar systems are microwave-based and do not depend on the cloud system, and are particularly useful during flooding events that usually occur in the presence of a cloud cover, allowing monitoring of the timing and spatial extent of flooding. Backscatter increases with biomass, and this makes it difficult to apply radar sensors in floodplain areas, which are usually characterized by very dense vegetation.

Satellite images provide information on the spatial variability of vegetation, but do not provide information about its vertical structure. LiDAR technology provides information on the three-dimensional structure of vegetation. Laser scanning (LS) is employed in terrestrial (TLS), airborne (ALS), and mobile (MLS) platforms. The airborne laser scanner (ALS) provides accurate information of forest canopy and ground elevations producing a digital terrain model and a digital surface model. The difference between the digital surface model and the digital terrain model gives the tree heights. Forzieri et al. [133] developed a model to identify individual tree positions, crown boundaries, and plant density using airborne LiDAR data. It needs an initial calibration phase based on a multiple attribute decision making simple additive weighting method. Jalonen et al. [134] employed multi-station TLS, both in field and laboratory conditions, to derive the total plant areas of herbaceous vegetation and the vertical distribution of the total plant area of foliated woody vegetation for different levels of submergence.

7. Flow Resistance and Vegetation Management

The different approaches and methods described above can be used in 1D, 2D, and 3D models.

These models can be useful in a multidisciplinary framework that includes hydrology, hydraulics, forestry, sediment transport, and ecology, which are the best vegetation management strategies. It is practically impossible to analyze the latter in the field, since one should see the flow behavior for the design discharge. To our knowledge, the only field experiments of different vegetation management is that of Errico et al. [135], who compared the effects of three different vegetation scenarios on flow velocity distribution, turbulence patterns, and flow resistance in drainage channel colonized by common reeds. The first scenario corresponded to the canopy in undisturbed conditions, the second and the third scenarios were obtained by clearing, respectively, the central part and the entire channel. The channel conveyance was obtained by clearing reeds in just the central part of the drainage channel and was comparable to that obtained by the total clearance, but with much less ecological impact and maintaining relatively high levels of turbulent intensities. Errico et al. [135] also evidenced that the most suitable models for representing natural reed canopies are those that quantify the blockage factor of the patch, rather than the effect of plant elements matrixes.

Phillips and Tadayon [90] were among the first to present some examples of vegetation management in both artificial and natural trapezoidal-shaped channels. Simulation results using HEC-RAS indicate that the design discharge for the channel for full-grown vegetation conditions would overtop channel banks and flood adjacent areas. Instead, simulations conducted for post-vegetation maintenance conditions, consisting in its partial removal, indicate that the design discharge would remain within the channel. A common element to some of the examples shown was verifying whether the design flow has the power to lay over the bushes. If this occurred, the associated roughness component was considered negligible and not used in the determination of the Manning coefficient. They also considered a minimum amount of 30 cm of freeboard above the design water-surface elevation. The purpose of this freeboard is to mitigate risk by providing a factor of safety. As part of the various vegetation maintenance schemes or scenarios, Phillips and Tadayon [90] have highlighted how leaving the vegetation randomly distributed is aesthetically more pleasant, while leaving the trees and bushes grouped together may present a better habitat environment for wildlife.

D'Ippolito and Veltri [136] analyzed the influence exerted by vegetation on the discharge with a return time of 200 years in the mountain reach of the Crati river (Calabria, Italy). The flow profiles, determined by the use of HEC-RAS, showed that, in the reach, there is an average increase in water heights of 5% with maximum peaks of 24%. The influence exerted by herbaceous vegetation is negligible compared to that due to woody vegetation.

Luhar and Nepf [11] assert that mowing patterns that produce less interfacial area per channel length (e.g., a single continuous cut on one side of the channel) are the most effective in reducing hydraulic resistance.

Van der Sande et al. [131] created a detailed land cover map of the villages of Itteren and Borgharen in the southern part of the Netherlands by using IKONOS-2 high spatial resolution satellite imagery and combining the use of spectral, spatial, and contextual information. It is one of the first applications of land cover maps used as inputs for flood simulation models. The plant cover is classified in three classes: Natural vegetation (n = 0.1), deciduous forest (n = 0.2), and mixed forest (n = 0.2). Using the above maps, the model results are not very different from those obtained using less accurate maps. Van der Sande et al. [131] conclude that their use provides better results (flow direction and water depth) for less extreme events.

Abu Aly et al. [88], with reference to a gravel-cobble river from California, analyzed the effects of vegetation on velocities, depths, and extent of the flooded areas for flows ranging from 0.2 to 20 times the bankfull discharge (Q_{BF}). They used a two-dimensional finite volume model that solves the vertical-mediated Reynolds equations (SRH-2D) and analyzed a stretch of the water course 28.3 km long with a mesh of 1–3 m. They obtained the height of the vegetation in each cell from LiDAR data and estimated the Manning coefficient based on the approach developed by Katul et al. [87] reported above. Compared to the case of absence of vegetation, they achieved an increase in the mean water depth of 7.4% and a mean velocity decrease of 17.5% for a flow of 4 Q_{BF}, values that rise, respectively, to 25% and 30% for a flow rate of 22 Q_{BF}. The model also shows how the vegetation has a strong channelization effect on the flow, in fact the flow is diverted away from densely vegetated areas and there is an increase in the difference between mid-channel and bank velocities.

Benifei et al. [137] investigate the effects of channel and riparian vegetation on flood event, with 200-year-return-period, really happened. Since the water levels were strongly influenced by riparian vegetation on the floodplain, they analyzed the effectiveness of different flood mitigation strategies employing the model Delft3D. They use the Baptist method [62]. Based on the results of the model, Benifei et al. [137] show that the most effective flood management strategy is obtained when the high vegetation (composed by trees) inside the flooding area with a return period of 2 years is removed, avoiding the growth of the bushes (5-years-old trees) as well.

8. Future Research and Conclusions

The paper is a review of a considerable number of studies on the flow resistance due to vegetation, based on experimental investigations in the presence of simulated or real vegetation; in addition, the paragraph on numerical methods also refers to phenomena on the local scale. It should be said, however, that flow resistance is a matter of detail because, according to Tsujimoto [138], flow, sediment transport, geomorphology, and vegetation constitute an interrelated system. The analysis of this interaction requires a multidisciplinary approach that can involve a set of disciplines such as hydraulics, hydrology, sediment transport, ecology, botany, and geotechnical engineering. It should also be noted that these phenomena vary greatly over time, in fact the climate and hydrology can alter growth patterns, rates of colonization, and vegetation density. The vegetation is simultaneously a dependent variable and an independent variable. Recent research has begun to investigate the interaction between sediment and vegetation in terms of sediment transport, erosion, and deposition, as stated in the introduction, but these interactions remain poorly quantified [16] and are new broad fields of research; the same is to be said for the interaction between vegetation and river morphodynamics [139,140].

With reference to flow resistance in the case of rigid vegetation many arrangements have been investigated taking into account different variables, so that a comparison of the proposed formulas is not immediate, even though possible. The first studies started from the use of basic equations, like momentum equation, and concepts, like the drag coefficient. Later, the studies developed towards the velocity and shear stress distribution. In the case of rigid vegetation, a few variables look to be able to derive a law for flow resistance. In the case of flexible vegetation, on the other hand, the various investigations refer to specific shrub and tree varieties, and the parameters of the different formulas are related to the specific varieties investigated, as well to the measuring range of the experimental tests. The wide variety of vegetation types and hydrodynamic conditions make comparing results and drawing general conclusions that are useful in practice difficult. Shields et al. [141] present a comparison between the Manning coefficients calculated, among others, with some of the formulas presented in the previous paragraphs both in the case of rigid vegetation and flexible vegetation. From it emerges how the formulas considered *predict values within the correct order of magnitude, however, clear guidelines for selecting one approach over another are difficult to provide* [141]. These indications could be obtained, as will be better specified later, seeing the ability of the different models to reproduce real situations. For operational purposes and for the correct use of formulas, it would be useful to establish a reference catalogue in which vegetation should be classified as rigid or flexible and in which ranges of biomechanical properties (flexural rigidity): Moreover, the geometric characteristics (shape of plants, foliage density, leaf area index) of the different species present in the river bed, on the banks, and in the floodplain areas, should be reported even better identifying how these properties vary with an appropriate parameter (age, height, or other). The vegetation, both in terms of species and density, varies considerably in space and in time. The hydraulic models that can best take into account this variability are the two-dimensional models. Remote sensing can be used to define the different roughness values, which is particularly useful in large areas and/or in areas not easily reachable. To accurately estimate the main vegetation properties, the simultaneous use of different remote sensing techniques could be used. Although methodologies have been proposed to determine height, flexural rigidity, density, stem diameter, crown base height, and diameter of vegetation, this constitutes a field with broad development perspectives. Very often it is difficult to classify the different types of vegetation, and remote sensing data require calibration through direct analysis in the field.

In order to identify which of the models previously seen are the most suitable to be used in the verification and design of interventions, it would be appropriate to apply them to real cases to reproduce flood events; an example of this is that of Benifei et al. [137], even if it is limited to the use of only one model. Considering the danger of flood events, reference could be made to controlled floods or indirect measures. At the same time, the same models can be used for the field verification of different riparian vegetation management strategies [135] and, subsequently, after calibration, for the design of the interventions.

Studies should be carried out on the dynamics of flood phenomena in the sense that high discharge can cause vegetation to bend on the bottom or break it or, due to the effect of localized erosion, can determine the uprooting of plants. Consequently, peak discharge or receding flow can have lower heights, even though this is valid only from a theoretical point of view because the material transported can accumulate on the bridge piles, causing dangerous backwater effects.

While the evaluation of flow resistance with reference to river reach is of considerable importance, it must be said that even the effect of vegetation patches exert considerable influence on flow resistance, on velocity distribution, and solid transport, and this is also a field of further research as witnessed by recent publications [34–38]. The kinetics of vegetation in terms of establishment, growth, and decay of riparian vegetation in response to dynamic hydraulic conditions are also beginning to be of interest to researchers [45,142].

While the present available results are mainly of global type, detailed numerical simulations can give insights on the characteristics of the flow fields when they interact with vegetation. This is also not a simple issue, due to the different properties that the vegetation can exhibit (rigid, submerged, etc.). Actually, the 3D numerical models can simulate very simple arrangements, and the currently required calculation resources go far beyond those used in professional practice. However, the continuous progresses in the numerical field and software development can guarantee that the detailed calculation of the flow–vegetation interaction will be soon possible. This issue involves both direct numerical simulations and numerical modeling of turbulence in this context, which may also benefit from the relevant theoretical developments and progress in observation techniques. So, in the future, they can provide more useful information for the conservation of habitats and biodiversity, maintenance of vegetation, and flooding protection.

Author Contributions: Conceptualization, F.C., A.D.; methodology, F.C., A.D.; writing—original draft preparation, F.C., A.D.; writing—review and editing, F.C., G.A., A.D., A.L.; supervision, F.C. All authors have read and agreed to the published version of the manuscript.

Funding: This research received no external funding.

Conflicts of Interest: The authors declare no conflict of interest.

Notation: The following symbols are used in this paper

A	area of water cross-section
a	projected plant area per unit volume
A_0	plant area projected onto a plan orthogonal to the flow direction when the plant is on free air
A_b	bed area related to a plant
A_c	characteristic area of plants
A_i	projected area of the ith plant on a plane normal to the streamwise direction
A_i^*	net submerged frontal area of the plant in the plane normal to the flow direction
A_L	one-sided leaf area
A_P	plant area projected onto a plan orthogonal to the flow direction when the plant is under the flow action
A_S	total cross-sectional area of all of the stems of an individual plant, measured at a quarter of the non-deflected height of the plant
b	Vogel exponent
B_x	fraction of channel cross section blocked by vegetation
C	Chézy coefficient
C_b	Chézy coefficient of the bed
C_D	drag coefficient
C_{DX}	species-specific drag coefficient
$C_{DX,F}$	species-specific drag coefficient of leaves
$C_{DX,S}$	species-specific drag coefficient of stems
C_f	Luhar-Nepf [11] friction coefficient
C_k	Chézy coefficient in the presence of emergent vegetation
C_r	Chézy coefficient in the presence of submerged rigid vegetation
C_u	coefficient in the Yang and Choi [76] model and in the Katul [87] model
C_0, C_1	coefficient in the Kouwen model
D	vegetation diameter

E	modulus of elasticity
f	Darcy Weisbach friction factor
f_c	function
F_D	drag force
f_v	friction factor due to vegetation
g	gravity acceleration
h	water depth or depth of the immersed part of the cylinder
h_v	vegetation height or vegetation height in the absence of flow
h_{vf}	bent vegetation height
h_*	representative length in Li et al. model [77]
i	bed slope
I	second moment of the cross-sectional area of the stems
J	energy line slope
K	Gauckler-Strickler velocity coefficient
L	length of river reach
LAI	Leaf area index
M	number of stems per unit bed area
m	number of cylinders per unit bed area
MEI	flexural rigidity
n	Manning roughness coefficient
n_b	soil Manning roughness coefficient
Q	discharge
R	hydraulic radius
Re_D	stem Reynolds number (=VD/ν)
Re_{D*}	vegetation Reynolds number calculate with the average pore velocity ($=V_vD/\nu$)
Re_v	vegetation Reynolds number ($=V_vr_v/\nu$)
r_v	vegetation-related hydraulic radius
r_v^*	dimensionless vegetation-related hydraulic radius
s	separation between individual resistance element
S	energy slope
SP	stream power
u	mean velocity along the vertical
u_s	mean velocity along the vertical in the surface layer in the case of submerged vegetation
u_v	mean velocity along the vertical in the vegetated layer in the case of submerged vegetation
u_z	local time-averaged velocity
u_*	shear velocity
u_{*c}	vegetal critical shear velocity
V	mean flow velocity or approach velocity
V_v	average pore velocity
V_χ	lowest velocity used in determining χ in Västilä et al. [104] model
x	streamwise coordinate
z	vertical coordinate
α_{0E}, α_{1E}	coefficients in the Ergun relationship for the drag coefficient
α_{KA}	characteristic eddy size coefficient in the Katul et al. [87] model
α_{KL}	characteristic turbulence length scale in the Klopstra et al. model [73]
β_1, β_2, β_3	numerical coefficients in the Carollo et al. [91] model
χ	vegetation parameter in Västilä et al. [104] model
χ_F	parameter in Västilä et al. [104] model relative to leaves
χ_S	parameter in Västilä et al. [104] model relative to stems
δ_e	depth of penetration of suspension layer in vegetation in Li et al. [77] model
γ	water specific weight
λ	density of vegetation
ν	water kinematic viscosity
ξ	parameter that takes into account the deformation of the plant
ξE	vegetation index
ρ	water density

References

1. Penman, H.L. *Vegetation and Hydrology*; Commonwealth Agricultural Bureau: Harpenden, UK, 1963; p. 124.
2. Peel, M.C. Hydrology: Catchment vegetation and runoff. *Prog. Phys. Geog.* **2009**, *33*, 837–844. [CrossRef]
3. D'Ippolito, A.; Ferrari, E.; Iovino, F.; Nicolaci, A.; Veltri, A. Reforestation and land use change in a drainage basin of Southern Italy. *iForest* **2013**, *6*, 175–182. [CrossRef]
4. Rowiński, P.M.; Västilä, K.; Aberle, J.; Järvelä, J.; Kalinowska, M. How vegetation can aid in coping with river management challenges: A brief review. *Ecohydrol. Hydrobiol.* **2018**, *18*, 345–354. [CrossRef]
5. Wynn, T.S.; Mostaghimi, S. The effects of vegetation and soil type on streambank erosion, southwestern Virginia, USA. *J. Am. Water Resour. Assoc.* **2006**, *42*, 69–82. [CrossRef]
6. Afzalimehr, H.; Dey, S. Influence of bank vegetation and gravel bed on velocity and Reynolds stress distributions. *Int. J. Sediment. Res.* **2009**, *24*, 236–246. [CrossRef]
7. Nepf, H.M. Hydrodynamics of vegetated channels. *J. Hydraul. Res.* **2012**, *50*, 262–279. [CrossRef]
8. Krzeminska, D.; Kerkhof, T.; Skaalsveen, K.; Stolte, J. Effect of riparian vegetation on stream bank stability in small agricultural catchments. *Catena* **2019**, *172*, 87–96. [CrossRef]
9. Julien, P.Y. *River Mechanics*; Cambridge University Press: Cambridge, UK, 2002.
10. Green, J.C. Comparison of blockage factors in modelling the resistance of channels containing submerged macrophytes. *River Res. Appl.* **2005**, *21*, 671–686. [CrossRef]
11. Luhar, M.; Nepf, H.M. From the blade scale to the reach scale: A characterization of aquatic vegetative drag. *Adv. Water Resour.* **2013**, *51*, 305–316. [CrossRef]
12. Anderson, B.G.; Rutherfurd, I.D.; Western, A.W. An analysis of the influence of riparian vegetation on the propagation of flood waves. *Environ. Modell. Softw.* **2006**, *21*, 1290–1296. [CrossRef]
13. Rutherfurd, I.D.; Anderson, B.; Ladson, A. Managing the effects of riparian vegetation on flooding. In *Principles for Riparian Lands Management*; Lovett, S., Price, P., Eds.; Land and Water Australia: Canberra, Australia, 2007; p. 190.
14. Dawson, F.H.; Charlton, F.G. *Bibliography on the Hydraulic Resistance or Roughness of Vegetated Watercourses*; Occasional Publication No. 25; Freshwater Biological Association: Ambleside, UK, 1998; p. 25. ISBN 0308-6739.
15. Curran, J.C.; Hession, W.C. Vegetative impacts on hydraulics and sediment processes across the fluvial system. *J. Hydrol.* **2013**, *505*, 364–376. [CrossRef]
16. Hession, W.C.; Curran, J.C. The impact of vegetation on roughness in fluvial systems. In *Treatise on Geomorphology*; Shroder, J.F., Butler, D.R., Hupp, C.R., Eds.; Ecogeomorphology; Academic Press: San Diego, CA, USA, 2013; Volume 12, pp. 75–93.
17. Marjoribanks, T.I.; Hardy, R.J.; Lane, S.N. The hydraulic description of vegetated river channels: The weaknesses of existing formulations and emerging alternatives. *WIREs Water* **2014**, *1*, 549–560. [CrossRef]
18. Wang, C.; Zheng, S.; Wang, P.; Hou, J. Interactions between vegetation, water flow and sediment transport: A review. *J. Hydrodyn.* **2015**, *27*, 24–37. [CrossRef]
19. Maji, S.; Hanmaiahgari, P.R.; Balachandar, R.; Pu, J.H.; Ricardo, A.M.; Ferreira, R.M. A Review on hydrodynamics of free surface flows in emergent vegetated channels. *Water* **2020**, *12*, 1218. [CrossRef]
20. Yen, B.C. Open channel flow resistance. *J. Hydraul. Eng.* **2002**, *128*, 20–39. [CrossRef]
21. Yang, J.Q.; Kerger, F.; Nepf, H.M. Estimation of the bed shear stress in vegetated and bare channels with smooth bed. *Water Resour. Res.* **2015**, *51*, 3647–3663. [CrossRef]
22. Liu, D.; Valyrakis, M.; Williams, R. Flow hydrodynamics across open channel flows with riparian zones: Implications for riverbank stability. *Water* **2017**, *9*, 720. [CrossRef]
23. Ben Meftah, M.; Mossa, M. Prediction of channel flow characteristics through square arrays of emergent cylinders. *Phys. Fluids* **2013**, *25*, 045102. [CrossRef]
24. Ben Meftah, M.; Mossa, M. A modified log-law of flow velocity distribution in partly obstructed open channels. *Environ. Fluid Mech.* **2016**, *16*, 453–479. [CrossRef]
25. Yang, J.Q.; Chung, H.; Nepf, H.M. The onset of sediment transport in vegetated channels predicted by turbulent kinetic energy. *Geophys. Res. Lett.* **2016**, *43*, 11261–11268. [CrossRef]
26. Tang, C.; Lei, J.; Nepf, H.M. Impact of vegetation-generated turbulence on the critical, near-bed, wave-velocity for sediment resuspension. *Water Resour. Res.* **2019**, *55*, 5904–5917. [CrossRef]
27. Västilä, K.; Järvelä, J. Characterizing natural riparian vegetation for modeling of flow and suspended sediment transport. *J. Soils Sediments* **2018**, *18*, 3114–3130. [CrossRef]
28. Box, W.; Västilä, K.; Järvelä, J. The interplay between flow field, suspended sediment concentration, and net deposition in a channel with flexible bank vegetation. *Water* **2019**, *11*, 2250. [CrossRef]
29. Yang, J.Q.; Nepf, H.M. Impact of vegetation on bed load transport rate and bedform characteristics. *Water Resour. Res.* **2019**, *55*, 6109–6124. [CrossRef]
30. Armanini, A.; Cavedon, V. Bed-load through emergent vegetation. *Adv. Water Resour.* **2019**, *129*, 250–259. [CrossRef]
31. Li, J.F.; Tfwala, S.S.; Chen, S.C. Effects of vegetation density and arrangement on sediment budget in a sediment-laden flow. *Water* **2018**, *10*, 1412. [CrossRef]
32. Sukhodolova, T.; Sukhodolov, A. Vegetated mixing layer around a finite-size patch of submerged plants: 1. Theory and field experiments. *Water Resour. Res.* **2012**, *48*, W10533. [CrossRef]

33. Sukhodolov, A.N.; Sukhodolova, T.A. Vegetated mixing layer around a finite-size patch of submerged plants: 2. Turbulence and coherent structures. *Water Resour. Res.* **2012**, *48*, W12506. [CrossRef]
34. Yilmazer, D.; Ozan, A.Y.; Cihan, K. Flow characteristics in the wake region of a finite-length vegetation patch in a partly vegetated channel. *Water* **2018**, *10*, 459. [CrossRef]
35. Li, W.; Wang, D.; Jiao, J.; Yang, K. Effects of vegetation patch density on flow velocity characteristics in an open channel. *J. Hydrodyn.* **2018**. [CrossRef]
36. Cheng, N.S.; Hui, C.L.; Wang, X.; Tan, S.K. Laboratory study of porosity effect on drag induced by circular vegetative patch. *J. Eng. Mech.* **2019**, *145*. [CrossRef]
37. Kitsikoudis, V.; Yagci, O.; Kirca, V.S.O. Experimental analysis of flow and turbulence in the wake of neighboring emergent vegetation patches with different densities. *Environ. Fluid Mech.* **2020**, 1417–1439. [CrossRef]
38. Yan, C.; Shan, Y.; Sun, W.; Liu, C.; Liu, X. Modeling the longitudinal profiles of streamwise velocity in an open channel with a model patch of vegetation. *Environ. Fluid Mech.* **2020**, 1441–1462. [CrossRef]
39. Liu, C.; Shan, Y. Analytical model for predicting the longitudinal profiles of velocities in a channel with a model vegetation patch. *J. Hydrol.* **2019**, *576*, 561–574. [CrossRef]
40. Mossa, M.; De Serio, F. Rethinking the process of detrainment: Jets in obstructed natural flows. *Sci. Rep.* **2016**, *6*. [CrossRef]
41. Mossa, M.; Ben Meftah, M.; De Serio, F.; Nepf, H.M. How vegetation in flows modifies the turbulent mixing and spreading of jets. *Sci. Rep.* **2017**, *7*. [CrossRef]
42. De Serio, F.; Ben Meftah, M.; Mossa, M.; Termini, D. Experimental investigation on dispersion mechanisms in rigid and flexible vegetated beds. *Adv. Water Resour.* **2017**, *120*, 98–113. [CrossRef]
43. Termini, D. Turbulent mixing and dispersion mechanisms over flexible and dense vegetation. *Acta Geophys.* **2019**, *67*, 961–970. [CrossRef]
44. Zhang, J.; Huai, W.X.; Shi, H.R.; Wang, W.J. Estimation of the longitudinal dispersion coefficient using a two-zone model in a channel partially covered with artificial emergent vegetation. *Environ. Fluid Mech.* **2020**. [CrossRef]
45. Yamasaki, T.N.; de Lima, P.H.; Silva, D.F.; Cristiane, G.D.A.; Janzen, J.G.; Johannes, G.; Nepf, H.M. From patch to channel scale: The evolution of emergent vegetation in a channel. *Adv. Water Res.* **2019**, *129*, 131–145. [CrossRef]
46. Mulahasan, S.; Stoesser, T. Flow resistance of in-line vegetation in open channel flow. *Int. J. River Basin Manag.* **2017**, *15*, 329–334. [CrossRef]
47. Chow, V.T. *Open Channel Hydraulics*; McGraw-Hill Book Co.: New York, NY, USA, 1959; p. 680.
48. Rouse, H. *Fluid Mechanics for Hydraulic Engineers*; McGraw-Hill Book Co.: New York, NY, USA, 1938; p. 422.
49. Cowan, W.L. Estimating hydraulic roughness coefficients. *J. Agric. Eng.* **1956**, *37*, 473–475.
50. Barnes, H.H., Jr. *Roughness Characteristics of Natural Channels*; U.S. Geological Survey Water-Supply Paper; Paper 1849; United States Government Printing Office: Washington, DC, USA, 1967; p. 213.
51. Phillips, J.V.; Ingersoll, J.V. *1998a—Verification of Roughness Coefficients for Selected Natural and Constructed Stream Channels in Arizona*; Professional Paper 1584; U.S. Geological Survey: Denver, CO, USA, 1998; p. 77.
52. Coon, W.F. *Estimation of Roughness Coefficients for Natural Stream Channels with Vegetated Banks*; U.S. Geological Survey Water-Supply Paper; Paper 2441; United States Government Printing Office: Denver, CO, USA, 1998; p. 133.
53. Land & Water Australia. *An Australian Handbook of Stream Roughness Coefficients*; Australian Government: Canberra, Australia, 2009; p. 29.
54. Arcement, G.J., Jr.; Schneider, V.R. *Guide for Selecting Manning's Roughness Coefficients for Natural Channels and Flood Plains*; U.S. Geological Survey Water-Supply Paper; Paper 2339; United States Government Printing Office: Denver, CO, USA, 1989; p. 38.
55. Petryk, S.; Bosmajian, G. Analysis of flow through vegetation. *J. Hydraul. Div.* **1975**, *101*, 871–884.
56. Li, R.M.; Shen, H.W. Effect of tall vegetations on flow and sediment. *J. Hydraul. Div.* **1973**, *99*, 793–814.
57. Nepf, H.M. Drag, turbulence and diffusion in flow through emergent vegetation. *Water Resour. Res.* **1999**, *35*, 479–489. [CrossRef]
58. Aberle, J.; Järvelä, J. Flow resistance of emergent rigid and flexible floodplain vegetation. *J. Hydraul. Res.* **2013**, *51*, 33–45. [CrossRef]
59. Liu, D.; Diplas, P.; Fairbanks, J.D.; Hodges, C.C. An experimental study of flow through rigid vegetation. *J. Geophys. Res.* **2008**, *113*, F04015. [CrossRef]
60. Petryk, S. Drag on Cylinders in Open Channel Flow. Ph.D. Thesis, Colorado State University, Fort Collins, CO, USA, 1969.
61. Stone, M.C.; Shen, H.T. Hydraulic resistance of flow in channel with cylindrical roughness. *J. Hydraul. Eng.* **2002**, *128*, 500–506. [CrossRef]
62. Baptist, M.J.; Babovic, V.; Rodríguez Uthurburu, J.; Keijzer, M.; Uittenbogaard, R.E.; Mynett, A.; Verwey, A. On inducing equations for vegetation resistance. *J. Hydraul. Res.* **2007**, *45*, 435–450. [CrossRef]
63. Ishikawa, Y.; Mizuhara, K.; Ashida, S. Effect of density of trees on drag exerted on trees in river channels. *J. For. Res.* **2000**, *5*, 271–279. [CrossRef]
64. Kothyari, U.C.; Hayashi, K.; Hashimoto, H. Drag coefficient of unsubmerged rigid vegetation stems in open channel flows. *J. Hydraul. Res.* **2009**, *47*, 691–699. [CrossRef]
65. D'Ippolito, A.; Lauria, A.; Alfonsi, G.; Calomino, F. Investigation of flow resistance exerted by rigid emergent vegetation in open channel. *Acta Geophys.* **2019**, *67*, 971–986. [CrossRef]

66. Tanino, Y.; Nepf, H.M. Laboratory investigation of mean drag in random array of rigid, emergent cylinders. *J. Hydraul. Eng.* **2008**, *134*, 34–41. [CrossRef]
67. Cheng, N.S.; Nguyen, H.T. Hydraulic radius for evaluating resistance induced by simulated emergent vegetation in open-channel flow. *J. Hydraul. Eng.* **2011**, *137*, 995–1004. [CrossRef]
68. Wang, H.; Tang, H.W.; Yuan, S.Y.; Lv, S.Q.; Zhao, X.Y. An experimental study of the incipient bed shear stress partition in mobile bed channels filled with emergent rigid vegetation. *Sci. China Technol. Sci.* **2014**, *57*, 1165–1174. [CrossRef]
69. Sonnenwald, F.; Stovin, V.; Guymer, I. Estimating drag coefficient for arrays of rigid cylinders representing emergent vegetation. *J. Hydraul. Res.* **2018**. [CrossRef]
70. Ergun, S. Fluid flow through packed columns. *Chem. Eng. Prog.* **1952**, *48*, 89–94.
71. Tinoco, R.O.; Cowen, E.A. The direct and indirect measurement of boundary stress and drag on individual and complex array of elements. *Exp. Fluids* **2013**, *54*, 1–16. [CrossRef]
72. Stoesser, T.; Kim, S.J.; Diplas, P. Turbulent flow through idealized emergent vegetation. *J. Hydraul. Eng.* **2010**, *136*, 1003–1017. [CrossRef]
73. Klopstra, D.; Barneveld, H.J.; van Noortwijk, J.; van Velzen, E. Analytical model for hydraulic roughness of submerged vegetation. In Proceedings of the 27th IAHR Congress, San Francisco, CA, USA, 10–15 August 1997; pp. 775–780.
74. Defina, A.; Bixio, A.C. Mean flow and turbulence in vegetated open channel flow. *Water Resour. Res.* **2005**, *41*, W07006. [CrossRef]
75. Huthoff, F.; Augustijn, D.; Hulscher, S. Analytical solution of the depth-averaged flow velocity in case of submerged rigid cylindrical vegetation. *Water Resour. Res.* **2007**, *43*, W06413. [CrossRef]
76. Yang, W.; Choi, S. A two-layer approach for depth-limited open-channel flows with submerged vegetation. *J. Hydraul. Res.* **2010**, *48*, 466–475. [CrossRef]
77. Li, S.; Shi, H.; Xiong, Z.; Huai, W.; Cheng, N. New formulation for the effective relative roughness height of open channel flows with submerged vegetation. *Adv. Water Resour.* **2015**, *86*, 46–57. [CrossRef]
78. Augustijn, D.C.M.; Galema, A.A.; Huthoff, F. Evaluation of flow formulas for submerged vegetation. In Proceedings of the EUROMECH Colloquium 523 Ecohydraulics: Linkages Between Hydraulics, Morphodynamics and Ecological Processes in University Blaise Pascal Clermont-Ferrand, Organiser Maison des Sciences de l'Homme (MSH), 4 rue Ledru, Clermont-Ferrand, France, 15–17 June 2011.
79. Vargas-Luna, A.; Crosato, A.; Uijttewaal, W.S.J. Effects of vegetation on flow and sediment transport: Comparative analyses and validation of predicting models. Earth Surf. Process. *Landform* **2015**, *40*, 157–176. [CrossRef]
80. Morri, M.; Soualmia, A.; Belleudy, P. Mean velocity modeling of open chnnel flow with submerged rigid vegetation. *Int. J. Mech. Mechatron. Eng.* **2015**, *9*, 302–307.
81. Pasquino, V.; Gualtieri, P. Flow resistance of submerged rigid vegetation: Focus and validation on two layer approach. In Proceedings of the 37th IAHR World Congress, Kuala Lampur, Malaysia, 13–18 August 2017; pp. 2502–2510.
82. Romdhane, H.; Soualmia, A.; Cassan, L.; Dartus, D. Flow over flexible vegetated bed: Evaluation of analytical models. *J. Appl. Fluid Mech.* **2019**, *12*, 351–359. [CrossRef]
83. Tang, X. Evaluating two-layer models for velocity profiles in open-channels with submerged vegetation. *J. Geosci. Environ. Prot.* **2019**, *7*, 68–80. [CrossRef]
84. Cheng, N.S. Representative roughness height of submerged vegetation. *Water Resour. Res.* **2011**, *47*, W08517. [CrossRef]
85. Van Velzen, E.; Jesse, P.; Cornelissen, P.; Coops, H. *Stromingsweerstand Vegetatie in Uiterwaarden; Handboek*; Part 1 and 2 Technical Report; RIZA Reports 2003.028 and 2003.029; Rijksinstituut voor Integraal Zoetwaterbeheer en Afvalwaterbehandeling—RIZA: Arnhem, The Netherlands, 2003.
86. López, F.; García, M. *Open Channel Flow Through Simulated Vegetation: Turbulence Modeling and Sediment Transport*; U.S. Army Corps of Engineers, Waterways Experiment Station; Wetland Research Technical Report WRP-CP-10; U.S. Army Corps of Engineers: Washington, DC, USA, 1997; p. 106.
87. Katul, G.; Wiberg, P.; Albertson, J.; Hornberger, G. A mixing layer theory for flow resistance in shallow streams. *Water Resour. Res.* **2002**, *38*, 1250. [CrossRef]
88. Abu-Aly, T.R.; Pasternack, G.B.; Wyrick, J.R.; Barker, R.; Massa, D.; Johnson, T. Effect of LiDAR-derived, spatially distributed vegetation roughness on two-dimensional hydraulics in a gravel-cobble river at flows of 0.2 to 20 times bankfull. *Geomorphology* **2014**, *206*, 468–482. [CrossRef]
89. Phillips, J.V.; McDoniel, D.; Capesius, J.P.; Asquith, W. *1998b—Method to Estimate Effects of Flow-Induced Vegetation Changes on Channel Conveyances of Streams in Central Arizona*; Water-Resources Investigations Report 98-4040; U.S. Geological Survey: Tucson, AZ, USA, 1998; p. 43.
90. Phillips, J.V.; Tadayon, S. *Selection of Manning's Roughness Coefficient for Natural and Constructed Vegetated and Non-Vegetated Channels, and Vegetation Maintenance Plan Guidelines for Vegetated Channels in Central Arizona*; Scientific Investigations Report 2006-5108; U.S. Geological Survey: Reston, VA, USA, 2007; p. 41.
91. Carollo, F.G.; Ferro, V.; Termini, D. Flow resistance law in channel with flexible submerged vegetation. *J. Hydraul. Eng.* **2005**, *131*, 554–564. [CrossRef]
92. Francalanci, S.; Paris, E.; Solari, L. On the vulnerability of woody riparian vegetation during flood events. *Environ. Fluid Mech.* **2020**, *20*, 635–661. [CrossRef]

93. Armanini, A.; Righetti, M.; Grisenti, P. Direct measurement of vegetation resistance in prototype scale. *J. Hydraul. Res.* **2005**, *42*, 481–487. [CrossRef]
94. Palmer, V.J. A method for designing vegetated waterways. *J. Agric. Eng.* **1945**, *26*, 516–520.
95. *Soil Conservation Service Handbook of Channel Design for Soil and Water Conservation*; SCS-TP-61; United States Department of Agriculture: Washington, DC, USA, 1966.
96. Ree, W.O.; Crow, F.R. *Friction Factors for Vegetated Waterways of Small Slope*; Pub. ARS-S-151; USDA Agricultural Research Service: Washington, DC, USA, 1977.
97. Gwinn, W.R.; Ree, W.O. Maintenance effects on the hydraulic properties of a vegetation lined channel. *Trans. of the ASAE* **1980**, *23*, 636–642. [CrossRef]
98. Temple, D.M.; Robinson, K.M.; Ahring, R.M.; Davis, A.G. *Stability Design of Grass-Lined Open Channels*; Agriculture Handbook Number 667; United States Department of Agriculture; United States Government Printing Office: Washington, DC, USA, 1987; p. 167.
99. Kouwen, N.; Li, R.M.; Simons, D.B. Flow resistance in vegetated waterways. *Trans. ASAE* **1981**, *24*, 684–698. [CrossRef]
100. Kouwen, N.; Unny, T.E. Flexible roughness in open channels. *J. Hydraul. Eng.* **1973**, *99*, 713–728.
101. Stephan, U.; Gutknecht, D. Hydraulic resistance of submerged flexible vegetation. *J. Hydrol.* **2002**, *269*, 27–43. [CrossRef]
102. Kouwen, N.; Fathi-Moghadam, M. Friction factors for coniferous trees along rivers. *J. Hydraul. Eng.* **2000**, *126*, 732–740. [CrossRef]
103. Freeman, G.E.; Rahmeyer, W.J.; Copeland, R. *Determination of Resistance Due to Shrubs and Woody Vegetation*; ERDC/CHL TR-00-25; Engineer Research and Development Center, US Army Corps of Engineers: Washington, DC, USA, 2000.
104. Västilä, K.; Järvelä, J.; Aberle, J. Characteristic reference areas for estimating flow resistance of natural foliated vegetation. *J. Hydrol.* **2013**, *492*, 49–60. [CrossRef]
105. Jalonen, J.; Järvelä, J. Estimation of drag forces caused by natural woody vegetation of different scales. *J. Hydrodyn.* **2014**, *26*, 608–623. [CrossRef]
106. Fathi-Moghadam, M.; Kowen, N. Non-rigid, non-submerged, vegetation roughness in flood plains. *J. Hydraul. Eng.* **1997**, *123*, 51–57. [CrossRef]
107. Ricardo, A.M.; Koll, K.; Franca, M.J.; Schleiss, A.J.; Ferreira, R.M.L. The terms of turbulent kinetic energy budget within random arrays of emergent cylinders. *Water Resour. Res.* **2014**, *50*, 4131–4148. [CrossRef]
108. Caroppi, G.; Gualtieri, P.; Fontana, N.; Giugni, M. Vegetated channel flows: Turbulence anisotropy at flow–rigid canopy interface. *Geosciences* **2018**, *8*, 259. [CrossRef]
109. Caroppi, G.; Västilä, K.; Järvelä, J.; Rowiński, P.M.; Giugni, M. Turbulence at water-vegetation interface in open channel flow: Experiments with natural-like plants. *Adv. Water Resour.* **2019**, *127*, 180–191. [CrossRef]
110. Penna, N.; Coscarella, F.; D'Ippolito, A.; Gaudio, R. Bed roughness effects on the turbulence characteristics of flows through emergent rigid vegetation. *Water* **2020**, *12*, 2401. [CrossRef]
111. Penna, N.; Coscarella, F.; D'Ippolito, A.; Gaudio, R. Anisotropy in the free stream region of turbulent flows through emergent rigid vegetation on rough beds. *Water* **2020**, *12*, 2464. [CrossRef]
112. Alfonsi, G. Reynolds Averaged Navier-Stokes equations for turbulence modeling. *Appl. Mech. Rev.* **2009**, *62*, Art. 040802. [CrossRef]
113. Lauria, A.; Calomino, F.; Alfonsi, G.; D'Ippolito, A. Discharge coefficients for sluice gates set in weirs at different upstream wall inclinations. *Water* **2020**, *12*, 245. [CrossRef]
114. Calomino, F.; Alfonsi, G.; Gaudio, R.; D'Ippolito, A.; Lauria, A.; Tafarojnoruz, A.; Artese, S. Experimental and numerical study of free-surface flows in a corrugated pipe. *Water* **2018**, *10*, 638. [CrossRef]
115. Alfonsi, G.; Ferraro, D.; Lauria, A.; Gaudio, R. Large-eddy simulation of turbulent natural-bed flow. *Phis. Fluids* **2019**, *31*, 085105. [CrossRef]
116. Stoesser, T.; Neary, V.; Wilson, C.A.M.E. Modelling vegetated channel flows: Challenges and opportunities. In Proceedings of the WSEAS (The World Scientific and Engineering Academy and Society) Conference on Fluid Mechanics, Corfu, Greece, 20–22 August 2005.
117. Cui, J.; Neary, V.S. Large eddy simulation (LES) of fully developed flow through vegetation. In Proceedings of the Fifth International Conference on Hydroinformatics, Cardiff, UK, 1–5 July 2002.
118. Alfonsi, G.; Lauria, A.; Primavera, L. Recent results from analysis of flow structures and energy modes induced by viscous wave around a surface-piercing cylinder. *Math. Probl. Eng.* **2017**, *10*, Art. 5875948. [CrossRef]
119. Alfonsi, G.; Lauria, A.; Primavera, L. A study of vortical structures past the lower portion of the Ahmed car model. *J. Flow Vis. Image Process.* **2012**, *19*, 81–95. [CrossRef]
120. Fischer-Antze, T.; Stoesser, T.; Bates, P.; Olsen, N.R.B. 3D numerical modelling of open-channel flow with submerged vegetation. *J. Hydraul. Res.* **2001**, *39*, 303–310. [CrossRef]
121. Shimizu, Y.; Tsujimoto, T. Numerical analysis of turbulent open channel flow over a vegetation layer using a $k-\varepsilon$ turbulence model. *J. Hydrosci. Hydraul. Eng.* **1994**, *11*, 57–67.
122. Choi, S.U.; Kang, H. Reynolds stress modeling of vegetated open-channel flows. *J. Hydraul. Res.* **2004**, *42*, 3–11. [CrossRef]
123. López, F.; García, M. Mean flow and turbulent structure of open channel flow through non-emergent vegetation. *J. Hydraul. Eng.* **2001**, *127*, 392–402. [CrossRef]

124. Kim, S.J.; Stoesser, T. Closure modeling and direct simulation of vegetation drag in flow through emergent vegetation. *Water Resour. Res.* **2011**, *47*, W10511. [CrossRef]
125. Stoesser, T.; Wilson, C.A.M.E.; Bates, P.D.; Dittrich, A. Application of a 3D numerical model to a river with vegetated floodplains. *J. Hydroinform.* **2003**, *5*, 99–112. [CrossRef]
126. Dudley, S.J.; Bonham, C.D.; Abt, S.R.; Fischenich, J.G. Comparison of methods for measuring woody riparian vegetation density. *J. Arid Environ.* **1998**, *38*, 77–86. [CrossRef]
127. Hohenthal, J.; Alho, P.; Hyyppa, J.; Hyyppa, H. 2011—Laser scanning applications in fluvial studies. *Prog. Phys. Geog.* **2011**, *35*, 782–809. [CrossRef]
128. Tomsett, C.; Leyland, J. Remote sensing of river corridors: A review of current trends and future directions. *River Res. Appl.* **2019**, *35*, 779–803. [CrossRef]
129. Huylenbroeck, L.; Laslier, M.; Dufour, S.; Georges, B.; Lejeune, P.; Michez, A. Using remote sensing to characterize riparian vegetation: A review of available tools and perspectives for managers. *J. Environ. Manag.* **2019**, *267*. [CrossRef]
130. Forzieri, G.; Castelli, F.; Preti, F. Advances in remote sensing of hydraulic roughness. *Int. J. Remote Sens.* **2012**, *33*, 630–654. [CrossRef]
131. Van der Sande, C.J.; De Jong, S.M.; De Roo, A.P.J. A segmentation and classification approach of IKONOS-2 imagery for land cover mapping to assist flood risk and flood damage assessment. *Int. J. Appl. Earth Obs.* **2003**, *4*, 217–229. [CrossRef]
132. Forzieri, G.; Degetto, M.; Righetti, M.; Castelli, F.; Preti, F. Satellite multispectral data for improved floodplain roughness modelling. *J. Hydrol.* **2011**, *407*, 41–57. [CrossRef]
133. Forzieri, G.; Guarnieri, L.; Vivoni, E.R.; Castelli, F.; Preti, F. Multiple attribute decision-making for individual tree detection using high-resolution laser scanning. *Forest Ecol. Manag.* **2009**, *258*, 2501–2510. [CrossRef]
134. Jalonen, J.; Järvelä, J.; Virtanen, J.P.; Vaaja, M.; Kurkela, M.; Hyyppa, H. Determining characteristic vegetation areas by terrestrial laser scanning for floodplain flow modeling. *Water* **2015**, *7*, 420–437. [CrossRef]
135. Errico, A.; Lama, G.F.C.; Francalanci, S.; Chirico, G.B.; Solari, L.; Preti, F. Flow dynamics and turbulence patterns in a drainage channel colonized by common reed (Phragmites australis) under different scenarios of vegetation management. *Ecol. Eng.* **2019**, *133*, 39–52. [CrossRef]
136. D'Ippolito, A.; Veltri, M. Influenza della vegetazione sulle portate al colmo in un tratto di un piccolo bacino. In Proceedings of the Atti del Terzo Congresso Nazionale di Selvicoltura Per il Miglioramento e la Conservazione Dei Boschi Italiani, Taormina (Messina), Italy, 16–19 October 2008; 2009; pp. 475–481. (In Italian).
137. Benifei, R.; Solari, L.; Vargas_Luna, A.; Geerling, G.; van Oorschot, M. Effects of vegetation on flooding: The study case of the Magra river. In Proceedings of the 36th IAHR World Congress, The Hague, The Netherlands, 28 June–3 July 2015.
138. Tsujimoto, T. Fluvial processes in streams with vegetation. *J. Hydraul. Res.* **1999**, *37*, 789–803. [CrossRef]
139. Camporeale, C.; Perucca, E.; Ridolfi, L.; Gurnell, A.M. Modeling the interactions between river morphodynamics and riparian vegetation. *Rev. Geophys.* **2013**, *51*, 379–414. [CrossRef]
140. Xu, Z.X.; Ye, C.; Zhang, Y.Y.; Wang, X.K.; Yan, X.F. 2D numerical analysis of the influence of near-bank vegetation patches on the bed morphological adjustment. *Environ. Fluid Mech.* **2020**, *20*, 707–738. [CrossRef]
141. Shields, D.F., Jr.; Coulton, K.G.; Nepf, H. Representation of vegetation in two-dimensional hydrodynamic models. *J. Hydraul. Eng.* **2017**, *143*, 1–9. [CrossRef]
142. Dombroski, D. *A deterministic Spatially-Distributed Ecohydraulic Model for Improved Riverine System Management*; Technical Rep. No. SRH-2014-26; Bureau of Reclamation: Denver, CO, USA, 2014.

MDPI
St. Alban-Anlage 66
4052 Basel
Switzerland
Tel. +41 61 683 77 34
Fax +41 61 302 89 18
www.mdpi.com

Water Editorial Office
E-mail: water@mdpi.com
www.mdpi.com/journal/water

www.ingramcontent.com/pod-product-compliance
Lightning Source LLC
Chambersburg PA
CBHW061844210926
51850CB00006B/967
9783036576633